Interpreting Technology

Philosophy, Technology, and Society

Series Editor: Sven Ove Hansson

Technological change has deep and often unexpected impacts on our societies. Sometimes new technologies liberate us and improve our quality of life, sometimes they bring severe social and environmental problems, sometimes they do both. This book series reflects philosophically on what new and emerging technologies do to our lives and how we can use them more wisely. It provides new insights on how technology continuously changes the basic conditions of human existence: relationships among ourselves, our relations to nature, the knowledge we can obtain, our thought patterns, our ethical difficulties, and our views of the world.

Titles in the Series
The Ethics of Technology: Methods and Approaches, edited by Sven Ove Hansson
Nanotechnology: Regulation and Public Discourse, edited by Iris Eisenberger, Angela Kallhoff, and Claudia Schwarz-Plaschg
Water Ethics: An Introduction, Neelke Doorn
Humans and Robots: Ethics, Agency, and Anthropomorphism, Sven Nyholm
Interpreting Technology: Ricoeur on Questions Concerning Ethics and Philosophy of Technology, edited by Wessel Reijers, Alberto Romele, and Mark Coeckelbergh
On the Morality of Urban Mobility, Shane Epting (forthcoming)
Test-Driving the Future: Autonomous Vehicles and the Ethics of Technological Change, edited by Diane Michelfelder (forthcoming)
The Ethics of Behaviour Change Technologies, edited by Joel Anderson, Lily Frank, and Andreas Spahn (forthcoming)

Interpreting Technology

Ricoeur on Questions Concerning Ethics and Philosophy of Technology

Edited by
Wessel Reijers, Alberto Romele,
and Mark Coeckelbergh

ROWMAN & LITTLEFIELD
Lanham • Boulder • New York • London

Published by Rowman & Littlefield
An imprint of The Rowman & Littlefield Publishing Group, Inc.
4501 Forbes Boulevard, Suite 200, Lanham, Maryland 20706
www.rowman.com

6 Tinworth Street, London SE11 5AL, United Kingdom

Selection and editorial matter © 2021 Wessel Reijers, Alberto Romele, and Mark Coeckelbergh

Copyright in individual chapters is held by the respective chapter authors.

All rights reserved. No part of this book may be reproduced in any form or by any electronic or mechanical means, including information storage and retrieval systems, without written permission from the publisher, except by a reviewer who may quote passages in a review.

British Library Cataloguing in Publication Information Available

Library of Congress Cataloging-in-Publication Data

Names: Reijers, Wessel, 1989- editor | Romele, Alberto, editor. | Coeckelbergh, Mark, editor.
Title: Interpreting technology : Ricœur on questions concerning ethics and philosophy of technology / edited by Wessel Reijers, Alberto Romele, and Mark Coeckelbergh.
Description: Lanham : Rowman & Littlefield, [2021] | Series: Philosophy, technology and society | Includes bibliographical references and index.
Identifiers: LCCN 2021016517 (print) | LCCN 2021016518 (ebook) | ISBN 9781538153468 (cloth) | ISBN 9781538153475 (epub)
Subjects: LCSH: Technology—Philosophy. | Technology—Moral and ethical aspects. | Hermeneutics. | Ricœur, Paul.
Classification: LCC T14 .I64 2021 (print) | LCC T14 (ebook) | DDC 601—dc23
LC record available at https://lccn.loc.gov/2021016517
LC ebook record available at https://lccn.loc.gov/2021016518

Contents

List of Figures and Tables vii

Introduction: Hermeneutic Philosophy of Technology:
A Research Program ix
Alberto Romele, Wessel Reijers, and Mark Coeckelbergh

PART I: RICOEUR AND THEORIES OF TECHNOLOGY

1 Ricoeur's Polysemy of Technology and Its Reception 3
  Ernst Wolff

2 Postphenomenology and the Hermeneutic Ambiguity
  of Technology 27
  Eoin Carney

3 Let's Narrate That Symmetry! Ricoeur and Latour 43
  Bas de Boer and Jonne Hoek

4 Ricoeur's Critical Theory of Technology 61
  David M. Kaplan

5 Free the Text! A Texture-Turn in Philosophy of Technology 75
  Bruno Gransche

PART II: RICOEUR'S ETHICS OF TECHNOLOGY

6 Narrative Self-Exposure on Social Media: From Ricoeur to
  Arendt in the Digital Age 99
  Annemie Halsema

7	Digital Hermeneutics: Will the Real Quantified Self Please Stand Up? *Noel Fitzpatrick*	117
8	The Pedagogical Relation in a Technological Age *David Lewin*	135
9	Prostheses as Narrative Technologies: Bioethical Considerations for Prosthetic Applications in Health Care *Geoffrey Dierckxsens*	153
10	Responsibility, Technology, and Innovation: The Recognition of a Capable Agent *Guido Gorgoni and Robert Gianni*	171

PART III: RICOEUR AND TWENTY-FIRST-CENTURY TECHNOLOGY

11	Ricoeur and E-health *Alain Loute*	189
12	The Force of Political Action in the Technological Polis *Todd S. Mei*	209
13	Software and Metaphors: The Hermeneutical Dimensions of Software Development *Eric Chown and Fernando Nascimento*	229
14	Narrating Artificial Intelligence: The Story of AlphaGo *Esther Keymolen*	249

Conclusion: Hermeneutic Responsible Innovation *Wessel Reijers, Alberto Romele, and Mark Coeckelbergh*	271
Index	285
About the Contributors	293

List of Figures and Tables

FIGURES

8.1	The Pedagogical Relation	142
8.2	The Technological Structure of Education	145

TABLES

5.1	Criteria of Textuality	81
5.2	Minimal Ontology of Understanding	87

Introduction

Hermeneutic Philosophy of Technology: A Research Program

Alberto Romele, Wessel Reijers, and Mark Coeckelbergh

Contemporary developments in technology and corresponding societal and cultural transformations urge us to turn to activities of interpretation. The burden of interpretation of these rapid and disruptive changes lies on the shoulders of many. Artificial intelligence (AI) experts need to interpret what it means that AlphaGo has defeated the human champion of the game GO. Philosophers need to interpret how digital objects impact metaphysics and philosophical theorizing. Policy makers need to interpret the possible impacts of apps they plan to introduce in society to track and trace people with COVID-19, to limit the damage of the current global pandemic. The public, in turn, needs to interpret technologically mediated knowledge about how to best behave in the times of a pandemic; whether to wear a face mask, to cancel face-to-face workshops (as the editors of this volume had to) and move activities online, or to hope for a vaccine. And not to speak of the impending catastrophe of climate change, which calls all of us to interpret the findings of the instruments of climate science, the technological changes toward a green and sustainable world, and the techno-political requirements for evading disaster. Interpretation of technology happens on many different levels, in many different domains, and involves basically all of us. This book is dedicated to this activity and takes as a major lead from Paul Ricoeur; a giant in contemporary philosophy who has thought and written more than almost anyone else on the topic of *interpretation*.

At first glance, it might look exaggerated to dedicate an entire book to Ricoeur and the philosophy of technology. Ricoeur is not a philosopher of technology, and his interest in technology has been rather marginal. Moreover, philosophers of technology have generally ignored Ricoeur's philosophy. So, one might think of this book as an eclectic attempt to bring yet another French voice in the field, and the chapters in it as mere exercises of style and eccentric scholarship. In these pages, we will not only show the potential of Ricoeur's philosophy for understanding technology in groundbreaking and original ways. We will also pave the way to a new research program in the philosophy of technology we call "Hermeneutic Philosophy of Technology" (HPT). At the end of the book, we will reflect on this program by also looking ahead, at a new horizon of Ricoeur-inspired questioning of technology, which takes the shape of "Hermeneutic Responsible Innovation" (HRI).

The use of the term "research program" is not casual. In Lakatos's terms (1978), a research program is made of a hard core of theoretical assumptions that cannot be abandoned without abandoning the program altogether, but it also tolerates more specific theories called auxiliary hypotheses. A program is considered progressive, whenever the auxiliary hypotheses contribute to improve the descriptive and predictive capacities of the hard core; it is considered degenerative, when the number of auxiliary hypotheses has become excessive, and their major function is to protect the hard core. We contend that many of the criticisms against the dominant programs in the philosophy of technology (in particular, the dual nature of technical artifacts program, and Postphenomenology) are so similar in their perspectives and scopes that they might represent different versions, each one with its own auxiliary hypotheses and methods, of a new—progressive and original—research program, namely HPT. Incidentally, we also wonder if the dominant programs in the philosophy of technology have not reached a degenerative status.

HPT aims at (1) mapping the multiple conditions of possibility in which technologies are embedded and (2) analyzing with different methods a certain number of these conditions of possibility. While a certain version of HPT might privilege language and narratives, and their concretizations in texts, documents, and monuments, we show that other grammars of technology exist and deserve to be explored. The transcendental—here to be understood in a very broad sense, as another way of saying "conditions of possibility" and "grammar"—of technology, like the Being for Aristotle, is said in many ways.

This introduction aims at sketching the outlines of HPT. It is structured in three parts. In the first part, we account for the reasons for resorting to Ricoeur's philosophy to understand technology. In particular, we study the role Ricoeur attributes to the linguistic and narrative dimensions in his thought and the way these dimensions are generally neglected in the contemporary, empirical philosophy of technology. We also highlight Ricoeur's constant effort to articulate truth and method. In the second part, we show that our perspective is part of an emerging movement in the philosophy of technology that focuses more and more on the conditions of possibility—or grammars—in which technology and technological mediations are always-already embedded. In the third part, we resume the different chapters included in this volume.

1. A RICOEURIAN PHILOSOPHY OF TECHNOLOGY

As already mentioned, Ricoeur is not a philosopher of technology, neither did he ever show a particular interest in technology. It seems fair to say that Ricoeur and the philosophy of technology have ignored one another. However, we contend that Ricoeur's philosophy might play a major role in the current debates and practices related to the philosophy of technology. In particular, we argue that his hermeneutics might help in correcting some exaggerations of the dominant empirical attitude in the philosophy of technology, but without losing sight of it. This book is less interested in Ricoeur's explicit philosophy of technology, that is, in what Ricoeur effectively said about technology, though this will certainly be touched upon, than in showing how Ricoeur's philosophy can be of inspiration to the contemporary philosophy of technology.

Ricoeur's hermeneutics is based on the following three pillars: (1) Human access to the world is always paired with or mediated by a symbolic and linguistic representation of the world; (2) if one wants to study the ways human beings access to the world, one has to employ hermeneutics as a method, that is, an *Auslegung*—an exegesis, an explication, an interpretation—of the symbolic and linguistic representations humans have of the world; and (3) the interpretation of texts represents the paradigm for every interpretation of the ways humans symbolically and linguistically have access to the world. This third pillar implies some corollaries, in particular (a) the centrality of the textual model, and more generally the ontological priority of writing over speech, (b) the principle of distanciation,[1] and (c) the overcoming of the distinction between truth and method, between understanding and explication, which is instead central in Heidegger and Gadamer's hermeneutics—on these pillars and corollaries, see in particular chapter 1 of Ricoeur (1991).

The merit of Ricoeur consists in having externalized and technicized language, without renouncing the ontological and anthropological dimension that characterizes hermeneutics after Heidegger and Gadamer. He speaks of a "long route" as opposed to Heidegger's—and Gadamer's—"short route." In the words of the French philosopher:

> I call such [Heidegger's] an ontology of understanding the "short route" because, breaking with any discussion of *method*, it carries itself directly to the level of an ontology of finite being in order there to recover *understanding*, no longer as a mode of knowledge, but rather as a mode of being. . . . The long

route which I propose also aspires to carry reflection to the level of an ontology, but it will do so by degrees. (Ricoeur 1974, 7)

He also speaks of truth and method not as an alternative, as it was the case in Gadamer's hermeneutics, but as a possible articulation. For him, the improvement of technical methods to approach texts and cultural productions does not reduce the possibility to grasp their truth, that is, the ways they reveal a world to us that concretely impact our modes of existence. Rather the contrary: "explain more in order to understand better" is a sentence often used to characterize Ricoeur's overall philosophical conviction.

First, we argue that there is familiarity between Ricoeur's hermeneutics and the contemporary philosophy of technology. As mentioned, Ricoeur externalized and technicized language by studying it in its crystallized forms: symbols first, and then signs, metaphors, and narratives. In the course of the years, he did not only plead for but also practiced the methods of semantics, semiotics, and the sciences of language in general. In other words, he assumed an empirical attitude toward his specific objects of study. In particular, Ricoeur's hermeneutics has an affinity with a branch of the contemporary philosophy of technology called postphenomenology, whose core is occupied by the notion of technological mediations. The idea is that the human access to the world is always technologically mediated.

In postphenomenology, which has also been called "material hermeneutics," technologies are hermeneutic "by nature," in the sense that in mediating between humans and the world, they magnify some aspects of it, while reducing some others. Technologies are hermeneutic also because they are multistable, that is, they can vary in uses and effects in different social and cultural contexts—the notions of "magnification-reduction" and multistability are introduced in Ihde (1990).[2] It is not by chance that Don Ihde, the founding father of postphenomenology, started his academic career as a Ricoeur's scholar, by editing the English translation of *The Conflict of Interpretations* (1974) and publishing the first monograph on Ricoeur in English, titled *Hermeneutic Phenomenology: The Philosophy of Paul Ricoeur* (1971). One could say that postphenomenology began as an expansion of Ricoeur's hermeneutics: texts and language, in general, become mediators between the humans and the world among many others.

Second, we contend that Ricoeur's hermeneutics can also be of support for overcoming some limits of the current debates in the philosophy of technology. In particular, we argue that Ricoeur's hermeneutics paves the way to a series of considerations about the linguistic and narrative conditions of possibility in which technologies are embedded. If postphenomenology would be our starting point, Ricoeur's hermeneutics would show that (1) there are other non-technological mediations next to the technological mediations, and

(2) technological mediations are wrapped up in non-technological, namely linguistic and narrative conditions of possibility.³

Philosophers of technology after the "empirical turn" (Achterhuis 2001) became increasingly attentive to the concrete inventions, innovations, and implementations of technology in society. Many of them started to engage in empirical research, and borrowed methods from disciplines such as science and technology studies and media studies. The empirical turn turned away from both the older monolithic philosophy of technology, and from the linguistic turn that dominated the philosophical debate of the twentieth century. However, we believe that philosophy of technology ended up throwing out the baby with the bathwater. Without negating the importance of the empirical turn and its manifestations, we believe that the moment has come to integrate it with a renovated—and empirically oriented—interest for the other dimensions in which technology are always embedded. Indeed, to rephrase Heidegger's famous sentence, it is our conviction that "the essence of technology is *not entirely* technological."⁴

2. THE TRANSCENDENTALS OF TECHNOLOGY

In Ihde's *Technology and the Lifeworld*, chapter 6 is titled "Cultural Hermeneutics." In this chapter, Ihde introduces the notion of technological multistability in the sense that technologies may be different in use in different cultural contexts. He also gives several examples, such as the case of the oval sardine cans left behind by the Australians after entering the New Guinean highlands for the first time, in the 1930s, in search of gold. These cans were immediately snatched by the New Guineans as treasured objects and made into centerpieces of the elaborate headwear they wore for special occasions. While in this case, a technology has been newly absorbed by the culture in which it found itself, things can go also the other way around: a technology can contribute to modifying an entire culture or "worldview" (Romele 2020a). Moreover, while Ihde considers the importance of culture in the *use* of technology, we rather suggest that culture, language, and narratives play an important role in the *entire* process of technological innovation.⁵

While Ricoeur has certainly given great importance to language, we believe that one cannot reduce Ricoeur's to a philosophy of language. Let us think, for instance, to his huge contribution to ethics and social philosophy. That means that by importing Ricoeur's thought in the philosophy of technology, we do not want to absolutize language or narratives as the sole conditions of possibility of technology. In this book, many other grammars of technology are considered. As a research program, HPT does not focus, indeed, exclusively on texts, narratives, and cultural productions. It rather wants to articulate, according to the principle of the hermeneutic circle, an interest for the empirical elements in the processes of technological innovation with a specific attention to the transcendentals in which these processes are always embedded.[6]

In this sense, HPT follows and gathers the insights of many scholars in philosophy of technology that recently felt a dissatisfaction with the empirical attitude that dominates the field. The fact is that the concern to "return to the things themselves" and to get a firmer grip on technological reality, an idea that is at the basis of the empirical philosophy of technology, has led in practice to ignoring those aspects of reality that do not yield to immediate intuition or perception because they transcend the technologies and the technological mediations, which they contribute to form.

Coeckelbergh (2017) argues that language and technology are both mediators between humans and the world, and conditions of possibility of linguistic and technological mediations, as well as of any kind of mediation.[7] He explicitly pleas for a "transcendental turn" in the philosophy of technology. Lemmens (2021) affirms that empirical philosophies of technology neglected the systemic character of technology. According to him, technology is the condition of possibility of technology itself, as well as of all sorts of human access to the world.[8] Romele (2020b) resorts to Pierre Bourdieu's social theory to discuss the way social dynamics of recognition, inclusion or exclusion, determine certain technological designs and uses. In particular, he introduces the concept of "technological capital" and its three states: objectified, institutionalized, and embodied.[9]

Lemmens et al. (2017) plea for a "terrestrial turn" in the philosophy of technology. With this expression, they mean both that technology has a planetary dimension, and that philosophy of technology should undertake an earth-oriented perspective. This implies to consider earth as one of the conditions of possibility of our technological being. In the words of the authors:

> Philosophy of technology needs to become capable—and urgently so—of facing the many new and unprecedented technological and ecological challenges that the emerging Anthropocenic and thoroughly "Earthbound" condition will present to a planetized humanity that is *threatened* by its own technology yet destined to start *remedying* this situation through this very same technology, by explicitly taking care of its increasingly instable, unreliable and precarious earthly habitat. (Lemmens et al. 2017, 124)[10]

While most of these authors have, implicitly or explicitly, outlined a hierarchy in the transcendentals of technology, we introduce here a principle of symmetry among the transcendentals of technology. Actually, Smith (2018) comes close to this perspective generalizing a transcendental approach to technology and proposing "mapping" as a specific method. First, he suggests understanding "transcendental" as an adjective rather than a noun, so one can account for multiple transcendentals, ontological *à la* Heidegger, epistemological *à la* Kant, and so on. Second, he criticizes "turning" as a method, and he introduces "mapping" instead:

> Picture a series of interactive and evolving maps, on which is possible to zoom in and out in terms of complexity, detail, and abstraction . . . . Imagine also that they have topological functionality: it is possible to simplify their elements in order to draw out relations between other maps and the elements on them. Imagine, crucially, that the limits of these maps are apparent . . . . This, I submit, is an alternative picture of method to which philosophy of technology might productively aspire today: as "mapping."

As far as we understand it, mapping has a twofold merit: (1) it allows zooming in and zooming out, that is, to focus on both the empirical dimension, and the a priori conditions of this dimension; (2) it allows linking, that is, going from one a priori to the other.[11] While very promising as a general attitude toward the conditions of possibility of technology, we believe that this approach is vague. Indeed, we believe that each condition of possibility must be described and empirically explored with methods specific to it. For this reason, we argue that the study of the grammars of technology represents more than a single research project. It is, indeed, an entire research program that should involve not only philosophers of technology but also scholars with different backgrounds.

In this book, our goal is limited to the presentation, and, in some cases, experimental explorations of how Ricoeur's philosophy can be used to interpret technology according to some of its technological and non-technological conditions of possibility. At the same time, the book engages with a promising new horizon of HPT, through engagement with ethics of technology and responsible innovation, which in the final chapter will be unfolded as "Hermeneutic Responsible Innovation" (HRI). In this way, the book not only envisions how a Ricoeur-inspired questioning of technology could lead to theoretical innovations, but also in a hermeneutic translation of theory to the practices of responsible innovation.

3. THE BOOK CHAPTERS

The fourteen main chapters of this book, written by a diverse group of seventeen authors from many different places (Belgium, Czech Republic, Germany, France, Ireland, Italy, the Netherlands, the United Kingdom, and the United States) illustrate more than only the value of Ricoeur's philosophy for thinking concerning technology. They also represent the transnational character of Ricoeur's legacy; its versatile nature—stretching between different disciplines—and its relevance, both for different generations of scholars and for cross-generational understanding of world affairs. The chapters contribute to the field of philosophy of technology in three different ways: (1) discussing the relation between Ricoeur's hermeneutics and the major theoretical approaches in the field; (2) exploring how Ricoeur's philosophy could contribute to practice ethics of technology; and (3) demonstrating how Ricoeur's hermeneutics is eminently relevant for understanding the most contemporary technological transformations that humankind is grappling with. These themes intersect in many of the chapters, but the book has been structured in accordance with the theme that each chapter focused on most.

Part I of the book discusses the first theme. It explores the place of Ricoeur's philosophy in the growing field of philosophy of technology, which is dominated by roughly three main approaches after the "empirical turn": Actor Network Theory (ANT), Critical Theory of Technology, and postphenomenology or material hermeneutics. In chapter 1, Ernst Wolff offers what might be considered as a bird's-eye perspective of Ricoeur's contribution to the thinking concerning technology. Through a fascinating exposition of the role of technology in Ricoeur's work, Wolff demonstrates the relevance of Ricoeur for understanding the technical dimension of the human capability. Whereas technology commonly understood as technical artifacts and systems is often absent from Ricoeur's writing, technology in the sense of *techne* figures prominently in many corners of his intellectual work.

This perspective opens up a dialogue between Ricoeur and philosophy of technology. First, Eoin Carney relates Ricoeur to postphenomenology in chapter 2, situating the first at the side of a hermeneutics of suspicion (Ricoeur was generally negative about modern technology) and the second at the side of a hermeneutics of trust. He demonstrates poignantly that Ricoeur can enrich postphenomenology by pointing at the ambiguity of technology, which constitutes conflicting interpretations. Second, Jonne Hoek and Bas de Boer open up a highly relevant dialogue between Latour and Ricoeur in chapter 3, through their common interlocutor, Greimas. Showing how Greimas's semiotics has inspired both Latour's material semiotics and Ricoeur's hermeneutics, Ricoeur's critique of Greimas opens up a "critique with a detour" of ANT, and through this, a discussion of how both perspectives can mutually benefit one another. Third, David Kaplan discusses how Ricoeur's view on technology relates to that of the Frankfurt Schule, in chapter 4, in particular of Feenberg's critical theory of technology. He shows how Ricoeur's distinction between utopia and ideology could be of great relevance for critical theorists of technology, for instance, through emphasizing the positive aspect of distanciation in emancipatory technical practices.

Part I concludes with a chapter by Bruno Gransche, who instead of relating Ricoeur to existing approaches explores in a very original way how Ricoeur could inspire a totally new understanding of technology, one that relates it back to one of its common ancestors: the text. In chapter 5, Gransche builds up toward a texture-hermeneutics, and proposes a number of distinctly new tools to understand our technological world.

Part II explores how Ricoeur's philosophy may contribute to the ethics of technology in different ways. While narrative and narrative identity have a prominent role in this part, the authors bring forward also other aspects of Ricoeur's social and political philosophy. In chapter 6, Annemie Halsema investigates what the use of digital technologies, in particular social networking sites, means for the notion of the self. According to her, there are several resemblances between narrative and digital identity. However, she also insists on their differences: while narrative identity implies expression of the self, digital identity aims at self-exposure. Noel Fitzpatrick sketches in chapter 7 the outlines of a "digital hermeneutics of the self." He considers Ricoeur's distinction between identity-*idem* (sameness) and identity-*ipse* (selfhood) to argue that the digital quantifications of the self concern the former rather than the latter. This causes major ethical and social issues. In his words, "the quantified self or data self acts as form of abstraction which does not possess the rights of the individual and yet the data sets are used to determine and impinge upon the rights of individuals within society."

In chapter 8, David Lewin explores the role of technology in education. The author opposes the "technological thinking," which is for him a way of seeing things and people only in terms of their apparent utility. Such thinking has ruinous consequences on the educational practices that are increasingly understood in terms of a neutral ensemble of techniques and technologies. It also causes major problems to the understanding of the subjects involved in the educational process that are seen as mere means or resources. The author proposes to raise narrative—understood in terms of processes of narration that allow narrative identity to form—as the horizon of understanding for every educational relation. Geoffrey Dierckxsens considers in chapter 9 the ways technologies, in particular biotechnologies and prostheses, are embedded in social imaginaries that have an impact on their uses—and design. He proposes to integrate the narrative approaches to technology with considerations about the social dimension in which the embodiment process of technologies like prostheses always take place. By looking at the concept of social imaginary as Ricoeur developed it, that is, as the ensemble of a community's social beliefs, values, and norms, we may gain better insight in ways patients ethically evaluate technological embodiments.

Par II concludes with chapter 10, in which Guido Gorgoni and Robert Gianni resort to Ricoeur's notion of responsibility to frame a new approach to responsibility in technological innovation. According to the authors, Ricoeur's contribution is important because he reconnects responsibility to the semantics of moral imputation, therefore regaining the sense of the connection between the agent and the action more than that of the attribution of its (typically negative) consequences. Responsibility in technological innovation is not just about a correct procedures and robust and trustworthy technological devices; it is also about personal commitment, toward oneself and the others.

Part III links Ricoeur's narrative hermeneutics and his work on political action and metaphor to contemporary technologies: e-health technologies, social media, software development, and artificial intelligence. It thus shows the continuing force of Ricoeur's thinking for addressing the problems of the twenty-first century, while also identifying some limitations of his work. In chapter 11, Alain Loute views e-health as a problem for Ricoeur's hermeneutics and defends the thesis that "e-health destabilizes hermeneutics on a threefold level, and that, as a result, it is difficult for hermeneutics to identify ethical and epistemological issues of e-health." Thus, rather than celebrating Ricoeur, Loute argues that it is philosophy of technology that has a heuristic power since it enables us to uncover and critically question some assumptions of Ricoeurian hermeneutics. At the end of his chapter, Loute suggests new and interesting lines of research, including a project that would discuss connections between space and narrativity. This is not only interesting for thinking about architecture, but also for philosophy of technology, which often neglects both topics.

Chapter 12 opens up a different conversation: one between Ricoeur's reading of Arendt and Austin's speech act theory. Todd Mei picks up Ricoeur's reading of Arendt on natality and J. L. Austin's speech act theory to consider the effects of social media as an instance of mediation in the technological polis. He defends the thesis that social media present a risk to "predicatory aims of political action," because political actions and utterances are divorced from their normative context, which leads to a minimization of the "rational or representational content that might inform a reasoned discussion." Mei's chapter thus opens up an original avenue for reflection on the political implications of Ricoeur's reading of Arendt for contemporary technologies. Eric Chown and Fernando Nascimento focus instead on the hermeneutic dimensions of software development. In chapter 13, they apply Ricoeur's concept of metaphor as semantic innovation to software development, casting software products as "redescriptions of interpreted action" and highlighting the significance of "productive imagination" in the process. Their chapter does not only enable us to theorize the interpretative dimension of software development but also invite developers to use their productive imagination to turn us away from what already exists in the world toward the creation of new meanings.

Finally, chapter 14 uses Ricoeur to shed light on the narrative dimension of AI. Esther Keymolen believes that such an approach could address some of the challenges AI poses to postphenomenological approaches in philosophy of technology. By adding a narrative approach inspired by Ricoeur to the "methodological toolbox," we may overcome the problem that currently AI is not much used yet, not very developed yet, or "blackboxed" away from direct, concrete experience, which means that current postphenomenology is not very helpful. The author unpacks three core elements of Ricoeur's account of narrative in turn: story, genre, and emplotment. Then she deploys this account of narrative in an analysis and close reading of the 2017 documentary *AlphaGo*, about the Google-developed computer program of the same name and its victories against the world's top players.

4. CONCLUSION

Via a reappropriation of Ricoeur's philosophy, the fourteen chapters of this book show that all empirical attitude toward technology is to be understood in the light of the not-just-technological conditions of possibility in which technologies and technological mediations are embedded. In other words, they represent a fundamental contribution to a general effort of thinking philosophy of technology *beyond* the limits—but also welcoming the strengths—of the empirical turn that dominate the discipline. Ricoeur's philosophy is characterized by the constant effort to articulate truth and method, two terms that we might translate here as a general attention to technology with the capital "T" and the concrete interest for technologies with lowercase "t." Responsibility in technological innovation must be seen in the light of personal commitment; technological mediations cannot be understood without an analysis of the symbolic, social, and cultural mediations that make technological mediations more or less effective. Even (partially) transcendental approaches to technology like Feenberg's critical theory of technology should be considered from a broader perspective, in which ideology is just one of the two poles of social imaginary; and so on. Each grammar of technology, be it linguistic, technological, social, cultural, and so on, deserves to be empirically explored, both in itself and in its consequences on the technological innovation. HPT announces a general attitude that in the chapters of this book is concretized via different research objects and methods. In the conclusion, we will contribute ourselves to the actualization of HPT by introducing a method in the ethical design and assessment of technological innovation called "Hermeneutic Responsible Innovation."

NOTES

1. According to Ricoeur (1991, 75), in Gadamer, there is an insolvable contradiction between "alienating distanciation," that is, the attitude that makes possible the objectivation that reigns in human and social sciences, and "appropriation," that is, "the fundamental and primordial relation whereby we belong to and participate in the historical reality that we claim to construct as an object." Ricoeur proposes to overcome this contradiction, arguing that the distance from our object of research—a distance that can be realized via the implementation of adequate methods—is the condition of possibility for both an adequate knowledge of the object itself, and a second-degree—that is, non-immediate and non-naive—existential appropriation of it.

2. In postphenomenology, all technologies are hermeneutic, but some technologies are more hermeneutic than others. Hermeneutic technologies in a strict sense are those representing a part of the world that must be correctly interpreted and understood in order to access the world. Hermeneutic technologies include of course texts, but also thermometers, airplane's cockpits, Google Maps, and many others.

3. Actually, language, and in particular narrative, and hence "emplotment" (*mise en intrigue*), might also be of inspiration to understand how technologies are articulated to each other and to other humans and nonhumans. While postphenomenology tends to privilege one mediation at the time, a narrative philosophy of technology might be able to account for the mutual dependencies among multiple technological and non-technological mediations.

4. Incidentally, it would be interesting to reconstruct the specific narrative in which the idea of empirical turn itself is embedded. Every community has its own origin myth. Similar narratives are part of what Ricoeur (1976) has called "cultural imagination," whose two poles are ideology and utopia. The empirical turn corresponds, for sure, to a series of concrete practices, but it is also a narrative in which a community of scholars recognize itself, and whose elements of representations have not been explored yet.

5. On this point, see Flichy (2007), who resorts to Ricoeur's distinction in the social imaginaire between utopia and ideology to describe the technological imaginaire and its function in the circular movement that goes from the technological project to the technological lock-in.

6. Moreover, it is obvious that every transcendental has its own "materiality." So one can, for instance, conduct an empirical study on the sociocultural representations of an emerging technology (nanotechnologies, AI, quantum computing, etc.) via an empirical analysis of their concretizations or actualizations in texts, and images, but also in institutions, behaviors, and so on.

7. In their criticism to the empirical philosophies of technology, some scholars insisted on the transcendentals, while some others on the necessity to think the

entanglement between technological and non-technological mediations. In his Ph.D. dissertation, Richard Lewis (2020) has distinguished, for instance, six types of mediations: technology, society and culture, body, mind, time, and space.

8. Lemmens's perspective is inspired by Bernard Stiegler. For Stiegler, technology is an "empirical transcendental," actually the first among the transcendentals, from four points of view at least: historical, anthropological, techno-evolutionary, and techno-phenomenological.

9. The objectified state corresponds to the technologies at disposal, or the technologies that are designed with that specific social actor's type in mind; the institutionalized state transcends already the empirical dimension, and corresponds to the ensemble of codified and written norms that allow some technological designs and uses, while prohibiting some others; the embodied state corresponds to the ways social actors 'allow themselves' or 'prohibit themselves' certain technological designs and uses. For a similar perspective, see also Rosenberger (2017), who proposed to integrate postphenomenology with a series of considerations on the ways technological mediations are wrapped up in social and political dynamics of power.

10. One can notice a certain technological optimism in this sentence. This is probably due to the prominent role the authors attribute to technology as transcendental. However, in the light of our perspective, we believe that one should be more cautious in presenting technology as *the* remedy (as well as *the* threat) to the ecological emergency. Indeed, we believe that an Earth-oriented philosophy of technology should consider the multitude of transcendentals (linguistic, social, cultural, etc.) that, along with technology, impact our relation with nature.

11. The use of a "digital" terminology in this context is not casual. Clearly, Smith has in mind a digital map accessible from the web as a model of mapping. On the zooming-in and zooming-out metaphor, and its constant use, especially in digital humanities and sociology, see for instance Latour (2017).

REFERENCES

Achterhuis, Hans. 2001. *American Philosophy of Technology: The Empirical Turn*. Bloomington: Indiana University Press.
Coeckelbergh, Mark. 2017. *Using Words and Things: Language and Philosophy of Technology*. New York and London: Routledge.
Flichy, Patrice. 2011. *The Internet Imaginaire*. Cambridge, MA: The MIT Press.
Ihde, Don. 1990. *Technology and the Lifeworld: From Planet to Earth*. Bloomington: Indiana University Press.
Lakatos, Imre. 1978. *The Methodology of Scientific Research Programmes*. Cambridge, MA: Cambridge University Press.
Latour, Bruno. 2017. *Anti-zoom*. http://www.bruno-latour.fr/sites/default/files/P-170-ELIASSON-GBpdf.pdf. Accessed October 1, 2020.
Lemmens, Pieter. 2021. "Thinking Technology Big Again. Reconsidering the Question of the Transcendental and 'Technology with a Capital T' in the Light of the Anthropocene." *Foundations of Science*.

Lemmens, Pieter, Vincent Block, and Jochem Zwier. 2017. "Toward a Terrestrial Turn in Philosophy of Technology." *Techné: Research in Philosophy and Technology* 21, no. 2–3: 114–126.

Lewis, Richard. 2020. *Relating Through Our Selves: Situating Media Literacy with Intrasubjective Mediation*. Ph.D. dissertation defended at the Vrije Universiteit Brussel Brussels, Belgium.

Ricoeur, Paul. 1974. *The Conflict of Interpretations*. Evanston: Northwestern University Press.

Ricoeur, Paul. 1976. "Ideology and Utopia as Cultural Imagination." *Philosophic Exchange* 7, no. 1: 17–28.

Ricoeur, Paul. 1991. *From Text to Action: Essays in Hermeneutics II*. Evanston: Northwestern University Press.

Romele, Alberto. 2020a. "Technological Capital: Bourdieu, Postphenomenology, and the Philosophy of Technology Beyond the Empirical Turn." *Philosophy & Technology*. https://doi.org/10.1007/s13347-020-00398-4.

Romele, Alberto. 2020b. "The Datafication of the Worldview." *AI & Society*. https://doi.org/10.1007/s00146-020-00989-x.

Rosenberger, Robert. 2017. *Callous Objects: Design Against the Homeless*. Minneapolis: Minnesota University Press.

Smith, D. 2018. *Exceptional Technologies: A Continental Philosophy of Technology*. London and New York: Bloomsbury.

Part I

RICOEUR AND THEORIES OF TECHNOLOGY

Chapter 1

Ricoeur's Polysemy of Technology and Its Reception

Ernst Wolff

Technology (*Technik* or *technique*) is not central as a theme in Paul Ricoeur's philosophy; he wrote no monograph in the field of philosophy of technology and even articles dedicated primarily to technical themes are rare in his oeuvre.[1] However, his thought on technology is not merely limited to occasional references. A first complex of thought is his reflection on the system of human-made artifacts and associated modes of usage, and often specifically as it developed in modernity (sections 1–4). A second theme is the ethical response to problems generated by technological progress (sections 5–6). Finally, technology is more or less directly thematized whenever Ricoeur explores different dimensions of human capability (section 7).

Ricoeur never developed his thought on technology into a neat system. I advance a broad synopsis of technology in Ricoeur by considering it from the point of view of a relatively "well-ordered polysemy" of *techné* in his work, which I will call his philosophy of "technology," for the sake of simplicity.

This encompassing view on Ricoeur's philosophy of technology has most often been neglected by the rapidly expanding reception of Ricoeur's philosophy in a range of technology-related fields of enquiry. His own thought on technology serves as one point of access, but not the only one, through which aspects of his work are received in these fields. In the last part of this chapter, I give an overview of various dimensions of this varied reception.

1. TECHNOLOGY, MODERNITY, AND CULTURAL PLURALITY

Ricoeur's views on modernity and his ways of approaching (or side-stepping) it in his thought are quite complex. During the early post–World War II years,

Ricoeur developed his initial perspective on a loosely defined contemporary era (rather than a neatly defined modernity).[2] It is within this framework that one has to approach Ricoeur's first inroads into the philosophy of technology.

The question of technology appears in "Christianity and the Meaning of History" (originally published in French in 1951),[3] as part of a generally constructed reflection on the meaning of history, which Ricoeur examines as consisting of three layers: progress, ambiguity, and hope. The idea of progress, which Ricoeur defines as the "accumulation of acquirements [*l'accumulation d'un acquis*]" (Ricoeur 1964, 93/1965, 81),[4] refers to the level at which one has to consider the history of "tools" [*outils*].[5] Since Ricoeur demarcates these as "material or cultural tools, tools of knowledge, and even tools of consciousness and of the spirit" (Ricoeur 1964, 1965, 93/1965, 81), it implies that he has the whole range of technology in mind, in as far as it can be analytically *abstracted* from human action and thought.

Accumulation allows for progress in tools and in their usage (in other words, labor in the widest sense), but also in knowledge and know-how. Moreover, accumulation allows for a sedimentation of technology, which transcends individual human beings' activities and lives. Writing and the printing of texts epitomize this point. But sedimentation has a further temporal dimension: it is irreversible (Ricoeur 1964, 95/1965, 82).

This brings Ricoeur to a curious conclusion (apparently derived from Pascal): humanity, understood as single entity, is the ultimate recipient of whatever is thus transmitted.[6] Ricoeur claims that this trend applies also to mental techniques (e.g., calculation and science) and spiritual techniques. This means that all cultural particularities of technological inventions belong to a specific group or culture only in a provisional way; in principle, cumulatively and in the long run, "tools," or technology, as a whole, belongs to humanity as a whole.

Only when the abstraction is lifted and technology's *embeddedness* in human interaction is reconsidered one witnesses again the drama of decisions and crises, of growth and decay (Ricoeur 1964, 94/1965, 84). Different people and different groups can be distinguished from one another by the ways in which they face history and the ways in which they let ethical demands and solutions weigh on their appropriation and use of technology. Hence, technology can never be separated from valuation (Ricoeur 1964, 101/1965, 87–88). Big historical tendencies in this respect allow us to identify civilizations, each with its own history of values and challenges, its own rise and fall—in any case, we are never looking simply at a history of progress, but have to consider complex processes of emergence, maintenance, invention, and decay. On the level of human interaction, technology is inseparable from the human history of power, in other words, of politics. Where value infuses the invention, use, and progress of technology, something is at risk: human survival, human flourishing, and harm to others. Ricoeur identifies this as the risk of guilt [*culpabilité*] (Ricoeur 1964, 105/1965, 91). On the whole, whereas technology in abstraction from action forms a history of progress, its integration in human action and evaluation reveals it as fundamentally ambiguous (Ricoeur 1964, 107/1965, 92 and see §1.2 below).

Subsequently, Ricoeur revises this view on technology in "Universal Civilization and National Cultures" (1961). Universal civilization refers precisely to those common aspects of the diversity of human cultures, which result from the spread of the "scientific spirit" (Ricoeur 1964, 322/1965, 271), and which are relayed in and incorporated into modern technology. Technological globalization transcends national and cultural boundaries, and, in turn, contributes to the gradually awakening awareness that people belong to one planetary humanity (Ricoeur 1964, 324/1965, 275). Other extensions of the scientific spirit—the modern state and its administration, rational economics, and aspects of everyday culture, such as housing and clothing—also have a technical dimension (Ricoeur 1964, 324–326/1965, 275–277). Ricoeur spells out the ambiguity of these developments. On the positive side, there is the spread of better material conditions for survival (food, safety, and medical treatment), mass education and the means to protect human dignity (which is not to claim that everything has already been realized!). On the negative side, there is the destruction of traditional cultures, the undermining of the normative core of these cultures—aside from all the well-known forms of abusive applications of new means for exploitation and oppression. The threat to inherited cultural traditions is also a threat to the human evaluative ability in matters technological and beyond.

But one could also approach the relation between technology and valuation from the angle of political intervention (already mentioned above), as Ricoeur does in "The Tasks of the Political Educator" (originally published in French in 1965). Here he identifies three domains of the effectiveness of education, corresponding with three levels of analysis of "civilization,"[7] namely industries (*outillages*), institutions, and values[8] (Ricoeur 1991, 242/1974, 272). Under *outillages*, Ricoeur understands "a very vast aspect of civilization which goes beyond the level of tools [*outils*], machines, and even of techniques [*la technique*]" (Ricoeur 1991, 242/1974, 272). He attributes similar characteristics to the term as in the 1951 article: accumulation, conservation, innovation, and acquisition (in principle) by the whole of humanity. The effect of accumulation ripples out into all forms of knowledge and the sciences, because it is captured in documents, monuments, libraries, and so on. Finally, the increasing reach of technological relations has a unifying effect on humanity, and this increasingly results in an awareness among the people of the earth that they belong to a single humanity.

However, as in the earlier discussion, this view on technology remains an abstraction. In reality, people appropriate technology through institutions and are guided by values in doing so (Ricoeur 1991, 243–244/1974, 274–275). Institutions make humanity plural, and they differentiate historically and geographically limited experiences. The analytical abstraction of technology from institutional life underscores the fact that, as dependent as politics may be on technology, it can never simply be reduced to technology or even to the economy that stimulates technological productivity (Ricoeur 1991, 245/1974, 276–277[9]). In addition, the temporality of technology is not that of progress, but of ambiguity. Hence, according to Ricoeur, technology experiences no real crisis: new inventions or new technical problems become crises only in so far as they are integrated in people's institutional life (Ricoeur 1991, 246/1974, 278).

This institutional life is, in turn, informed by common acts of valuation (which is Ricoeur's third level of analysis of "civilization"). Values, now firmly defined as forms of practice, represent the most specific detail of people's action in history: Ricoeur argues that "it is the industry [*outillage*] which is abstract and value which is concrete" (Ricoeur 1991, 247/1974, 279). Knowing how to use other people's physical power against their will is a technology; abolishing slavery is a historically concrete way of discarding that technology. Hence, this mediation, which started with technology, opens up the whole hermeneutics of symbols and traditions (by which valuation is exercised) on which Ricoeur worked in the same period.[10]

Finally, Ricoeur claims that there is no totalizing vision of the three levels of analysis of civilization. There remains a tension between the three levels of analysis, which, in turn, intensifies the questions about the tasks of political educators. For if in modernity, more than ever before, technological development is subject to planning, the need for collective decision-making about such developments is also increased (Ricoeur 1991, 250–251/1974, 283–284). Political education in the broadest sense should equip the citizenry to participate responsibly in such decision-making. Ricoeur is ultimately concerned with the democratization of decision-making in the economy, and thereby also the development and use of technology.[11] Moreover, collective responsibility has to assume that there is a tension between two ethics—one dominated by realism about the means to be used and the uncertainty of some outcomes, the other dominated by context-independent principles. Understanding responsibility as the prudent arbitration in practice between these two moments is a consistent aspect of Ricoeur's view of responsibility (see section 2) and it is derived, among others, from Max Weber.[12] For each context, the appropriate midway between moralism and cynical realism has to be sought. Hence, reflection on technology and means is integrated into a larger three-tier project on the "struggle for a democratic economy, the proposal of a project both for all people together and for the single person, and the reinterpretation of the traditional past in the face of the rise of the consumer society" (Ricoeur 1991, 257/1974, 293, translation modified).

To summarize, Ricoeur reflects on technology, not for itself, but as an indispensable part of affirming the unity of humanity, the irreducibility of politics, and the significance of valuation.

2. THE AMBIGUITY OF TECHNOLOGY

The contradictory possibilities opened up by new technologies are thematized in "The Adventure of Technology and Its Planetary Horizon" (originally published in French in 1958), a rare essay exclusively devoted to technology. In the face of nuclear (and other) destructive power, people ask themselves: "What happens to humanity [*qu'advient-il de l'homme*]—and to the meaning of humanity [*l'homme*]—when it acquires such power? To what extent does humanity change itself by changing its power over things?" (Ricoeur 1958, 67). Instead of siding with military realism or a moralist rejection of modern (military) technology, Ricoeur underscores the ambiguity of modern technological development, situating it within the longer history of humanity.

With the industrial revolution, technology became a core aspect of how people define themselves and interpret their destiny. Only since the industrial revolution does technology touch on "the 'sacred' of the human—what is essential, necessary, primordial to it" (Ricoeur 1958, 68)[13] and that there emerges a "technological civilization." The essay explores three dimensions of this turn: work, consumption and self-understanding.

First, the domain of work has seen numerous changes: generally, technology has brought gradual relief regarding the arduousness and danger of work, the fragmentation of traditional arts into small, repetitive jobs with a negative impact on the experience of work as meaningful, and the laborer's self-image. But changing technologies also led to the emergence of new forms of work, ranging from the maintenance of machines to administrative services. Whereas the steadily increasing technological capabilities change people's view of their place in the universe, their working conditions are determined by very mundane interactions.

Second, Ricoeur considers the improvement in people's living conditions, as far as nutrition and medical treatment, housing and basic education, and so on are concerned. But this improvement of living conditions is the flip side of increasing dependency on "machines," that is, industrial productivity, which henceforth reshapes people's transport and housing, infiltrates into household chores, modifies the transmission of culture, and so on (Ricoeur 1958, 71). And at this point, the ambiguity of technology has an impact on people as consumers (in the largest sense of the word): widespread accessibility, coupled with a lowering of standards in mass culture.

Third, before the industrial revolution, work was often considered unfree, in contrast to speech, be that in the service of contemplation or of political praxis. Now, thanks to new means of action, work has become part of human nature, a means to pursue freedom, a conquest.

Although technology has given an unparalleled impetus to globalization and the unification of humanity, this process will be completed only when (if ever) there will be one humanity with one politics and one culture (Ricoeur 1958, 74). Even the view, or rather the significance, of the heavens has changed: "Heaven is becoming a domain of action, an object of covetousness, domination, possession" (Ricoeur 1958, 75). Once again, one sees how Ricoeur's meditation on technology opens up the question of meaning and nihilism, which informed his early hermeneutics. But this meditation also has a geopolitical dimension: Ricoeur denounces the fact that space technology had little to do with service to people and much more with propaganda and the strategies of the two big geopolitical blocs of the world at that time. Indeed, the budget spent on the fantastic projects each time deprives the destitute of the earth of an equal amount of support.

3. TECHNOLOGY, WORK, AND POLITICS OF A LABOR SOCIETY

Work is a recurrent theme in Ricoeur's early post–World War II practical philosophy and political thought. Since this theme is closely related to that of technology, I review a representative text in this respect: "Work and Language" or "Work and the Word" of 1953, an article that responds to a variety of interpretations of what was then called the "civilization of work" (Dosse 2008, 171).

Ricoeur is concerned that the social specificities of work will get lost in an overly general deployment of the term "work" and also that the centrality of work will lead to neglect of that which cannot be reduced to work itself. Consequently, he attempts to undermine an understanding of work that excludes the spheres of activity of speech and prefers to integrate them into a dialectic of work and word (Ricoeur 1953, 238–223/1964, 197–198). The conclusion of his article reads: "Hence, every human civilization will be both a civilization of work *and* a civilization of the word" (Ricoeur 1953, 263/1964, 219). I would like to focus on the context-specific sociopolitical diagnosis that Ricoeur makes on the basis of this demonstration, which is of a more general anthropological nature.

He notes a number of contemporary tendencies in the relations between work and the word. A first tendency is the worker's alienation[14] from his or her wage-paying labor and the humiliation and under-evaluation of manual labor (Ricoeur 1953, 253–254/1964, 210–211). A second tendency is introduced by the technological form of work: the fragmentation of the arts, but also of office work and science (hence Ricoeur's reluctance to celebrate the self-realizing potential of work, cf. Ricoeur 1953, 255–256/1964, 212–213) and consequently attempts to compensate for the loss of meaning in work through leisure and consumption. Thus, culture (in the broadest sense) and work both experience difficulty in adapting to technological modernity. At the same time, culture contains the possibility of non-adaptation to dehumanization, openness to the future, and the quest for a realistic trade-off between the new conditions and the aspiration for more humane alternatives, a quest that is firmly rooted in the task of education (Ricoeur 1953, 257/1964, 213–214).

Next, Ricoeur insists on the political participation by which people should be allowed to help steer the technical environment in which they work. However, he advocates a decentralized model of participation (instead of a socialist centralist model). But the democracy of work would have to spiral out and also have an impact on the management of the working environment and industries. Furthermore, a civilization of work has to be an economy of work, which implies that a fair redistribution of wealth and a policy of full employment should be pursued. This wide vision of the requirements for a fair labor society also has implications for culture—not ideological management of culture in the form of a naïve celebration of industrial work, but a creative critique of labor practices and poetic reflection on solidarity between people (Ricoeur 1953, 259–260/1964, 215–216).

Lastly, language has a set of roles to play in the context of work. First, the range of linguistic functions can act as a corrective to the division and fragmentation of work. It can express frustration, open this division to a plurality of views, give access to a total view on an industry's activity, and even serve as distraction. Articulation is the first step to political mobilization. Second, Ricoeur thinks that forms of speech can serve as a (partial) compensation for difficult experiences at work through leisure. Third, innovation in technology, and thus the invention of new forms of work, passes through the practice of theory, specifically in the form of the sciences. Fourth, Ricoeur counts on the creative function of language to contribute to the discovery and invention of meaning, which will always be needed in all forms of work.

Hence, one gets an impression of the interrelation of two major families of human ability—work and the word, neither of which suffices on its own as a form of the pursuit of human well-being.

4. URBANIZATION

Finally, two further traits of the modern world are dealt with, in combination, in an aptly titled article "Urbanisation and Secularization" (originally published in French in 1967). I consider it here only in as far as it completes Ricoeur's view on modern technology.[15]

The four main traits of city life that drew Ricoeur's attention can be understood only if one considers the city as a technical construct. First, a city has to be understood as a colossal instrument for connecting people and enabling communication between them. The city multiplies the number of links among people and augments the possibilities of exchange over longer distances and with greater frequency. Consequently, many exchanges are more abstract or anonymous, requiring a renegotiation of the public-private relation. Also, the sheer mass of information implies intensification of the decisions to be made.

Second, the city is a place of multiple and accelerated mobility[16] and "internal migration," which can be seen in the act of commuting. The movement between a person's home and workplace corresponds with a more marked differentiation of people's social roles and, in turn, with the impact of psychological stress or opportunities. This requires tolerance and adaptability.

Third, the city is a place of concentrated technology and organization, a technopolis, as Cox called it. It concentrates health systems, bureaucracies, commerce, and so on.[17]

Fourth, the city is also an image to people of what a "city" really is. Such images range from the religious construct of the "City of God" to the political "polis." In recent history, the changing material city has changed people's view of themselves as urbanites. Now the city becomes an image of modernity, energy, and initiative.

Whereas the city opens up many opportunities and freedoms, it is also a space of specific suffering (Ricoeur 2003, 116/1974, 180). Each of the four points of urbanization corresponds to a number of urban pathologies. Anonymity, traffic congestion, and information saturation figure among the pathologies corresponding to the first point mentioned earlier. Being uprooted, a multiplication of conflict and the degeneration of inner cities correspond to the feature of mobility. Between under-administration and over-bureaucratization, the concentration or organization is also subject to pathologies. Even the view of the city can succumb to pathologies: directedness at a vague, distant future, which may lead it to be experienced as a locality dominated by a senseless technological logic.

And thus, the city too reflects that essential characteristic of all modern technological developments: their two-sidedness, their ambiguity.

Just as Ricoeur follows Cox in his presentation of the modern city, Ricoeur follows other authors: Weber on the nexus between rule and technology

(Ricoeur 1986b, chapter 12), Habermas on technology as ideology (Ricoeur 1986b, chapter 13), and Taylor on the role of technology in the malaise of modernity (Ricoeur 2000b, 410–441/2004b, 312–313).

5. ETHICS WITH AND AMONG CHANGING TECHNOLOGICAL MEANS

In Ricoeur's work, philosophical reflection on the ethics of technology is concentrated in his later texts and is focused mostly on dilemmas in the biomedical professions. Yet, the link between ethics and technology in Ricoeur's work is quite old, as I will now first demonstrate.

Initially, this link emerges in Ricoeur's political thought in relation to the question of the *efficacy* of one's ethical stance. A first example of this point is Ricoeur's reflection on Gandhi's anticolonial politics in the article "Non-Violent Man and His Presence in History" (originally published in French in 1949). Ricoeur attributes the efficacy of Gandhi's movement to the fact that it is a good combination of a "technology [*technique*]" and a "spirituality" (Ricoeur 1964, 273/1965, 230); in other words, of operative strategy and a legitimate normative stance. A second example is found a decade later in the first formulation Ricoeur gave of the political paradox. People aspire to relations of political self-realization but can ultimately achieve this only through a state. But the state, understood as the ultimate instrument of power (following Weber), can instead of realizing the will of the people, turn against people. Hence the need "to devise institutional *techniques* [*techniques*] especially designed to render possible the exercise of power and render its abuse impossible" (Ricoeur 1964, 311, similarly 314–315/1964, 261–262, similarly 264–265). Thus, Ricoeur refers to technology in many of his political essays, but on a more profound level, the question of technologies of action emerges from reflection on ethico-political efficacy.

Furthermore, these two briefly presented cases[18] contain, in outline, the structure of Ricoeur's later understanding of responsibility as a practical choice between two opposing ethical claims. This is best seen in the way Ricoeur explains responsibility in the ethical chapters of *Oneself as Another*. Any responsible or practically wise action is here presented as the outcome of a difficult, *prudent* trade-off between two equally valid, but contradictory claims: the *ethical* claim of pursuing the good life with and for others in just institutions, and the *moral* claim, which seeks to filter out all non-universalizable principles of action. Hence, Ricoeur's view of responsibility is also constructed on the tension between points of view that may be held simultaneously when one considers them *theoretically* but calls for final arbitration *in practice*.[19]

6. NEW TECHNOLOGICAL CONDITIONS, NEW ETHICS?

From the perspective of this broad understanding of ethics, changes in the technological environment do not alter the essential structure of responsibility but rather change the specific ways in which people may want to strive for the good life; the series of new practical possibilities, which they may want to prohibit from the point of view of a moral principle; and the kinds of trade-off that may be relevant in practice. All of these changes are represented in the preface Ricoeur wrote for Frédéric Lenoir's volume *Le temps de la responsabilité* (Ricoeur 1991, 271–294). As one might expect, technological changes are seen as inducing a shift in what it means for people to act. These changes are played out in six domains: the life sciences, the environment, economics, industry and development, the media and communication, and politics—for all of which Ricoeur catalogues salient aspects and the concomitant dilemmas as they appear in this book.[20] Moreover, in this book, Ricoeur observes that the catalogued dilemmas can be approached either through an appeal to stable commonly held convictions from the normative tradition (which have to be applied to new problems) or through new forms of ethics, of which responsibility would be an example (cf. 1991, 282).

In *Oneself as Another*, Ricoeur clarifies his understanding of responsibility with reference to a set of exemplary ethical dilemmas in biomedical ethics (cf. Ricoeur 1990, 313–318/1992, 269–273). The increasing expansion of technological capabilities in the domain of biomedical ethics makes it ideal material for reflecting on ethical problems. Ricoeur's aim is, however, rather to suggest a general approach to such dilemmas than to make a contribution to specific points of debate.

Let us summarize the main points of this approach. First, one should have a sound understanding of the dilemma *as* a dilemma (i.e., one needs to admit that there is no solution that can make its problematic character disappear). One's most intimate convictions and professional ethics may inform one's interpretation of these dilemmas. Second, a good technical and scientific grasp of the problem at hand is required. The specifics in each case depend on the problem itself—end of life, pre-natal life, and so on. In both points, the contradictions (between different ethical claims and between different expert opinions) are an integral part of ethical decision-making. Finally, it is more prudent to come to a collective decision than to rely on the judgment of a single individual (Ricoeur 1990, 318/1992, 273/2001, 252–253/2007, 219–220).

This interest in biomedical issues is taken up again in the "exercises" in practical philosophy, indeed, in practical wisdom in *Reflections on the Just* (Ricoeur 2001, 40–46, 215–256 and 289–297/2007, 31–36, 187–197 and 249–256, but also Ricoeur 1991, 399–404).

A last point to observe is that the general way in which Ricoeur approaches dilemmas in biomedical ethics could be redeployed in any other field of human action, in any technological context.

7. CAPABILITIES

In the first two sections of this exposition, I have mostly spoken about technology in the way it is most often used in everyday parlance, namely as the evolving set of artifacts and procedures. A third point of this exploration of technology in Ricoeur needs to take us to a much older and wider use of the idea of "technology," namely as an aspect of human abilities to act.

Let us go to the heart of Ricoeur's understanding of action, as he elaborates on it in the first volume of his *Philosophy of the Will* (originally published in French in 1950). Every action has to be understood by means of a complement (the *pragma*) (Ricoeur 2009, 263–264/1979, 209–210). Different *pragmata* can be identified: to be distinguished, to be used, to be manipulated, each time directing the inceptive action to that which is to be done. In fact, there is a plethora of such forms of actional directedness:

> This eminently technical character of the human milieu and of human action depends, as we know, on the fact that humans *work* with tools to produce the "artificial" objects of their civilizational needs and even of their vital needs. This is why humans' action is typically "artificial": it is *techné*, mother of arts and techniques. (Ricoeur 2009, 266/1979, 211–212, translation modified)

This means that the whole of human actional directedness at fields of practice is shaped by the technical environment and the corresponding technical formation of the agent. The world as a world of action, and the agent itself, are inseparable from the technical milieu.

Once one has granted this basic point, a number of consequences can be derived from it. First, the body of the agent is a shaped "organ" through which action is realized. Through the body, the use of tools and instruments is mediated; indeed, the body and the tool may even become a single practical unit. But Ricoeur insists rather on the ambiguity in the chain of action: will-organ-tool-work—an ambiguity that consists in the fact that the chain lends itself to a phenomenology of the will (seen from the start) or a study in physics (seen from the end) (Ricoeur 2009, 268/1979, 213).

But to act, the body must act to maintain its ability to act—it must be fed. Feeding inserts the human organism into a natural environment whence technological processes also draw their energy (Ricoeur 2009, 119/1979, 87). The same body also has inborn instincts to defend or attack but is part of the wider acquisition of bodily techniques (*techniques du corps*). Thus emerge the automatisms of action (Ricoeur 2009, 380/1979, 302), which are clearly instantiated in some actions performed by a good sportsperson, artisan, or worker. Rather than a relaxation of the hold of the will to action, Ricoeur sees in the exercise of automatisms an achievement of the will, which, according to him, never ceases to survey such actions. A derivative of automatisms is machine-like actions, which, formerly acquired and mastered, are somewhat of a misfit with a new context of action (Ricoeur 2009, 38/1979, 304). What has been said thus far holds for bodily actions, but also for thought. Thought technologies cover knowledge learned by heart, forms of grammar and syntax, association of ideas, etc. (Ricoeur 2009, 368/1979, 292).

Ricoeur deepened this understanding of the technical dimension of action when he passed from a pure phenomenology of the will to an exploration of human fallibility in the second volume of his *Philosophy of the Will* (originally published in French in 1960). Nothing is as revealing of the conflictual nature of human fallibility than feelings ("sentiment," *thumos*) (Ricoeur 2009, 153/1986c, 106). Three kinds of feeling involved in inter-human relations have to be explored to clarify this point: the desire for possession, for domination, and for worth (Ricoeur 2009, 158/1986c, 111). Each of these leads to a positive quest, but risks being distorted into an obstinate, perverted pursuit. Each of these quests and pursuits is in turn realized in a different part of social life: economics, politics, and culture, by which relations between people are mediated by object systems (Ricoeur 2009, 160–161/1986c, 113). Strictly speaking, possession, domination, and worth all three apply to human technical ability and to the human relation to technology. But I single out domination for closer scrutiny, since Ricoeur touches directly on technology here (Ricoeur 2009, 164–169/1986c, 116–120).

The quest for domination is rooted in the human desire to exercise power, which is ingrained in action as such. Working is a first way by which people exercise their power over things, and labor relations is a form of the exercise of power over other human beings. Already this simple observation opens up a wide view on technological realities: from the energy of human labor and the natural resources on and with which people work, to the organization of labor and the pursuit of efficiency. The latter point also indicates the intersection of the economic sphere with that of labor, in the form of management.[21] Labor relations are maintainable as power relations, because they are instituted, which makes them dependent on another kind of power relation, namely that of politics (Ricoeur 2009, 166/1986c, 118). Ultimately, political power cannot be reduced to the availability of the powerful instruments of the state, but it does require the use of the mechanisms of the state.[22] What needs to be highlighted here is how Ricoeur integrates, in principle, the whole breadth of technological artifacts and systems in an encompassing view of human action, from the most intimate affects through mediated interaction to the largest political scale.

As one may expect, technology recedes as a theme in Ricoeur's hermeneutic writings from the 1960s onward. However, it would be an error to think that it is absent from that point on. One way to highlight its presence would be to insist on the Diltheyan aspect of Ricoeur's hermeneutics, with its emphasis on the (material) object as the starting point of hermeneutic work. Another would be to recall that Ricoeur understood his hermeneutics as a generalization of the age-old capability of interpretation, also called *techné hermeneutiké* (Ricoeur 1974b, 4). Once we realize this, it is worthwhile to point out numerous variations of "technique" spread throughout Ricoeur's work. There is a rhetorical technique (Ricoeur 1975b, 14–17, 41–44/1994, 9–12, 31–33/1985, 289–290/1988, 160–161) just as there is a narrative technique (throughout Ricoeur 1984/1988b). Somewhat different is the psychoanalytic technique (see Ricoeur 2009, 498/1979, 398, taken up again in 1974b/1974, 177–195 and in 1965b, 410–416 and 426–439/1970, 166–168, 390–396, 406–418) and the *ars memoriae* (Ricoeur 2000b, 73–82/2004b, 61–68). In the case of narration, its technical counterpart is the narrative, for writing in general, there is the text (which is utterings transmitted in autonomous form),[23] and the work of historians is supported by archives—in all three cases, the technologies that constitute the objects are indispensable for understanding Ricoeur's argument (however, one notes with frustration that in general he does not examine these objects *as* technological).

This does not mean that one could not develop this theme—Ricoeur himself gives an example when in "Architecture and Narrativity" (1998) he takes up central tenets of his *Time and Narrative 3* and argues for a strong parallel between architecture and narration. Construction does for space what narration does for time. It allows human beings, who, in whichever way, already inhabit the world, to reconfigure their relation to the environment (or come to a new grasp of their situatedness in the world) in interaction with constructed initiatives of special arrangement. Construction is evidently here meant in the widest sense of the entire formal and informal urbanization and also includes destruction. Each reconfiguration has a range of social, economic, and political dimensions.

In *Oneself as Another* (1990), technology does not feature as a major theme. However, it is worthwhile to consider the ways in which this hermeneutics of human capabilities—and hence most of Ricoeur's philosophy up to *The Course of Recognition* (2004)—constantly presupposed the technical constitution of action and the exercise of action in a technical milieu. Thus, each of the four key categories of action has a technical dimension. Speaking involves all communication technologies. Doing involves all the mediations of action by technical artifacts and the steering of sources of energy. Narrating is integrally linked with one's grasp of the unity of one's life, which, in turn, includes practices and professions, which are unthinkable without their relevant instruments and know-how. Finally, normative imputation (or responsibility) applies to all action, but one may consider the fact that all action is mediated by institutions, which, in turn, depend on technical support. The hermeneutics of human capabilities could be understood as the groundwork for further reflection on human-technology relations, but Ricoeur does not elaborate on it in this direction.[24]

If the technical constitution of human capabilities and action is basic to both Ricoeur's explicit discussions of technology and to his understanding of responsibility in the face of technologically generated dilemmas, one might well ask why I did not start the discussion with this point. Logically, this would have been a sound approach, but doing so may have created the impression of a greater coherence and systematicity than there really is in the theme of technology in Ricoeur's work.

8. RECEPTION IN PHILOSOPHY AND ETHICS OF TECHNOLOGY

I have now given an overview of the theme of technology in Ricoeur's work. This has never been a central theme in Ricoeur scholarship, but there has been a steady increase in interest in the potential Ricoeur's thought for thinking through technology-related problems.

The first explorations of Ricoeur and technology approached this relation from a very broad perspective. David Kaplan (2003, 2006) offers an overview of technology in Ricoeur's oeuvre, which he embeds in his own reconstruction of Ricoeur's "critical theory" to offer a view on the contributions that Ricoeur's work and the philosophy of technology can have for each other. From a more interdisciplinary perspective, Ernst Wolff (2006, 2007, 2012, 2013) examines the possibility of deploying Ricoeur's textual and narrative hermeneutics to clarify the relation between human agents and technology in general. Quite different is David Lewin's (2012) attempt to demonstrate that Ricoeur contributes to the question of the nexus of technology and being (in the sense of Heidegger's later philosophy).

More recent studies tend to focus on specific areas of technology or specific questions, but at the same time, they open up the range of uses made of, and deepens the engagement with, Ricoeur's work in technology-related domains.

A new line of enquiry is the deployment of Ricoeur's hermeneutics in the domain of information technology. Jos de Mul (2008) has already argued for an extension of Ricoeur's understanding of narrative identity in the face of the massive social changes in the use of new media. Much impetus to this new enquiry has subsequently been given by Alberto Romele (2013, 2015, 2018). He explores the interpretive value of Ricoeur's narrative hermeneutics for processes of identity formation in the use of social networking sites and investigates the implications of the use of such media for a Ricoeurian narrative hermeneutics of identity. Taking further Ricoeur's interest in traces and externalizations of meaning in general, Romele also examines the ethical and juridical implications of digital traceability and the right to be forgotten.[25] Elsewhere, Romele argues that emerging media technologies are characterized by new forms of imagination, drawing on Ricoeur's understanding of productive imagination through narrative mimesis (Romele 2018b).[26] Mark Coeckelberg and Wessel Reijers (2016a, 2016b) argue for the "narrative capacity" of technology in general, and specifically how technology mediates of language and people's self-understanding. Ricoeur's narrative hermeneutics plays a key role in this study. From the same philosophical background, the same authors also advanced an interpretation of block chain technologies, in particular of cryptocurrencies. Johan Fornäs (2016) also draws insights from *Time and Narrative* but develops insights on "third-time" technologies, which mediate the lived experience of time and objective time in historically and culturally specific ways.

Again, drawing on Ricoeur's philosophy or narrative, a couple of authors have worked on the heuristics and simulation of the possible future impact of technological inventions and interventions. This is the case in Murray and Wolff's study (2015) on the introduction of mobile communication for managed health care in low-income sections of society. Bruno Gransche (2015, 2017) gives a more developed theoretical view on a narrative hermeneutics of future technological changes, and explores a specific application in computer simulations.

A whole series of studies follow directly from Ricoeur's own writings on biomedical ethics. Some examples are the studies by Byung-Hye Kong (2005), Jérôme Porée (2012), Gaëlle Fiasse (2013), Theo Hettema (2014), Eoin Carney (2015), Corine Mouton Dorey (2016), and Philippe Svandra (2016). While they demonstrate the value of Ricoeur's understanding of ethics and responsibility in a sensitive and changing technological milieu, these essays rarely work on the technological dimension of the ethical dilemma and therefore mainly contribute more directly to Ricoeur's ethics.

A more surprising development is an ethics and hermeneutics of the environment by means of Ricoeurian resources. By looking at technologically advanced ways to destroy the natural environment, these studies are directly or indirectly relevant to the field of technology studies. The texts of David Utsler (2009); Nathan Bell (2014); Jean-Philippe Pierron (2016); and Thiago

Souza Silva, Nádia Sampaio, and Monique de Jesus Bezerra Dos Santos (2019) exemplify this direction in research.

Focusing more directly on the human-made and built environment, in particular, city life and dwelling, there are several studies that explore and elaborate on Ricoeur's texts on the city, such as the studies by Franco Riva (2013), Anna Borisenkova (2015), Jean-Philippe Pierron (2016b), and Jérôme Porée (2017). Another side of this line of exploration can be found in the studies of Breviglieri (2012) and of Wolff (2014), who rather engage with Ricoeur's understanding of capabilities in relation to dwelling.

A number of scholars have taken up one of Ricoeur's earliest concerns, namely work. Todd Mei (2006, 2019) elaborates on Ricoeur's coordination of the word and work in an examination of the poetic moment in each. This double poetics holds on to the productive and literal meaning of work, while seeking to open up new possibilities through the figurative dimension of such a poetics. In a follow-up study, he reflects on meaningful work as analogous to meaningful speech acts. Marcel Hénaff (2015) tries to show numerous ways to reengage with Ricoeur on the problem of work, in its relation to social justice, through a critical reading of Ricoeur's evolving stance on work. Likewise, Nicolas Smith (2017) attempts to consider the contemporary relevance of Ricoeur's project in "Work and the Word," but only through a critical engagement with the tension between the anthropological and phenomenological commitments at work in that text.

I know of three cases, to date, where the scholarly projects discussed earlier have been worked out to a full monograph. They subject Ricoeur to a critical reading, while giving a significant place to core aspects of his philosophy.[27] Alberto Romele (2020) offers a hermeneutic investigation of the digital technical milieu and the corresponding digital culture. He examines the changing relations between the virtual and the real and expands his earlier study on imaginative machines. He also uses his understanding of digital hermeneutics to oppose the anthropocentrism of traditional hermeneutics. Wessel Reijers and Mark Coeckelberg (2020), advance a narrative theory and ethics of technology. Their approach to artifacts and practices, research and innovation, develops insights from Ricoeur's narrative hermeneutics to a broader view on technological mediation. Furthermore, the virtue ethical orientation of their ethics also extends aspect of Ricoeur's later ethics. Ernst Wolff (2021) develops an interpretive theory of the technical dimension of all human action, that is, of embodied capabilities and the mediation of action by technical means. The technicity of action, he argues, is found in individual and collective action, over the whole range of practical contexts, from everyday practices and in ethico-political engagement. All three of these projects integrate a critical appropriation of parts of Ricoeur's work in a broader interdisciplinary framework and are concerned with the sociopolitical implications of these studies. The present volume adds to this expanding field of research.

9. CONCLUSION

In Ricoeur's work, there is an ordered polysemy of *techné*. He examines the anthropological fact of capabilities, the socio-historical complexities of changing human-technology relations and the ethical dilemmas of technological progress. Recent receptions of Ricoeur in the philosophy of technology have made selective use of Ricoeur's own contributions, while also developing other insights from his work for the philosophy of technology.[28]

NOTES

1. His name is absent from compendia such as that by Hubig et al. (2013).
2. Cf. Wolff (under review a), chapter 4, §4.1b. In the subsequent discussion, I draw from this longer discussion.
3. On this point, see already Dauenhauer (1998, 30–32).
4. Where two page numbers are separated by a backslash, the first refers to the French version, the second to the English translation.
5. This understanding of progress can probably be upheld even if one considers the limitations imposed by the destruction of the ecology and climate (a fact which Ricoeur does not seem to have in mind at that period). These limitations come into play when the abstraction of technology from politics of values is lifted.
6. Elsewhere, in an exposition on Eric Weil, Ricoeur claims: "The human of technology, of calculation, of mechanism is the first human who lives universally and understands himself/herself by this universal rationality" (Ricoeur 1991, 100, my translation).
7. Here, as elsewhere, Ricoeur's use of the term "civilization" is rather like that of Norbert Elias, in the sense that it refers to the most encompassing category of human forms of existence—in this sense, all people are civilized.
8. See in the earlier text, "L'aventure technique et son horizon planétaire" (1958) (discussed further): "Tool, sign and institution imply each other: for the tool, ultimately, proceeds from the power to transform things by speech, and according to a prescribed order. These three notions can be turned around as one wishes, each one refers to the others" (Ricoeur 2003, 68, my translation).
9. This view on the specificity of the political is a core component of Ricoeur's political philosophy. See, for example, Ricoeur "The Political Paradox," (Ricoeur 1964/1965, 247–270) and "Ethics and Politics" (Ricoeur 1986/1991b, 325–338).
10. Cf. Wolff (under review a), chapter 5.
11. This echoes Ricoeur's earlier insistence on laborers' say in the management of industry and the economy, in "Work and the Word" (Ricoeur 1964/1965,

197–222—this chapter is discussed in section 1.3 here), and of the citizenry's limitation of state power in "The Political Paradox" (Ricoeur 1964/1965, 247–270).

12. Cf. Ricoeur (1991), and Wolff (under review b).

13. "Le 'sacré' de l'homme—son essentiel, son nécessaire, son primordial," in Ricoeur (1958, 68), my translation.

14. Elsewhere Ricoeur develops his view on alienation and objectification, cf. Ricoeur (1968, 1975).

15. I will not discuss the theological implications discerned by Ricoeur, nor the way he borrows (extensively) from Cox (2013, chapter 2).

16. See more recently Rosa (2005).

17. This point is taken up, albeit only briefly, as late as Ricoeur (1995, 36/2000, 6–7/2004, 318/2005, 204)—both times with reference to Ferry (1991).

18. I have studied them in more detail elsewhere—see Wolff (2021), Introduction, section 3.

19. For a discussion of the conflictual constitution of responsibility in Ricoeur, see Wolff (2021), chapter 8, section 1 and Contreras Tasso (2020).

20. The discussion derives from Lenoir's systematization and spans Ricoeur (1991c, 272–278).

21. I have already touched on this point earlier in relation to Ricoeur's "Travail et parole" / "Work and the Word."

22. Just as Ricoeur argued in "The Political Paradox," see section 2.1 earlier in the chapter.

23. Cf. Wolff (2021), chapter 1. Ricoeur does not seem to be interested in texts as technical artifacts although his understanding of the objectivation of meaning into a work and of the text as the result of *techné* and *praxis* (Ricoeur 1986, 120–121/1991b, 80–81) calls for an elaboration in this direction.

24. This claim is developed in Wolff (2021), chapters 3 and 4.

25. See also Romele and Severo (2016), which is applied to the Ricoeur reception.

26. Since Romele's essay publications were followed by a recent monograph (see further), I will not cite all of them here.

27. I report also the doctoral dissertation of Carney (2018).

28. This chapter is an extended version of my article "Ricoeur and the philosophy of technology," forthcoming in *Studia z Historii Filozofii*.

REFERENCES

Beatriz Contreras Tasso. 2020. "The Vicissitudes of Justice in the Ethical Triad of Desire, Norm and Prudential Judgment, According to Paul Ricoeur." In *The Ambiguity of Justice: New Perspectives on Paul Ricoeur's Approach to Justice*, edited by Geoffrey Dierckxsens. Leiden: Brill, 13–34.

Bell, Nathan. 2014. "Environmental Hermeneutics with and for Others. Ricoeur's Ethics and the Ecological Self." In *Interpreting Nature: The Emerging Field of Environmental Hermeneutics*, edited by Forrest Clingerman, Brian Treanor, Martin Drenthen, and David Utsler. New York: Fordham, 141–159.

Borisenkova, Anna. 2015. "Reading the City: From the Inhabitant to the Flâneur." In *Poetics, Praxis, and Critique: Paul Ricoeur in the Age of Hermeneutical Reason*, edited by Roger Savage. Lanham, MD: Lexington Publishers, 86–98.

Breviglieri, Marc. 2012. "L'espace habité que réclame l'assurance intime de pouvoir. Un essai d'approfondissement sociologique de l'anthropologie capacitaire de Paul Ricoeur." *Études Ricoeuriennes / Ricoeur Studies* 3, no. 1: 34–52.

Carney, Eoin. 2015. "Depending on Practice: Paul Ricoeur and the Ethics of Care." *Les ateliers de l'éthique/The Ethics Forum* 10, no. 3: 29–48.

Carney, Eoin. 2018. *Technologies in Practice: Paul Ricoeur and the Hermeneutics of Technique*. Doctoral dissertation in Philosophy, University of Dundee.

Coeckelberg, Mark, and Wessel Reijers. "Narrative Technologies: A Philosophical Investigation of the Narrative Capacities of Technologies by Using Ricoeur's Narrative Theory." *Human Studies* 39, no. 3: 325–346.

Cox, Harvey. [1965] 2013. *The Secular City: Secularization and Urbanization in Theological Perspective*. Princeton: Princeton University Press.

Dauenhauer, Bernard. 1998. *Paul Ricoeur: The Promise and Risk of Politics*. Lanham, MD: Rowman and Littlefield.

de Mul, Jos. 2008. "Von der narrativen zur hypermedialen Identität. Dilthey und Ricoeur gelesen im hypermedialen Zeitalter." In *Dilthey und die hermeneutische Wende in der Philosophie. Wirkungsgeschichtliche Aspekte seines Werkes*, edited by Frithjof Rodi and Gudrun Bertam-Kuhne. Göttingen: Vandenhoeck & Ruprecht, 313–331.

Dorey, Corine Mouton. 2016. "Rethinking the Ethical Approach to Health Information Management Through Narration: Pertinence of Ricoeur's 'Little Ethics.'" *Medicine, Healthcare & Philosophy* 19, no. 4: 531–543.

Ferry, Jean-Marc. 1991. *Les puissances de l'expérience. Essai sur l'identité contemporaine*. Paris: Cerf.

Fiasse, Gaëlle. 2013. "Ricoeur's Medical Ethics: The Encounter Between the Physician and the Patient." In *Reconceiving Medical Ethics*, edited by Christopher Cowley. New York: Continuum Press, 30–42.

Fornäs, Johan. 2016. "The Mediatization of Third-Time Tools: Culturalizing and Historicizing Temporality." *International Journal of Communication* 10: 5213–5232.

Gransche, Bruno. 2015. "Narrative Hermeneutik." In *Vorausschauendes Denken. Philosophie und Zukunftsforschung jenseits von Statistik und Kalkül*. Bielefeld: Transcript, 241–311.

Gransche, Bruno. 2017. "The Art of Staging Simulations: Mise-en-scène, Social Impact, and Simulation Literacy." In *The Science and Art of Simulation I*, edited by Michael Resch, Andreas Kaminski, and Petra Gehring. Cham: Springer.

Hénaff, Marcel. 2015. "Labor, Social Justice, and Recognition: Around Paul Ricoeur." In *Poetics, Praxis, and Critique: Paul Ricoeur in the Age of Hermeneutical Reason*, edited by Roger Savage. Lanham, MD: Lexington Publishers, 21–33.

Hettema, Theo. 2014. "Autonomy and Its Vulnerability: Ricoeur's View on Justice as a Contribution to Care Ethics." *Medical Health Care Philosophy* 17, no. 4: 493–498.

Hubig, Christoph, Alois Huning, and Günther Ropohl, eds. 2013. *Nachdenken über Technik. Die Klassiker der Technikphilosophie und neuere Entwicklungen*, 3rd edition. Berlin: Sigma.
Jonas, Hans. [1979] 1984. *The Imperative of Responsibility: In Search of an Ethics for the Technological Age*. Chicago: University of Chicago Press.
Kaplan, David. 2003. *Ricoeur's Critical Theory*. New York: State University of New York Press, 164–173.
Kaplan, David. 2006. "Paul Ricoeur and the Philosophy of Technology." *Journal of French Philosophy* 16, no. 1/2: 42–56.
Kong, Byung-Hyu. 2005. "A Philosophical Inquiry into Caring in Nursing Based on Ricoeur's Narrative Ethics." *Taehan Kanho Hakhoe Chi* 35, no. 7: 1333–1342.
Lewin, David. 2012. "Ricoeur and the Capability of Modern Technology." In *From Ricoeur to Action: The Socio-Political Significance of Ricoeur's Thinking*, edited by Todd S. Mei and David Lewin. London and New York: Continuum, 54–71.
Mei, Todd. 2006. "Form and Figure: Paul Ricoeur and the Rehabilitation of Human Work." *Journal of French Philosophy* 16, no. 1 and 2: 57–70.
Mei, Todd. 2019. "The Poetics of Meaningful Work: An Analogy to Speech Acts." *Philosophy & Social Criticism* 45, no. 1: 50–70.
Murray, Montagu, and Ernst Wolff. 2015. "A Hermeneutic Framework for Responsible Technical Interventions in Low-Income Households: Mobile Phones for Improved Managed Health Care as Test Case." *Journal for Transdisciplinary Research in Southern Africa* 11, no. 3: 171–185.
Pierron, Jean-Philippe. 2016a. "Ecologie et herméneutique. Le franciscanisme de Paul Ricoeur." In *Ricoeur: Philosopher à son école*. Paris: Vrin, 187–208.
Pierron, Jean-Philippe. 2016b. "Entre habiter et politiser: urbaniser." In *Ricoeur: Philosopher à son école*. Paris: Vrin, 171–186.
Porée, Jérôme. 2012. "Expliquer, comprendre et vivre la maladie." *Revue d'Histoire et de Philosophie Religieuses* 92, no. 1: 5–20.
Porée, Jérôme. 2017. "Le philosophe, l'architecte et la cite." In *L'existence vive. Douze études sur la philosophie de Paul Ricoeur*. Strasbourg: Presses Universitaires De Strasbourg, 147–163.
Reijers, Wessel, and Mark Coeckelbergh. 2016. "The Blockchain as a Narrative Technology: Investigating the Social Ontology and Normative Configurations of Cryptocurrency." *Philosophy & Technology* 31, no. 1: 103–130.
Reijers, Wessel, and Mark Coeckelberg. 2020. *Narrative and Technology Ethics*. Cham: Palgrave.
Ricoeur, Paul. 1950. *Philosophie de la volonté 1. Le volontaire et l'involontaire*. Paris: Point.
Ricoeur, Paul. 1958. "L'aventure technique et son horizon planétaire." Republished in *Autres Temps. Cahiers d'éthique sociale et politique* 76–77 (2003): 67–78.
Ricoeur, Paul. 1960. *L'homme faillible*. In *Philosophie de la volonté 2. Finitude et culpabilité*. Paris: Edition Points.
Ricoeur, Paul. 1964. *Histoire et verité*. Paris: Seuil.
Ricoeur, Paul. 1965a. *History and Truth*. Translated by Charles A. Kelbley. Evanston: Northwestern University Press.

Ricoeur, Paul. 1965b. *De l'interpétation. Essai sur Freud.* Paris: Editions du Seuil.
Ricoeur, Paul. 1967. "Urbanisation et sécularisation." Republished in *Autres Temps. Cahiers d'éthique sociale et politique* 76–77 (2003): 113–126.
Ricoeur, Paul. 1968. "Aliénation." In *Encyclopaedia Universalis I*. Paris: Encyclopaedia Universalis France, 660–664.
Ricoeur, Paul. 1969. *The Symbolism of Evil*. Translated by Emerson Buchanan. Boston: Beacon.
Ricoeur, Paul. 1970. *Freud and Philosophy. An Essay on Interpretation*. New Haven: Yale University Press.
Ricoeur, Paul. 1974a. *Political and Social Essays*. Edited by David Steward and Joseph Bien. Athens: Ohio University Press.
Ricoeur, Paul. 1974b. *The Conflict of Interpretations: Essays in Hermeneutics*. Translated by Don Ihde et al. Evanston: Northwestern University Press.
Ricoeur, Paul. 1975a. "Objectivation et aliénation dans l'expérience historique." *Archivio di filosofia* 45, no. 2–3: 27–38.
Ricoeur, Paul. 1975b. *La métaphore vive*. Paris: Seuil.
Ricoeur, Paul. 1979. *Freedom and Nature: The Voluntary and the Involuntary*. Translated by Erazim Kohák. Evanston: Northwestern University Press.
Ricoeur, Paul. 1984. *Temps et récit 2. La configuration du temps dans le récit de fiction*. Paris: Seuil.
Ricoeur, Paul. 1985. *Temps et récit 3. Le temps raconté*. Paris: Seuil.
Ricoeur, Paul. 1986a. *Du texte à l'action. Essais d'herméneutique*. Paris: Seuil.
Ricoeur, Paul. 1986b. *Lectures on Ideology and Utopia*. Edited by Georg H. Taylor. New York and London: Columbia University Press.
Ricoeur, Paul. 1986c. *Fallible Man*. Translated by Charles A. Kelbley. New York: Fordham University Press.
Ricoeur, Paul. 1988a. *Time and Narrative. Volume 3*. Translated by Kathleen Blamey and David Pellauer. Chicago and London: University of Chicago Press.
Ricoeur, Paul. 1988b. *Time and Narrative. Volume 2*. Translated by Kathleen Blamey and David Pellauer. Chicago and London: University of Chicago Press.
Ricoeur, Paul. 1990. *Soi-même comme un autre*. Paris: Seuil.
Ricoeur, Paul. 1991a. *Lectures 1. Autour du politique*. Paris: Seuil.
Ricoeur, Paul. 1991b. *From Text to Action. Essays in Hermeneutics II*. Translated by Kathleen Blamey and John Thompson. London: Athlone Press.
Ricoeur, Paul. 1992a. *Oneself as Another*. Translated by Kathleen Blamey. Chicago: University of Chicago Press.
Ricoeur, Paul. 1992b. *Lectures 2: La contrée des philosophes*. Paris: Seuil.
Ricoeur, Paul. 1994. *The Rule of Metaphor: Multi-disciplinary Studies of the Creation of Meaning in Language*. London: Routledge.
Ricoeur, Paul. 1995. *Le juste 1*. Paris: Esprit.
Ricoeur, Paul. 2000a. *The Just*. Translated by David Pellauer. Chicago: University of Chicago Press.
Ricoeur, Paul. 2000b. *La mémoire, l'histoire, l'oubli*. Paris: Seuil.
Ricoeur, Paul. 2001. *Le juste 2*. Paris: Esprit.
Ricoeur, Paul. 2004a. *Parcours de la reconnaissance*. Paris: Gallimard.

Ricoeur, Paul. 2004b. *Memory, History, Forgetting*. Translated by Kathleen Blamey and David Pellauer. Chicago: University of Chicago Press.

Ricoeur, Paul. 2005. *The Course of Recognition*. Translated by David Pellauer. Cambridge, MA: Harvard University Press.

Ricoeur, Paul. 2007. *Reflections on the Just*. Translated by David Pellauer. Chicago: University of Chicago Press.

Ricoeur, Paul. 2016. "Architecture and Narrativity." *Études Ricoeuriennes/Ricoeur Studies* 7, no. 2: 31–42.

Riva, Franco. 2013. "L'angoscia dell'abitare. Ricoeur, Lyotard e la città postmoderna." In *Leggere la città. Quattro testi di Paul Ricoeur*. Roma: Castelvecchi, 7–49.

Romele, Alberto. 2013. "Narrative Identity and Social Networking Sites." *Études Ricoeuriennes/Ricoeur Studies* 4, no. 2: 108–122.

Romele, Alberto. 2015. "Digital Traceability and the Right to be Forgotten: Ricoeurian Perspectives." *Tropos: Journal of Hermeneutics and Philosophical Criticism* 8, no. 2: 105–118.

Romele, Alberto. 2018a. "Imaginative Machines." *Techné: Research in Philosophy and Technology* 22, no. 1: 98–125.

Romele, Alberto. 2018b. "From Registration to Emagination." In *Towards a Philosophy of Digital Media*, edited by Alberto Romele and Enrico Terrone. London: Palgrave Macmillan, 257–273.

Romele, Alberto. 2020. *Digital Hermeneutics: Philosophical Investigations in New Media and Technologies*. New York: Routledge.

Romele, Alberto, and Marta Severo. 2016. "From Philosopher to Network: Using Digital Traces for Understanding Paul Ricoeur's Legacy." *Azimuth* 4, no. 7: 113–128.

Rosa, Hartmut. 2005. *Beschleunigung. Die Veränderung der Zeitstrukturen in der Moderne*. Frankfurt-am-Main: Suhrkamp.

Silva, Thiago Souza, Nádia Sampaio, and Monique de Jesus Bezerra Dos Santos. 2019. "Environmental Ethics and Theory of Responsibility in Paul Ricoeur and Hans Jonas: An Approach on Sustainability in Postmodernity." *International Journal of Development Research* 9, no. 12: 32396–32400.

Smith, Nicholas H. 2017. "Between Philosophical Anthropology and Phenomenology: On Paul Ricoeur's Philosophy of Work." *Revue internationale de philosophie* 278, no. 4: 513–534.

Svandra, Philippe. 2016. "Rethinking Ethics with Paul Ricoeur. Nursing: Between Responsibility, Care and Justice." *Recherche en soins infirmiers* 124: 19–27.

Utsler, David. 2009. "Paul Ricoeur's Hermeneutics as a Model for Environmental Philosophy." *Philosophy Today* 53, no. 2: 173–178.

Wolff, Ernst. 2006. "Transmettre et interpréter." *Médium* 6: 30–47.

Wolff, Ernst. 2007. "Mediologie en hermeneutiek." *Tydskrif vir Geesteswetenskappe* 47, no. 1: 81–94.

Wolff, Ernst. 2012. "Habitus—Means—Worldliness. Technics and the Formation of 'Civilisations.'" In *Shaping a Humane World: Civilizations, Axial Times, Modernities, Humanisms*, edited by Oliver Kozlarek, Jörn Rüsen, and Ernst Wolff. Bielefeld: Transcript, 25–53.

Wolff, Ernst. 2013. "Compétences et moyens de l'homme capable à la lumière de l'incapacité." *Études Ricoeuriennes/Ricoeur Studies* 4, no. 2: 50–63.

Wolff, Ernst. 2014. "Hermeneutics and the Capabilities Approach: A Thick Heuristic Tool for a Thin Normative Standard of Well-Being." *South African Journal of Philosophy* 33, no. 4: 487–500.

Wolff, Ernst. 2021. *Between Daily Routine and Violent Protest: Interpreting the Technicity of Action*. Berlin: De Gruyter.

Wolff, Ernst. under review a. *Lire Ricoeur depuis la périphérie. Décolonisation, modernité, herméneutique*.

Wolff, Ernst. under review b. *Ricoeur Reading Weber: Responsibility, Ideology, Explanation*.

Chapter 2

Postphenomenology and the Hermeneutic Ambiguity of Technology

Eoin Carney

Postphenomenology, a highly influential strand within the philosophy of technology, is often sold as offering a more pragmatic, evidence-based account of technologies than earlier strands within the field. One of the major representatives of the earlier, more speculative approaches is Martin Heidegger. Paul Ricoeur is also, to a certain extent, representative of this generation of philosophies of technology, even though technology plays a less central role in his philosophy. For example, David Kaplan writes that Ricoeur, when he occasionally discussed technology, "generally agreed with Heidegger, Marcuse, and Habermas, each of whom contrasts the dehumanizing characteristics or technology and technological reasoning with more human forms of experience and action" (Kaplan 2006, 42). As with several other discourses that emerged after the so-called empirical turn (Achterhuis 2001), Postphenomenology advocates an entirely different way of thinking about technologies. The question of Technology (with a capital "T") is set aside, in favor of localized investigations into concrete technologies and their associated practices and meanings. Furthermore, the question of Technology, as such, cannot appear as a metaphysical or speculative question at all, since technologies are simply inescapable *facts* of human existence, a point summarized by Ihde's quip that "Heideggerians love their laptops" (Scharff 2006, 132). Given these fundamental conflicts, between a postphenomenological outlook and earlier, hermeneutically informed reflections on technology, it is relevant to ask whether Postphenomenology puts an "end" to hermeneutic questions about technology, and by extension, a Ricoeurean, hermeneutic approach to technology.

In this chapter, I will examine both perspectives in more detail. The first, hermeneutic perspective, will be read as representative of a *hermeneutics of suspicion*. That is, as an attempt to unmask the modes of presentation associated with modern technologies. The key point here will be that hermeneutic approaches such as those of Ricoeur and Heidegger aim to situate hermeneutic reflection around the plurality and ambiguity of human action and understanding. It is this plurality that they see as threatened by Technology. Postphenomenology, as I will read it, is a hermeneutic response to these suspicions. It re-asserts the ambiguity and plurality of technologies and their mediating processes. In Ricoeur's terms, it asserts the *symbolic* nature of technologies themselves, that is, technologies become a locus for reflection and self-understanding. In this regard, and against some of its own claims to be neutral and "non-essentialist," I will read Postphenomenology as a corollary of a hermeneutic of suspicion, that is, as a *hermeneutics of trust.*

In this sense, it is not simply that earlier approaches to technology were not pragmatic enough, rather, they were not *hermeneutic* enough. That is, they failed to see that technology as a general phenomenon invites both attitudes of trust and suspicion. Like other symbolic phenomena, the meaning of technology cannot be encompassed by a single discourse or methodology, but only through a persistent conflict of interpretations. Similarly, postphenomenological approaches ought to acknowledge the role "pragmatic" and phenomenological description plays in constituting technology in a particular light. While Postphenomenology does succeed in highlighting the hermeneutic character of individual technologies, it arguably misses a chance to articulate the hermeneutic character of technology-as-a-whole and its own relation to it.

With these points in mind, the first part of this chapter will examine the hermeneutic critique of technology. In particular, I will focus on Ricoeur's understanding of the symbol as ambiguous and the way in which modern technology is sometimes set in opposition to this ambiguity. The second part will touch on some key aspects of Postphenomenology, which, I argue, respond to these earlier characterizations of technology by demonstrating that technological designs and artifacts are, in fact, pluralistic and multivocal. Finally, the third part will suggest that reading Postphenomenology as a hermeneutic response to previous perspectives on technology, as opposed to reading it as a simple alternative, allows for us to conceive technology itself as a symbolic phenomenon whose meaning is uncertain and, therefore, invites perpetual hermeneutic reflection.

1. HERMENEUTIC SUSPICION

It is often pointed out by postphenomenologists that earlier attempts to define and understand the nature of technology tended toward hyperbole and alarmism. This was due, it is claimed, to a fault in methodology. Earlier approaches were more speculative, while later approaches aimed to ground their theories in empirical observations and analysis. Here, however, I aim to challenge this narrative and suggest that, instead of reading this transition along the lines of an empirical turn, we should read it instead as a *hermeneutic* turn, that is, as a move from interpretative strategies that are guided by *suspicion* toward a set of strategies guided by *trust* or "faith." Reading the transition in this sense, as a conflict between interpretive strategies, allows for an appreciation of the continuity and dialectical unity of both approaches.

This hermeneutic unity, between suspicion and trust, is articulated in Ricoeur's early work on the hermeneutics of the symbol (Ricoeur 1967, 1970, 1974). Unfortunately, Ricoeur himself excludes technology from this kind of discourse, writing:

> Essentially, it was in adopting science and technology, not just as a form of knowledge, but as a means of dominating nature, that we left behind the logic of correspondences [logic of the sacred]. Because of this we no longer participate in a cosmos, but we now have a universe as an object of thought and as matter to be exploited. This mutation is not something we could control; if the sacred constitutes a world of being, this mutation reveals a history of being. And we could follow Heidegger in his writings on technology in discussing that history of being. (Ricoeur 1995, 61–62)

In contrast to a "means"-oriented development of knowledge, which Ricoeur here associates with "science and technology," hermeneutic approaches adopt a different strategy. Interpretation begins from the symbol as opposed to the sign. Whereas signs, for Ricoeur, are abstract, symbols are concrete; they are modes of language or expression that are reflective of their contexts. They are expressive of *lived* experience. In this sense, they call for interpretation because their literal or sign meaning cannot be separated from their figurative, contextual expressions. Classic hermeneutic examples of these kinds of symbolic phenomena are biblical or legal forms of language. In the case of biblical language, what is being expressed is not simply a literal meaning but also a deeper, anthropological relation between the human and the sacred.

For example, in his analysis of the "masters of suspicion"—Marx, Nietzsche and Freud—Ricoeur stresses the difference between direct forms of knowledge/mastery (e.g., as in science and technology), and the interpretive art of *unmasking,* which defines the enterprises of the masters of suspicion. To suggest that a subject matter is necessarily opaque or obscured, as in the case of material exchanges in a capitalist society (Marx), the unconscious (Freud), and the will to power (Nietzsche), means that understanding it will always involve an *interpretive detour.* The masters of suspicion were not trying to directly explain their subject matters, but were trying to deconstruct them, by revealing them as phenomena that have two ambiguous levels: a literal, direct meaning *and* a figurative or indirect form of presentation. These two sides cannot be uncoupled, and therefore can only be grasped through a hermeneutic strategy that recognizes the fundamental relation between the literal and the figurative. This, of course, has implications for modes of knowledge, which rely on more direct forms of data, for example, empirically grounded approaches, since the claim by the three masters of suspicion is that direct knowledge is always accompanied by an obscured or hidden context, whether it be class struggle, unconscious sexual drives, or the will to power.

With regard to technology, a hermeneutic of suspicion would claim that, as in the Ricoeur quote earlier, whereas technologies are presented in a particular light (as modes of knowledge, as neutral and unmediated ways of encoding/representing the world, etc.), their deeper meaning is hidden or obscured by this presentation. Their deeper or contextual/living meaning would be, as Ricoeur suggests, the relation between technology and the history of being and, in particular, a history of the "domination of nature" by technical means.

Ricoeur's advocates a hermeneutics of suspicion, however, with an important caveat: it needs to recognize the meaningfulness of its *own* enterprise. A deconstructive analysis is lacking if it remains solely within the critical sphere. Instead, it must be prepared, also, to take up the task of *appropriation*. Appropriation occurs through what Ricoeur sometimes calls a hermeneutics of trust or faith. That is, a hermeneutic strategy of listening again to what has been unmasked or critically evaluated, in order to find new meanings or new configurations. For example, in the case of the essential critical analysis of religions that we get through Marx, Nietzsche, and Freud, Ricoeur argues that we must then embrace a "second naivety" and return again to the question of the sacred. In this sense, a crucial feature of the symbolic for Ricoeur is also its inexhaustibility. It will continually invite both suspicion and trust; understanding the nature of symbolic phenomena is more akin to an ongoing dialogue, where the subject matter is continually evolving through the activities of questioning and listening. Although Ricoeur himself seems to side with Heidegger and exclude the sphere of science and technology from the question of the sacred or the symbol, I will suggest later that this framework ought also to be applied to our understanding of technologies.

Hermeneutic appropriation, the guiding thread of a hermeneutics of trust, is similar to what Gadamer calls hermeneutic *application* (Gadamer 1975, 306). As in the case of Ricoeur's contrasting of the hermeneutics of suspicion with the "direct" methods of science, Gadamer contrasts the direct forms of application found in the sciences with the hermeneutic art of applying/appropriating in a meaningful, context-sensitive way. As with a hermeneutically suspicious analysis, hermeneutic application does not operate along the lines of a subject-object dichotomy. Instead, interpreter and subject matter are mutually implicated in application. In the sciences, by contrast, the problem of application:

> already presupposes that science as such possesses its self-certain and autonomous existence prior to all application and free from all reference to possible application; but thanks to just this freedom from purpose, its knowledge is available for any application whatsoever, precisely because science has no competence to preside over its application. (Gadamer 1972, 8)

It is on these same grounds that the hermeneutic critique of technology is founded. In the case of technological applications, their "universality" eschews the process of hermeneutic appropriation/application—a process that is dialogical. For example, David Lewin (2012), drawing on Ricoeur's work, has analyzed what he terms, the "technical interface" along similar lines. The logic of the interface is simple: it works to reduce complicated, difficult to negotiate technical process to a series of easily navigable buttons, icons, functions, and actions. It is designed to reduce misunderstandings, unexpected uses, delays, detours, and so on. In other words, it works to reduce hermeneutic activity. Even as technological processes become more complex, and the capability of modern technology grows and impacts many aspects of everyday life, actual user engagement with technologies remains confined within the dictates of the interface.

Furthermore, this "technical language" remains relatively common and consistent across all technological phenomena that are mediated by interfaces. The result is a "utopia of functionality"; underlying technological processes are explicated and translated into a user-friendly form of presentation, progressively eradicating any uncertainty or need for interpretation (in the sense of meaningful application):

> This tendency towards circumscribed functionality gives us the impression that device does not exist with an interpretive context, but within a decontextualized utopia of functionality. As attractive as this utopia seems to be—it is, after all, exactly what the user wants the device to achieve—it negates the hermeneutic dimension of existence by its presentation of unmediated function. (Lewin 2012, 65)

A key contrast is established here; between, in Ricoeurean terms, the "univocity" of the interface and the "equivocity" of the hermeneutic experiences of uncertainty and ambiguity. Lewin reinforces this point by echoing Ricoeur's famous dictum that "the symbol gives rise to thought" (Ricoeur 1967, 247–257), when he writes: "the device only ever does what we ask of it. In so doing it cannot give rise to genuine thought" (Lewin 2012, 66).

Furthermore, ethical considerations accompany the hermeneutic and existential implications of the tendencies of the interface. By concealing uncertainty, the interface also conceals fragility and alterity, and threatens the human capacity to recognize otherness and for *phronēsis*:

> The whole point of the interface is to stabilize what discloses itself. We might say that it fixes and closes, and thereby opposes disclosure. By its attempt to conceal complex (that is, fragile or insecure) interaction and deliberation, the interface denigrates and excludes the human faculty of practical reason, named by Aristotle as *phronēsis*. (Lewin 2012, 65)

Here, it is claimed that whereas modern hermeneutics places emphasis on the value of ambiguity, incomplete dialogue, and the necessity of meaningful self-transformation in the process of understanding, the tendencies underlying the development of modern technologies seek to circumvent these experiences.

However, I am reading this type of approach as representative of a hermeneutics of suspicion, that is, as not necessarily making essentialist claims about technologies as such, but attempting to unmask or disenchant a certain way that technologies are presented (e.g., through the interface) and our contemporary modes of comportment toward things. In short, whereas Technology tends to present the process of application as straight-forward, a hermeneutic of suspicion is an interpretive strategy that gestures instead toward the uncertainties and pluralities inherent in every application, whether we acknowledge them or not. In this regard, Lewin's approach is instructive; technological process *are* complex and manifold, it is simply that our relation to these processes is mediated in a deficient way. This, as I see it, is also the point at which postphenomenology arises, as an attempt to build a new conceptual interface or way of interacting with technologies, one which acknowledges, and harnesses, the complexities and pluralities inherent in every mediation.

2. POSTPHENOMENOLOGY AND TECHNOLOGICAL MEDIATION

A key starting point for postphenomenological approaches is a rejection of earlier claims about the alienating effects of contemporary technologies. Technologies do not stand against the human, as a separate object, rather, they are a medium through which our embodied and hermeneutic existence is extended. Questions about whether this or that technology ought or ought not to be adopted, for example, mistakenly presuppose a vision of human existence, which is independent of technological mediation. Instead, technologies are taken as a factual element of human experience; any epistemological, practical or moral questions must arise alongside, and with technologies. However, although technological mediation is taken as a necessary feature of human experience, the nature of this mediation is open and dynamic. Technological designs are multifaceted; they often perform in unanticipated ways, depending on concrete contexts. Questions of hermeneutic appropriation or application do not disappear in a society shaped by technological forces.

In *Technology and the Lifeworld* (1990), Don Ihde outlines an influential framework for conceptualizing technological mediation. The framework is centered around three terms; the human, technology, and the world, and there are multiple variations on how these terms can relate to one another. The first, phenomenological, variant is labeled "embodiment relations":

(Human-technology)-world

In these cases, a technology has been calibrated to match, seamlessly, with human embodiment. The classic example here is of eyeglasses. When I am wearing glasses, they act as an extension of my embodied perception of the world. A separation remains, which I may occasionally notice when I need to wipe the lenses, but for the most part, these technologies "work" by remaining invisible.

The second variant, which is the important one for the purposes of this chapter, is labeled "hermeneutic relations:"

Human-(technology-world)

Here, technologies are active mediators; they translate or decode aspects of reality that may otherwise remain imperceptible or inaccessible. In these cases, the "extension" provided by a technology moves beyond embodied perception. Now, my linguistic or hermeneutic capacities are also called upon. Indeed, the most originary examples of hermeneutic technics for Ihde are systems of writing. The presence produced by these kinds of technologies differs from embodied technics, it is a hermeneutic presence; the world or reality rendered through hermeneutic relations is always an in-between world, distanced from both the context of what is being represented and the context of the "reader."

The final variant is labeled "Alterity relations,"

Human-technology-(world)

This variant stands at the extremis of hermeneutic relations, referring to cases where what is being represented or mediated is no longer an "object." For example, in cases in which a technology itself seems to take on an "autonomous" status. This can range from contemporary cases of machine learning or neural networks to more subtle instances, like the experience of playing with a spinning top, "Note that once the top has been set to spinning, what was imparted through and embodiment relation now exceeds it. What makes it fascinating is this property of quasi-animation, the life of its own" (Ibid, 100).

This simple framework already indicates a fundamental shift in thinking about technologies. Technologies, especially as they become more advanced, can incorporate multiple combinations of these relations at once, meaning that concrete encounters with technologies produce a range of hermeneutic experiences, from immediate, integrated modes of presencing (embodied relations), to interpretive, encoded modes of interaction (hermeneutic relations), to encounters with otherness and uncertainty (alterity relations). In other words, in place of the "metaphysical closure" that is suggested by earlier, speculative accounts of technology, there are instead multiple modes of mediation that work through a play of presence and absence, concealing and revealing.

In particular, the aforementioned outline of hermeneutic relations with technology fits well with Ricoeur's notion of *distanciation*. The act of translating a living ("world") phenomena into a technological form is akin to the act of transcribing speech as writing. As Ihde writes, "There is a partial opacity between the machine and the world and thus the machine is something like a text. I may read an author, but the author is only indirectly present in the text" (Ihde 1974, 275–276). In the case of scientific instruments, for example, a device may relay information back and forth between the living, material world and the operator of the device, just as a text relays information across historical and cultural distances. There is always a hermeneutic cost to the achievement of these kinds of transferences; the difference between one system (writing, the scientific instrument) and the other (living speech, the material being studied), means that certain features are left untranslatable. Yet, at the same time, this difference also allows for the emergence of a new perspective, which can, in turn, project further avenues or worlds of its own. Speech or cultural milieus, for example, are governed by a set of customs and patterns, which produce meaning. Writing attempts to provide formal parameters for recording these patterns and preserving them over time. However, writing is not *solely* a recording instrument, its systems and rules hold their own intricacies and wonders, leading to linguistic innovations at the level of writing itself, which in turn feed back into living speech and culture. Furthermore, the act of transcribing also provides an opportunity to consider something from a new, distanciated perspective. For example, the act of recording daily activities or events in a journal does not only serve the purpose of leaving a record for posterity, it also provokes self-reflection and self-examination. Immediate experiences and the distanciated reflections on these experiences, then, co-exist in a recursive relationship, each feeds-back onto the other resulting in a productive hermeneutic spiral.

Similarly, when it comes to technologies, it may seem at first that they serve a merely utilitarian purpose—executing a task more efficiently or rapidly, for example—but, as with writing, their "ends" lie beyond a mere encoding of a living practice in technological form. Technological designs can produce unexpected outcomes in practice and their development is often the product of an interaction between technical innovation and practical engagement. That is, just as in the case of writing, technological design is not one-sided or solely instrumental. There are countless historical and contemporary examples of this point in the phenomenological literature that are grouped, broadly, by Ihde using the phenomenological category of *intentionality*. A device may be designed initially to fulfill a certain purpose, and the relation between the design and this intended end is one of intentionality. For example, a microwave fulfills the purpose of heating food in a fast and more efficient way. However, due to, what Ihde terms, the *multistability* of designs, they can also give rise to unintended, unexpected consequences. Multistability refers to the way that a design can take on several different stabilities (intentional relations), depending on its concrete context. In other words, technological intentionality is hermeneutically conditioned. As Peter-Paul Verbeek explains:

> The insight that technologies cannot be separated from their use contexts implies that they have no "essence;" they are what they are only in their use. A technology can receive an identity only within a concrete context of use, and this identity is determined not only by the technology in question but also by the way in which it becomes interpreted. (Verbeek 2005, 117)

As with the hermeneutic notion of *application*, "applying" technological solutions to real-world contexts is not a straightforward process. For example, Verbeek has analyzed the ultrasound scan from this perspective. This technology, which is noninvasive in the physical sense, is far from noninvasive in the moral sense (Verbeek 2008, 14). Verbeek charts the various intersections of new meanings and questions introduced by the ultrasound scan, ranging from the personal level to the medical. For example, expecting couples may now include an ultrasound image as their "baby's first picture" (Ibid, 16) and the role of the father is affected; they "appear to feel more involved because of the new visual contact with their unborn" (Ibid, 17). Aside from these personal and cultural effects, new moral questions are raised: the image of the foetus not only represents a "person," but also a "patient" (Ibid, 15-17). A host of new ethical decisions arise alongside this technology, decisions that are shaped and guided by the information being presented through the device. In this regard, Verbeek claims that we ought to consider more deeply the types of moral *agency* that can be ascribed to devices like ultrasound scanners since "they appear to actively help answer our moral questions" (Ibid, 18).

The example of the ultrasound scan is illustrative of the postphenomenological approach as a whole. Ultrasound is a technology that could easily be criticized using a hermeneutics of suspicion; the device "enframes" the body of the foetus, artificially separating it from its actual context (continuous with the body of the mother), and therefore presenting it as something that is malleable, separable, and so on. However, these kinds of perspectives, as Verbeek demonstrates, miss a range of tangential and unexpected phenomena produced by the device, phenomena that can only be disclosed concretely, and through a "hermeneutics of trust" of sorts. Just as with any other hermeneutic object, this device puts in play a variety of intersecting questions, related not only to technical considerations but also to medical, cultural, and political. The postphenomenological wager involves bracketing speculative conclusions and turning instead to the things themselves, armed with the conviction that our fundamental relationship to artifacts and devices is not one-sided or utility based.

Verbeek pushes this conviction even further than Ihde, arguing that we are undergoing another Copernican revolution, this time in relation to human agency and morality. Just as Copernicus demonstrated we are not the "center" of the universe, and Darwin that we are not the center of our environmental processes, and, finally, Freud that we are not the center of our conscious knowing and acting, technological designs demonstrate that we are not at the center of our moral decision-making. Not only do we rely on technological instruments to gather and reveal the knowledge that leads to moral decisions (e.g., as in the case of the ultrasound), but we are also immersed in what Thaler and Sunstein famously call a "choice architecture" (Thaler and Sunstein 2009, 81–100). That is, our moral decisions and orientations are constantly shaped and mediated, in both conscious and unconscious ways, by the devices, designs and systems that we engage with daily. This fact can be harnessed in cynical and manipulative ways but also in productive ways, for example, by switching organ donor systems to be opt out as opposed to opt in. The point is that we ought to shift our thinking about morality; from the model of the Kantian subject who has dominion over their moral agency by virtue of reason, toward a postphenomenological understanding, whereby morality, like other forms of human agency and perception, is a *product* of intersecting processes—human agents, technological designs, scientific knowledge, cultural backgrounds, and so on. Once we recognize this fact, the task becomes one of "technological ascesis," that is, working with technologies to cultivate the self,

> Technological *ascesis* . . . consists in *using* technology, but in a deliberate and responsible way, such that the "self" that results from it—including its relations

to other people—acquires a deliberate shape. Not the moral acceptability, then, is central in ethical reflection on technology use, but the quality of the *practices* that result from it, and the *subjects* that are constituted in it. (Verbeek 2008, 23)

This unfolding of Postphenomenology toward questions of moral orientation and human flourishing by Verbeek is crucial for understanding its central meaning. Whereas Ihde's accounts of technology tend toward grasping Postphenomenology as a more "neutral," or at least "tempered," representation of technological artifacts than the earlier accounts offered by hermeneutic philosophers, Verbeek recognizes the transformational core of Postphenomenology: that it is not simply an application of phenomenological description to the question of technology, but rather a practical mode of interpretation, one which allows for a productive working-with the question of the meaning of technology.

3. POSTPHENOMENOLOGY AS A HERMENEUTIC OF TRUST

Comparing the two approaches to technology, it would seem there is a stark contrast. One approach (the hermeneutic) claims that essential features of human flourishing and understanding are being eclipsed by the tendencies of modern, global technics. The other (Postphenomenology) claims that technologies are, and have always been, an essential feature of our patterns of knowledge and action, and therefore cannot be simply "rejected" or overcome. Ricoeur's work, which is mostly silent on the question of technology, is arguably open to both approaches. On the one hand, a transformative understanding of interpretation and of "working-through" meaning is central to his philosophy and would seem to be threatened by any attempts to "outsource" interpretive, epistemological, and moral decisions to nonhuman devices and systems. On the other hand, Ricoeur, arguably in distinction to both Heidegger and Gadamer, always accords a large role to the function of *distanciation* in hermeneutic understanding. Ihde's outline of technological mediation allows us to think of devices and artifacts as modes of distanciation, and therefore as hermeneutically productive. As I have already suggested, however, the proper Ricoeurean move is not to resolve the tension between the two perspectives, but to recognize the unity of both suspicion and trust, of alienation and belonging.

Such a perspective can only be developed, however, once the hermeneutic nature of Postphenomenology is grasped, a fact that is arguably revealed more so through Verbeek's developments of Postphenomenology. From a hermeneutic perspective, the question of methodology or of an "empirical" versus a speculative approach to technology is only one portion of the story. Central to a hermeneutic account of interpretation is a recognition of the interaction that occurs between localized interpretations or encounters and general understandings. As Gadamer famously states, philosophical hermeneutics is not concerned with our wanting or doing, but what happens "over and above" our wanting or doing (Ihde 1975, xxviii). Again, this is why hermeneutics is not solely interested in "explication" (e.g., an interpretation, which reveals a particular nuance about a technological design), but also with "application," that is, the active transformation of an understanding or worldview that is affected by a particular interpretation. In this regard, it is important to ask: Amid the postphenomenological analysis of this or that device, what kind of relationship or general understanding of technology is being asserted?

Although, when placed within the context of the history of the philosophy of technology, postphenomenology does indeed appear to be less "romantic" than earlier approaches (Ihde 2010, 13), taken on the basis of its own claims, it appears, in fact, to exhibit its own, if not romantic viewpoint, then at least a "second naivety," in Ricoeur's productive sense of the term. In lieu of a hermeneutic deconstruction of "Technology" (e.g., vis-à-vis the history of being), it offers a reconstructive hermeneutics of trust.

This point is not always acknowledged directly within Postphenomenology. For example, Robert C. Scharff (2006) details some of the ways that, in Ihde's writings, there is a tacit reliance on a dualism between embodiment/perception, on the one hand, and language/hermeneutics/culture, on the other, with hermeneutic/cultural questions being merely "additive,"

> Ihde's body-perceptual/ cultural-linguistic cut would seem to reflect his inheritance of a Husserlian dualism that was in its day even less successful in "adding" history and culture to perception. And this, it seems to me, would indeed be an albatross-like feature in any "phenomenology." (Scharff 2006, 137)

As a side note, this is an important point when thinking about the relation between Postphenomenology and Ricoeur's work, since, in contrast, Ricoeur is far more "Heideggerian"; continually emphasizing the role language plays as the central medium for all modes of understanding. Scharff does acknowledge the importance and value of Ihde's core recognition, the pervasive presence of technological mediation in human life. However, for Scharff, this recognition brings Ihde close to Heidegger again. Both are asserting a global or general sense of technoscientific existence, albeit with a different emphasis:

> Ihde and Heidegger both understand themselves to be thinking and acting "in the midst of" the pervasive technoscientific character of life in the democratic

capitalist West. They differ mainly in the fact that Heidegger thinks a "free relation" to technoscience has to be reflectively won, and Ihde thinks we already have it. Either way, we are speaking of Technology, but without reifying it. (Ibid, 139)

Here, Scharff is correcting the picture of Heidegger that is depicted in Postphenomenology; like Ihde, Heidegger is simply recognizing the determining character of technology on everyday life, and, crucially, this viewpoint is achieved by "thinking and acting 'in the midst of'" an epoch. That is, both are thinking *hermeneutically* about technology. Comparing this point with the aforementioned criticism of Ihde's dualism, we can see that Scharff is also suggesting that Heidegger is thinking more *clearly* "in the midst" of technology, since he is acknowledging, more fully, a properly hermeneutic perspective (the perspective disavowed, at times, by Ihde's emphasis on pragmatism and embodiment). Or, in other words, Heidegger is less *naïve* about technology than Ihde.

This point is reinforced toward the end of the chapter, when Scharff proposes: "let Ihde ask: Who am I, and for whom do I speak, as philosopher of technology? As hermeneut of technology? Of science? Of technoscience? What sort of 'phenomenologist' is able to declare, and do it with knowing irony, 'Heideggerians love their laptops?'" (Ibid, p. 141). On the one hand, Scharff is right to suggest that Postphenomenology needs to clarify and reflect upon its own historical embeddedness more, but on the other, perhaps, he is too quick to dismiss the "naivety" of the position. Looked at another way, Postphenomenology need not be read as simply a naïve return to Husserl and a misunderstanding of Heidegger. Instead, we can potentially read it as "willful" or "second" naiveite, in Ricoeur's sense, that is, as a hermeneutic strategy of learning to *listen* again to areas of existence that have been demythologized, but which we nevertheless remain dependent on. Ihde's point about "all Heideggerians loving their laptops" then is asserting something akin to the *symbolic* persistence of technology. We can deconstruct it, demythologize it and trace its origin in the history of being, but afterwards, we are still faced with the inescapable presence of tools and devices. In this case, the strategy that remains is one of accepting this technical condition and learning to listen again for new ways that it can manifest itself. Importantly, Scharff also draws attention to the ways that both Heidegger and Ihde are, in spite of the protestations of the latter, attempting to think-through the same relation, albeit from different standpoints. In this sense, Postphenomenology is not an "overcoming" of earlier hermeneutic suspicions, but a hermeneutic unfolding of these positions in new directions, via an alternative interpretive strategy.

A Ricoeurean, hermeneutic approach to the question of technology, then, ought to aim to incorporate a postphenomenological perspective, whilst also remaining attentive to the existential questions of the technology-human relation introduced by thinkers like Heidegger. This perspective, I claim, would constitute a properly hermeneutic approach to technology, taking hermeneutics as defined by Ricoeur as animated by a "double motivation: willingness to suspect, willingness to listen; vow of rigor, vow of obedience" (Ricoeur 1970, 27). It means, also, recognizing the "symbolic" nature of "Technology" (with a capital "T"), in Ricoeur's sense of the symbolic as ambiguous and open to conflicting interpretations. The symbolic also possesses a degree of mystery or unknowability, which renders hermeneutic activity an existential task, not simply a quasi-scientific or "pragmatic" enterprise. In this sense, Postphenomenology, at least in terms of its claims about its own methodology, can be criticized for neglecting the ontological dimension of technology-human relations (e.g., see Zwier et al. 2016). However, reading its methodology instead as a hermeneutic strategy of trust and appropriation allows us to consider the ontological depth of its approach.

Accepting this, Ricoeur's work on symbols becomes instructive. The hermeneutic task is not to come to a final conclusion about the meaning of technology, just as we cannot do so with the meaning of evil, for example. Instead, the task is return again and again to the question of technology, and to be prepared to have both our faith and our suspicions thwarted by what we find. We need to recognize, along with Postphenomenology, the potential for technical inventions and artifacts to project new, untried possibilities for human understanding, but also, along with hermeneutic thought, we need to recognize the existential implications of interpretation and application, that is, to recognize technology also as a "threat" or potential "end," in the sense that an interpreting self before a symbol is a vulnerable thing.

4. CONCLUSION

Postphenomenology, as with other discourses that, in principle, accept the inextricable relation between human-technology-world, is an invaluable research tool for discovering new meanings and understandings produced by technological designs and inventions. However, the central questions concerning technology and the possibilities of developing a "free relation" to it, posed by Heidegger and also core to many features of hermeneutic thinking, continue to be relevant and operative. I have suggested in this chapter that Postphenomenology can be understood as presupposing a hermeneutics of trust. A hermeneutic account of technology, following Ricoeur, would aim to reflect further on what this means as a speculative position, not just how it may enrich this or that understanding of a particular technological device.

In this regard, Ricoeur's greatest contribution to the debate on technology might not be in demonstrating the value of either empirical or speculative approaches to technology, but in allowing us to trace the ways that, just as with other culturally significant phenomena, technologies, in their ambiguity and undecidability, give rise to conflicting discourses and hermeneutic strategies. The point here is not to suggest that each discourse is equivalent to the other, or that we should never come to a final decision about the meaning of technology. Rather, the point is to recognize, along with both Heidegger and Ihde, the importance of the question of technology *as a question*. As contemporary societies become more dependent on technological forms of mediation, it does not seem like a question that will decrease in importance.

REFERENCES

Achterhuis, Herman Johan. 2001. *American Philosophy of Technology: The Empirical Turn*. Bloomington: Indiana University Press.

Gadamer, Hans-Georg. 1972. "Welt Ohne Geschichte?" In *Truth and Historicity/ Vérité et Historicité*, edited by Hans-Georg Gadamer, 1–8. The Hague: Martinus Nijhoff.

Gadamer, Hans-Georg. 1977. *Philosophical Hermeneutics*. Translated by David E. Linge. Berkeley: University of California Press.

Gadamer, Hans-Georg, Joel Weinsheimer, and Donald G. Marshall. 1975. *Truth and Method*. London and New York: Continuum.

Heidegger, Martin, John Macquarrie, and Edward Robinson. 1962. *Being and Time*. Oxford: Basil Blackwell.

Ihde, Don. 1975. "The Experience of Technology: Human-Machine Relations." *Cultural Hermeneutics* 2, no. 3: 267–279.

Ihde, Don. 1990. *Technology and the Lifeworld: From Garden to Earth*. Bloomington: Indiana University Press.

Ihde, Don. 2010. *Heidegger's Technologies: Postphenomenological Perspectives*. New York: Fordham University Press.

Kaplan, David M. 2006. "Paul Ricoeur and the Philosophy of Technology." *Journal of French and Francophone Philosophy* 16, no. 1/2: 42–56.

Lewin, David. 2012. "Ricoeur and the Capability of Modern Technology." In *From Ricoeur to Action: The Socio-Political Significance of Ricoeur's Thinking*, edited by Todd S. Mei and David Lewin, 54–74. London and New York: Bloomsbury.

Ricoeur, Paul. 1967. *The Symbolism of Evil*. Translated by Emerson Buchanan. Boston: Beacon Press.

Ricoeur, Paul. 1970. *Freud and Philosophy: An Essay on Interpretation*. Translated by Denis Savage. New Haven and London: Yale University Press.

Ricoeur, Paul. 1974. *The Conflict of Interpretations: Essays in Hermeneutics*. Edited by Don Ihde. Illinois: Northwestern University Press.

Ricoeur, Paul. 1995 *Figuring the Sacred: Religion, Narrative, and Imagination.* Translated by David Pellauer. Minneapolis: Fortress Press.
Scharff, Robert C. 2006. "Ihde's Albatross: Sticking to a 'Phenomenology' of Technoscientific Experience." In *Postphenomenology: A Critical Companion to Ihde*, edited by Evan Selinger, 131–144. New York: SUNY Press.
Stiegler, Bernard. 1998. *Technics and Time: The Fault of Epimetheus* (Vol. 1). California: Stanford University Press.
Thaler, Richard H., and Cass R. Sunstein. 2009. *Nudge: Improving Decisions About Health, Wealth, and Happiness.* New York: Penguin.
Verbeek, Peter-Paul. 2005. *What Things Do: Philosophical Reflections on Technology, Agency, and Design.* Pennsylvania: Pennsylvania State Press.
Verbeek, Peter-Paul. 2008. "Obstetric Ultrasound and the Technological Mediation of Morality: A Postphenomenological Analysis." *Human Studies* 31, no. 1: 11–26.
Zwier, Jochem, Vincent Blok, and Pieter Lemmens. 2016. "Phenomenology and the Empirical Turn: A Phenomenological Analysis of Postphenomenology." *Philosophy & Technology* 29, no. 4: 313–333.

Chapter 3

Let's Narrate That Symmetry!

Ricoeur and Latour

Bas de Boer and Jonne Hoek

The French anthropologist and philosopher of science and technology Bruno Latour is a prominent figure in the development of philosophy of technology. Among other things, Latour has been a major voice in the advance of the science and technology studies (STS), recognizing the intertwinement of societal, political, scientific, and technological domains. Furthermore, he is one of the driving forces behind what later has been called the "empirical turn" in the philosophy of technology (e.g., Achterhuis 2001). Latour formulated a methodological outlook for this empirical reorientation: the symmetry principle, or the principle of generalized symmetry. With it, he aimed to counter a set of preconceptions he ascribed to Modernity that unjustly keep apart the human and the nonhuman, the subject and object, society and nature. Entrenched in such oppositions, so Latour argued, we remain blind for the intertwinements, mediations, and co-operations of humans and technologies by which our reality is actually shaped. Therefore, we should instead approach both human and nonhuman symmetrically, in terms of what they do in concord, that is: as actants.

It is well-established that Latour's theoretical advances in this direction were significantly inspired by the semiotics of Algirdas Greimas (e.g., Beetz 2013; Høstaker 2005; McGee 2014). And it is here, so we will show in this chapter, that an interesting point of comparison with Ricoeur appears. For also Ricoeur's engagement with semiotic theory was in critical discussion with Greimas. Ricoeur, moreover, tried to complement and improve upon Greimas's semiotic approach by positioning his hermeneutics of narrative understanding in a productive dialectic with it. We will here draw out some inferences from this association, establishing how Ricoeur's appreciation and critique of Greimas might apply *mutatis mutandis* to Latour. Subsequently, we will also look at this association the other way around and see how

Latour's symmetry principle of explanation extends the scope and significance of Ricoeur's thinking into the philosophy of technology.

Our comparison thereby rests upon an analogy drawn between the explanation of texts on the one hand, and the interpretation of technologically mediated practices on the other hand that can be found in the work of Latour. Various scholars connecting Ricoeur to the philosophy of technology have highlighted the importance of this association, as well as the crucial importance of Latour breaking new ground in this respect. David M. Kaplan, for instance, states that "Ricoeur believes that if human action can be read and interpreted like written works then the methods and practices of textual interpretation can function as a paradigm for the interpretation of action for the social sciences" (Kaplan 2006, 43). However, since for Ricoeur the technological is not evidently a subject of the social sciences, Latour can function here as a bridge toward the technological. This is why Kaplan states that "if Latour is right, then we must give up the illusion that technological understanding is somehow different from social understanding." It is via Latour, therefore, that one can "extend [Ricoeur's] notion of narrative understanding beyond human action to include the natural and technical worlds as well" (Kaplan 2003, 170).

Coeckelbergh and Reijers have recently taken up this challenge and argue similarly that "[Ricoeur's] theory of narrative configuration can go beyond the works of literary fiction and history and can even include visual objects like paintings and knowledge in the scientific field of biology" (Coeckelbergh and Reijers 2016, 335). In this setting, they note that even though Latour (together with Madeleine Akrich) already spoke metaphorically of "scripts" that artifacts carry along in their programs of action, Latour on his part failed to see the genuine linguistic dimension, let alone the "wider social-linguistic environment (prescriptions, discourse, narratives)" in which such scripts play out in practice (Ibid. 328).

This latter observation ties in with a more fundamental critique that is sometimes directed at Latour's symmetry principle. For in approaching humans and nonhumans alike, it is feared that Latour loses appreciation for what also sets apart the human subject from its socio-material context. For example, Krarup and Blok argue that "Latour seems in practice to be 'heterogenizing' human subjectivities into a background of materially stabilised, and technologically shaped, assemblages. . . . His analyses leave us short of understanding, let alone describing, how bodies, symbols, and subjective desires simultaneously contribute to the process of forging socio-technical effects" (2011, 57). Put differently: though it might be that Latour has found a fine-grained description devise to analyze vast networks of human-technology relations with, one wonders: "will not something crucial be lost in our view of inter-subjective social life if we just 'go on describing'" (Ibid. 43)?

It is here, at this point, so we will argue, that Ricoeur's hermeneutic appreciation and critique of Greimasian semiotics appears as relevant for today's philosophy of technology. For also Latour recognizes that, instead of describing objects in the world as simple matters of fact, which are supposedly there "whether we like it or not," we should instead approach them as matters of concern, which means that "they have to be [continuously] kept up, cared for, accompanied, restored, duplicated, saved" (Latour 2008, 48/50). We will contend that Ricoeur's narrative hermeneutics can acknowledge how things—as matters of concern—point to their constitutive role for (inter-) subjectivity, though without giving up the radical symmetry that has been the semiotic backbone of Latour's philosophical approach.

1. RICOEUR'S CRITICAL APPRECIATION OF GREIMASSIAN SEMIOTICS

Ricoeur discusses the work of Greimas in several texts. He is particularly interested in Greimas's semiotics as it presents a challenge to his own hermeneutic account of understanding narrativity and time (Ricoeur 1984b, 29). Ricoeur's intuition is that human time is inherently wound up with a narrative mode of understanding and that "irreducible factors of temporality" are captured in the plot of any story (Ricoeur 2016, 247). By contrast, Greimas presents his semiotic approach as offering a method to study the composition of meaning structurally, a method that de-chronologizes a story by looking at features immanent to the logical composition of its signs. This presupposes a focus on what Greimas calls "deep structures": structures that are present—not in the plot of a story, not evidently in the character of the personages, not in the singular actions described, and not even in the particular order of words the text presents us with, but in the various *actants* that stand contrasted logically and call forth one-another in semiotic relations.

An *actant* is therefore different from an *actor*, as the latter is usually a human character, and an actant can be anything whatsoever. Whatever thing an actant in a story happens to be (a knight, a king, the sky, a forbidden love, death, a rose, the color red, a gust of wind), its true significance for the story is only to be understood from the semiotic relations it entertains in the text. For this reason, actants appear always paired, as any "object" implicates a "subject" that desires it, any message of communication ties together "sender" with "receiver," and any "helper" will pragmatically oppose an "adversary." By looking at such structures, Greimas states in discussion with Ricoeur, we see "the production of meaning" in detail, which is "the production of difference, the production of oppositions" (Greimas and Ricoeur 1989, 559). These are the deep structures of narrative.

According to Ricoeur, Greimas not only reverses the primacy in order of explanation between "depth" and "surface" structures of narrative, but entirely removes the crucial difference between a hermeneutic *understanding* of a narrative and its semiotic *explanation*. Thereby, the hermeneutic prefigurations of a story (its emplotments) are reduced to a convenient clothing, "taken to be a [mere] surface effect of explanation" (Ricoeur 1990, 119). To this, Ricoeur objects. Greimas's schematization of actants offers a structure of connections maybe, but not real composition, and hence no means for real comprehension. Despite its merit for detailed scientific explanations, therefore, Ricoeur argues that Greimas's logic of actants considered alone does not generate understanding. For understanding, Ricoeur believes, we simultaneously should take recourse to more comprehensive sense of hermeneutical pre-understanding: a plot, a moral of the story that orients the temporal, teleological composition of the whole. What are Ricoeur's arguments for this contention?

Firstly, Ricoeur maintains that Greimas's semiotics cannot illuminate relations between actants without taking implicit recourse to phenomenology. For instance, Greimas holds that "the deprivation of a value/object a subject undergoes is a modification that affects this subject as a victim" (Ricoeur 1984b, 59). A rose (subject), for example, can lose its scent (value/object). Now, if only put in contrast, Ricoeur would argue that the actants do not readily provide an enriched sense of meaning and significance in a narrative. Only when understood in sensuous, temporal terms we can speak here about the rose as a victim, as the fragrance as something being lost or violated enduringly. But in that case, Ricoeur holds, "the implicit recourse to phenomenology is flagrant" (Ibid. 59). Also in Greimas's work, a phenomenological pre-understanding must therefore already be involved.

Secondly, a text might follow the logic of actants explained, but this does not yet produce a story, so Ricoeur holds. For according to this logic, "all the subsequent operations would also be foreseeable and calculable" (Greimas 1983, 166). But then nothing would happen. There could be no event, no surprise. "There would be nothing to tell" (Ricoeur 1984b, 56–57). In connecting the elements only logically, no chronological comprehension is produced. This is why the most basic and encompassing narrative opposition, that between a quest and its fulfillment, cannot, according to Ricoeur, be understood logically and a-historically. When doing so, no reference is made to a structuring of time that we are already involved with and which is necessary for understanding: "What is missing from the operations of conjunction and opposition is precisely the *diastasis* [literally: standing apart] of time. Thus, the mediation effected by the narrative cannot be of a mere logical order: the transformation of the terms is properly historical" (Ricoeur 2016, 248). Time is not an epiphenomenon of the logical, but is the staging of its actants *as* development, *as* a story. Accordingly, emplotment, so Ricoeur holds, cannot be a mere surface manifestation, as if a mere clothing of the actants for convenience of discourse (Ibid. 248). In a discussion with Greimas, Ricoeur puts it as follows: Narrative "figures are much more than a garment. . . . Precisely what is productive is that you cannot have spatialisation, temporalisation, and actorialisation without plot" (Greimas and Ricoeur 1989, 556).

Ricoeur's criticism of Greimas targets its reductionism, but it does not lead into the other extreme. Impressed with the Greimassian actor model, Ricoeur sees a productive tension or dialectic existing between a more detailed scientific *explanation* of elements, and a hermeneutic *understanding* comprehending those elements into a whole. "To explain more," so Ricoeur's adage follows, "is to understand better" (Ricoeur 1984a, 32; Ricoeur and Greimas 1989) That is to say: our understanding calls forth semiotic explanation, not as its "adversary, but as its complement and its mediator" (Ricoeur 1990, 125).

2. THE SEMIOTIC LEGACY OF LATOUR'S SYMMETRICAL TREATMENT OF HUMANS AND NONHUMANS

One of the main goals of Latour's work is to counter what he calls the *Modern Constitution*. The Modern Constitution provides us with a number of flawed beliefs regarding the nature of reality, and Latour aims to develop an alternative to it that does justice to the complexity of the processes that constitute the world in which we live. The most important legacy of the Modern Constitution is the presence of the subject-object dichotomy, from which the dichotomies between nature and society, human and nonhuman, and rational and social can be derived (e.g., Latour 1993b, 35–38). Semiotics is for Latour a way of analyzing that evades, and even dismantles the vocabulary of the Modern Constitution. In this section, we show how semiotics is constitutive of Latour's symmetric treatment of humans and nonhumans, thereby informing his analyses of technologies.

A common-sense understanding of the relation between human beings and artifacts would conceptualize the former as having the capacity to act upon their beliefs and desires, while the latter are considered "mute" in the sense that their being is fully dependent on the intentions and actions of humans. In strong opposition to such a view, Latour proposes to overcome this distinction by understanding agency as a capacity of *hybrid* entities that are compositions of *actants*—a term Latour adopts from Greimas, and uses to refer to both humans and nonhumans indiscriminately. This has come to be known as the *symmetry principle* by which neither human beings nor artifacts a priori have a privileged position in the constitution of the agency of hybrid entities. A gunman is a good example of a hybrid entity: "you are another subject because you hold the gun; the gun is another object because it has entered in a relationship with you.... They become 'someone, something else'" (Latour 1999, 180).

Several authors have criticized the symmetry principle for unjustifiably eliminating central capacities of human actors (e.g., experience and cognition) that strongly distinguish them from nonhuman actors. Latour's notion of actant then is, so it is argued, insufficiently capable of addressing important differences between human beings and artifacts (e.g., Bloor 1999; Feenberg 2017; Krarup and Blok 2011). While such criticisms may raise a valid point, they do not explicitly engage with Latour's reasons for proposing to analyze humans and nonhumans in symmetrical terms. When looking closer at how Latour comes to understand a gunman as a hybrid entity, it becomes clear that it is derived from a specific, semiotic *reading* of the situation like a text being written: "If we study the gun and the citizen as [two] *propositions*, however, we realize that neither subject nor object (nor their goals) is fixed. When the *propositions* are articulated, they join into a new *proposition*" (Latour 1999, 180, our emphasis). When approaching guns and citizens as *propositions*, a specific perspective is adopted toward their role in the constitution of the gunman. We make clear Latour's semiotic perspective by showing: (a) how Latour uses it in his analyses of the coming into existence of scientific facts to move from texts to laboratories, and (b) how he subsequently uses semiotics for analyzing technological artifacts in everyday life.

3. LATOUR'S APPROACH TO SCIENTIFIC PRACTICE: USING SEMIOTICS TO MOVE FROM THE TEXT TO THE FLESH

Initially, Latour's use of semiotics was restricted to the study of scientific facts. In *Laboratory Life*, Latour's first anthropological study of scientific practice conducted at the Salk Institute published in 1979, one of his first observations is that all the work conducted in the laboratory is done for the purpose of the writing of scientific texts, which he considers the laboratory's end-product. These texts, then, so he observes, are organized in such a way that they contain truths about a reality external to them. A semiotic approach toward scientific texts, so Latour holds, allows to investigate how the statements they contain are made to be facts, without referring to a scientific rationality that mysteriously endows these statements with truth and certainty. The latter would explain factuality of scientific statements with reference to an external nature that is described and is reached by employing a scientific method, this is what Latour calls the vocabulary of *ready-made science* (Latour 1987, 4).

Adopting the vocabulary of *ready-made science* for Latour implies to buy into a particular narrative about science accepting its unique rationality uncritically. However, so he holds, another type of explanation arises when scientific facts and technical artifacts are encountered when looking deeper into the structure of scientific activity: when investigating *science-in-the-making*. Accordingly, he suggests to:

> enter facts and machines while they are in the making; we will carry with us no preconception of what constitutes knowledge; we will watch the closure of the black boxes and be careful to distinguish between to contradictory explanations of this closure, one uttered when it is finished, the other while it is being attempted (Ibid. 13-15).

To this purpose, Latour organizes his analysis by staging a character he calls "the dissenter," which allows him to show what world of science opens up when a naive outsider decides not to believe a sentence written in a scientific text.

Disagreeing with a factual statement in a scientific paper is never to disagree with something isolated. Authors of a scientific paper include semiotic characters that affect the character of disagreement (Ibid. 54): scientific statements make use of other scientific statements that are cited and referred to and are therefore embedded in a larger network of truth claims. Accordingly, when one wishes to disbelieve a specific sentence, it is implied that the reader disagrees with many other authors. Even more, scientific texts are full of visual displays presented as evidence that what is written down can be observed right away. When the dissenter is serious about her disagreement, she has to move from the text to the world it refers to and find possibilities for reasonable disbelief in there. Doing so is to take seriously the idea that there "were a set of semiotic actors presented in the text but not *present* in the flesh; they were alluded to as if they existed independently from the text; they could have been invented" (Ibid. 64). Latour thereby uses semiotics to migrate from text to laboratory, and vice versa.

Now, what is found in the laboratory where semiotic actors are present in the flesh? In the laboratory, the dissenter can see with her own eyes that these inscriptions are not simply invented when being confronted with the instruments responsible for the creation of the diagrams and graphs present in scientific texts. Latour calls these instruments *inscription devices* because they inscribe visual displays into texts that increase the reality of the statements that are written down (Ibid. 69). However, in isolation, also visual displays are not indicators of the truth of a statement: they need someone that speaks for them in order to become meaningful. When being spoken for, it becomes clear what visual inscriptions do. They visualize what happens in a carefully managed experimental set-up in which effects of something on something else remains present when tested and manipulated to the extreme. Eventually, this is what is visualized by scientific instruments and inscribed into scientific texts.

At this point, Latour introduces the term *actant* to describe what is spoken for by the statement that the dissenter first decided to disbelieve. Actants are "behind the texts, behind the instruments, inside the laboratory. . . . [They are] an array allowing next extreme constraints to be imposed on 'something.' This 'something' is progressively shaped by its re-actions to these conditions" (Ibid. 89). Actants are both the things described and the things describing it, ranging from the scientists, the experimental set-ups and the scientific articles in which the actant figures and is experimented upon over and over again.

Latour credits his semiotic approach to scientific texts as allowing him to move from the world of written statements to the world of the laboratory, taking seriously actants as constitutive of scientific statements. Semiotics helps doing so because it does not assume an "a priori distinction . . . between an anthropomorphic actor and a 'physimorphic' or 'zoomorphic' one" (Latour 1993a, 130–131). Accordingly, Latour's semiotic approach to analyzing the coming into being of scientific facts forces him to challenge the constraints imposed on reality by the Modern Constitution, because he is confronted with the important role of nonhuman actants that needs to be articulated.

4. SEMIOTICS AND THE AGENCY OF TECHNOLOGICAL ARTIFACTS

Later in his career, Latour starts turning his attention to the technological artifacts that structure how society is organized. According to him, these artifacts are remarkably absent in philosophical and sociological theories about modern society. Because of this absence, Latour calls technological artifacts the "missing masses," and uses his semiotic approach to give them a voice.

One of the technological artifacts that Latour attempts to give a voice is the automatic door-closer (or in French: the groom). Of course, there is no door-closer without a door that shapes what to expect from a door-closer, namely, to allow to be used by an individual that intends to enter a building, and to close immediately after the individual has done so. The door-closer thus prescribes a certain set of actions. Because of this, so Latour holds, non-human actants such as the door-closer have a clear moral component: without them, visitors of the building might well leave the door wide open after entering. Latour proposes to analyze such prescriptions (and give technological artifacts a voice accordingly) by replacing the mechanisms embodied in an artifact "by strings of sentences (often in the imperative) that are uttered (silently and continuously) by the mechanisms for the benefit of those who are mechanized: do this, do that, behave this way, don't go that way, you may do so, be allowed to go there" (Latour 1992, 157). In other words, technological artifacts such as an automatic door-closer can be given a voice by reading them as a set of interrelated *propositions*. When read in this way, it becomes clearly visible that an action is never executed by a human being alone: the particular way in which a visitor enters a building can never be reduced to human intentions, but necessarily involves the specific capacities of the door closer as well.

However, ascribing agency to the door-closer remains indebted to the Modern Constitution because it tacitly assumes that agency can be ascribed to isolated entities. This would simply be a mirror-case of reserving the ascription of agency for human subjects: "To speak of 'humans' and 'non-humans' allows only a rough approximation that still borrows from modern philosophy the stupefying idea that there exist humans and non-humans, whereas there are only trajectories and dispatches, paths and trails" (Latour 2000, 12). Agency is thus constituted within particular trajectories and is a capacity of those as a whole, rather than the actants that participate in these. The notion of trajectory thus denotes the process through which action takes place: actants mutually shape one another in a specific way such that if a trajectory were organized differently, a different type of action would be constituted. Hence, when Latour interprets agency as a capacity of *hybrids*, he uses this term to refer to the entire trajectory within which actions are generated. When approached in this way, the automated door-closer is an actant because it is a being "that participate[s] in processes in any form whatsoever, be it only a walk-on part and in the most passive way" (Tesnière, op. cit. Greimas and Courtes 1982, 5). Articulating the role of the automated door-closer (as an actant) is accordingly to articulate how the process in which it participates would be organized differently without it. We can ascribe to the actant a specific role or figure because of this articulation.

The fact that humans and nonhumans have symmetrical capacities as actants does not necessarily imply that their role within trajectories cannot be articulated differently. The crux for Latour is that there is no a priori asymmetry between humans and nonhumans, they are symmetric as entities "that ac[t] in a plot *until* the attribution of a figurative or nonfigurative role" (Latour 1994, 33, our emphasis). The attribution of either a figurative or a nonfigurative role is thus a game changer for how we conceive of the role of particular actants within certain trajectories. Let us illustrate this with

Latour's analysis of how a speed bump functions within a trajectory giving rise to the slowing down of car-drivers: in principle, a trajectory within which car-drivers slow down can be realized by a multiplicity of different actants, ranging from placing a police officer at the side of the road from the placing of speed bumps. In the case of the police officer, it is clear that the mechanism responsible for the slowing down is authorized because the police officer is a *figurated* character making clearly visible the intervention necessary for the slowing down. In the case of a speed bump, which is an unfigurated character, it is suggested, precisely because the speed bump is not figurated, it becomes less obvious that the setting in which car-drivers slowdown is one crucially dependent on active manipulation.

According to Latour, labeling the speed bump "inhuman" is a way to overlook the translation mechanisms responsible for creating the incentive to slow down. When articulating which action plans are delegated to the speed bump, it becomes clear that the way the speed bump acts is, just as any technical action, "a form of delegation that allows us to mobilize, during interactions, moves made elsewhere, earlier, by other actants" (Ibid. 52).

Just as statements made in scientific texts, technical artifacts fold acts conducted at another time and place into the here and now. Yet, the crucial difference between written texts and technical artifacts is that in the former case "I am here *and* elsewhere . . . I am myself *and* someone else, [while in the case of technical artifacts], an action, long past, of an actor, long disappeared, is still active here, today, on me—I live in the midst of technical delegates" (Ibid. 40). This implies that human behavior is structured by other actions that are hardly detectable in the present, because it is perfectly possible "to position no *figurated character* at all as the author *in* the scripts *of* our scripts (in semiotic parlance there would be no *narrator*)" (Latour 1992, 165). When sidestepping the trajectories mobilized in technological artifacts (i.e., by conceptualizing them as mute nonhumans), they become black-boxed, thereby making incomprehensible the role of other actants in how our lives are shaped in the midst of technical delegates. Latour's semiotic analyses of technical artifacts can therefore be seen as the re-introduction of their narrator(s). In taking up the role of the narrator of the trajectories within which technical artifacts function, Latour draws attention to the politics of artifacts and the forms of morality that they embody.

5. THE NARRATIVE TELEOLOGY IN LATOUR'S SEMIOTIC EXPLANATIONS: A RICOEURIAN CRITIQUE

Let us recapitulate. Greimas argues that semiotic explication underlies our understanding of plot and narrative, rather than the other way around. This thesis is challenged by Ricoeur's phenomenological hermeneutics of narrative and time. Latour, in turn, argues that also technology, like texts and everything else in the world, is semiotically structured. He takes it that a generalized "semiotics is a method to describe the generative path of any narration" (Latour 1990, 9). Neither Ricoeur nor Greimas addresses this latter option explicitly. However, staying within bounds of the associations established, a Ricoeurian critique directed at Latour will entail the following: that when explicating technologies semiotically, *as if* a text with narrative, a phenomenological sense of pre-understanding remains both indispensable and primary. This is the critique we will draw out here.

In order to pursue this argument convincingly, we need (for now) to accept Latour's premises and hold back any preconceptions about what differentiates the narration of literary texts from that of technologies. Even though Ricoeur has addressed his appreciation of Greimas' semiotics specifically with regard to fictional narratives and has in fact much to say about other genres too (e.g., fictional, historic, and mythical), introducing such distinctions here would make the argument question-begging. After all, it is the endeavor of semiotics to explicate things without imposing such structures of meaning from the outset. For the truly scientific analyst, it is "forbidden to tell the actants what they should do" (Latour 1990, 9). Both Greimas and Latour aim to provide a point of access for the researcher that looks "beneath" the surface of narrative expectations and leading away from a Modern Constitution that plots to keep the human separated from the nonhuman artificially. For this reason, Latour and Greimas explain all things as describing their own narrative trails; as actants.

A Ricoeurian critique of Latour might start—similarly as in the case of Greimas—by showing how the mere connectivity of actants describing "trajectories and dispatches, paths and trails" (Latour 2000, 12) does not in and of itself amounts to understanding yet. Or, to put this more precise: explicating an actor-network semiotically does not yet translate into what Ricoeur would consider a *narrative understanding*. A narrative involves time, not as a number of motion, but as it "is meaningful to the extent that it portrays the features of temporal experience" (Ricoeur 1984a, 3). We can illustrate this argument with an example. In his article on the door-closer, Latour describes an ordinary door as being complex already, it is a hybrid, namely a "hole-wall," which, mediated by a linchpin, allows a manifold of characters (humans, mice, dossiers, fresh air, and noise) entering (hole), or not-entering (wall) a room that "would be mausoleum or tomb" without it (Latour and Johnson 1988, 298). So, Latour invites us to write down the extensive and diverse efforts needed for these same actants to go in- or out of the room without doors (e.g., breaking the wall and building it up again). Doing so, we soon start to appreciate the semiotic "*translation* or *delegation*" enacted by the hybrid actor-network of the door (Latour and Johnson 1988, 298).

With Ricoeur, we can argue that the actant of the door does not appear meaningfully without us drawing implicitly—and we might add: rather extensively—on our phenomenological preunderstanding of the matter. Surely, we can think of a wooden board being connected to a frame via a linchpin. But such a state of affairs does not yet mediate my *narrative* understanding of the door as a *hybrid* connection, as a composition that actually *does* something in the larger constellation of things. The semiotic logic remains static here; it does not narrate anything without a phenomenological grasp of the opposition "hole-wall" by which we see that a hole "allows" for movement, whereas a wall "hinders" it. Hence, when the door is "open," the walled room is exposed, available, or inviting. "Closed," the door restricts movements—accordingly, the room is secured, inaccessible, shut. We side with Ricoeur here by noting that "the preunderstanding that we have of these distinctions on the phenomenological plane seems ... to exercise an irreplaceable guiding role" (Ricoeur 1990, 131).

Moreover, the semiotics of the door-hybrid become infused with a hermeneutic teleology when investigating the *translation* or *delegation* that is invested in the door, which is explored by listing its counterfactuals, as Latour makes us do: "you simply have to imagine that every time you want to get in or out of the building you have to do the same work as a prisoner trying to escape or a gangster trying to rob a bank, plus the work of those who rebuild either the prison's or the bank's walls" (Latour 1988, 299). Besides actual experiment (e.g. putting a lock on the door and throwing away the key), there is no way to open-up this black-box of hybridity other than by recourse to our memory, imagination, and expectation of common sense. What is more: these counterfactuals themselves do not immediately make out part of the composition of this actor-network—they do so only through my association, meaning they are traces of my comprehension in extending the actor-network of the door. In making this a narrative, I connect this door with my own comprehending why what seems contingent here at first is contingently *thus* rather than *otherwise*.

Making this argument with Ricoeur, we break ground for a minimal, but very crucial difference (which, so we will argue, is a difference without distinction), namely that between the semiotic explanation of the actor-network itself, and our narrative activity that "guides" the semiotic explanation into structures of understanding. This ties in directly with Ricoeur's criticisms of structuralism, of which semiotics can be considered a branch. Against the semioticians, Ricoeur holds that a text is not a closed-off reality. Even though its meaning seems semiotically composed by mutual, *internal* reference of its actants only, the text always also "speaks" to something *outside* of it, namely the reader that interprets it (Ricoeur 2016, 107–126). By virtue of this, a text allows for different and changing interpretations over time, and harbors so always an irreducible surplus of meaning.

Now Latour, quite differently, considers it the "major achievement" of the "semiotic turn" to do away with any such references outside of the text (be that "nature" as the thing described, or "society" as providing words with meaning) (Latour 1990, 8). This is why he also extends semiotics beyond the text, making him "elevate things to the dignity of texts or"—what is the same at that point—"to elevate texts to the ontological status of things" (Ibid. 10). In that case, "Every network surrounds itself with its own frame of reference, its own definition of growth, of referring, of framing, of explaining" (Ibid. 11). All things, as actor-networks, now "provide explanations of themselves" when connecting (Ibid. 11). However, we argue—with Ricoeur—that there is still a difference to be made between semiotic *explanation* and narrative *understanding*, for the trajectory of an actant, which becomes self-explanatory in Latour's semiotic approach, does not coincide with our activity of tracing that actor-network; understanding it, interpreting it. Understanding requires at least one extra actant to connect to the self-explanatory network: the one who comprehends the semiotics into the narration of time.

It seems like we are making a very minimal claim here with Ricoeur, yet Latour would likely object to it inasmuch as it could be reminiscent of a Modern, dualist narrative in which the human subject will again start describing the nonhuman object. As if we are saying with Ricoeur that only the human breathes true meaning and life into matters such as doors and hinges; as if we are claiming there is an added intersubjective reality beyond the complexity of actor-networks already irreducibly rich in meaning and consequences. But does Ricoeur's appeal to narrative understanding as different from semiotic explanation necessarily reinforce such a Modern Constitution? We believe this not to be the case. In itself, Ricoeur's notion of narrative might need not entail any a priori commitment to humans being the only possible readers, or to nonhumans being unable to have a sense of narrative time. It certainly also does not entail (as Latour might fear) that a reader "jumps outside of the network, and then adds explanation" to it (Ibid. 11). For we can uphold a *difference* between a text and its reader without buying into ontological *distinction* that is often attached to it; and it is the distinction that Latour considers a Modern delusion.

Making a semiotics universal, as Latour does, means that we as readers are now part of the text. And since Latour holds that all actor-networks "provide explanations of themselves" when connecting, then so do we as readers, like any other actant, according to our "own frame of reference" (Ibid. 11). Such a frame of reference is what Ricoeur calls a narrative, or sometimes also: human time. But here comes the punchline: readers can never be wholly reduced to the disinterested, neutral scientists that study the semiotics of other actants according to *their* own logic only (i.e., as a dissenting, observing, attesting, translating, and semiotician putting things to trial). Contrary to Latour, a Ricoeurian critique points to the fact that as readers in the text we do not only encounter things as if it is "forbidden to tell the actants what they should do" (Ibid. 9). Quite the opposite is the case, as in order to arrive at understanding, a reader employs a sense of narrative time, meaning we are an anticipating actor in the text: the reader is a "who," historically and socially situated, pursuing its own goals, bringing along its own frame of reference, by virtue of which it is affective—vulnerable even—to the things that it is presented with. To put this point differently: a reader is not external to text, such that (s)he can become either a passive recipient of the narrator's intentions or an active formative element of the text's meaning. Instead, a text is only a text insofar the reader's frame of reference inscribed into it.

6. CONCLUSION

As stated in the opening of this chapter, Latour is often credited for drawing attention to the importance of technologies in our lives but criticized for having a blind-spot with regard to how also intersubjective relations and subjective desires are constitutive of human action and understanding (Coeckelberg and Reijers 2016; Krarup and Blok 2011). Specifically referring to Ricoeur, Coeckelberg and Reijers propose a movement "from Text to Technology" and use Ricoeur's hermeneutics "to understand how technologies configure intersubjective relations and how these relations can be traced back to the subjective realm of human action" (2016, 343). But using Ricoeur to re-introduce a focus on social relations and intersubjectivity—now as mediated by technologies—comes with a danger, as it could lead us back into the Modern Constitution. Are we not abandoning Latour's symmetry principle and re-introduce a dichotomy between entities capable of engaging into social relations (i.e., subjects) and entities that are incapable of doing so (i.e., objects)? Or is there a way to give Latour's missing masses a voice such that it is also *their* voice, and not just the echo (or mediation) of the human subject relating to them?

According to Latour, blowing life into technological artifacts should not be considered an attempt to remove subjectivity from our view entirely, but rather a way to take seriously the basic observation that when you "consider humans . . . you are by that very act interested in things" (Latour 2000, 14). Human subjectivity is therefore never something given, but instead the outcome of the particular trajectories within which it arises—like anything else is. Because of this, to be a subject always implies to be formed as a particular kind of subject dependent on the constraints other actants put upon it. When the human subject is drawn out of its "splendid isolation," it becomes clear that subjects "depend on a flood of entities allowing them to exist." When analyzing how subjectivity arises from within a flood of entities, the possibility opens up to turn subjectivity into "a fully artificial and fully traceable gathering" (Latour 2005, 208). Because of this, so Latour holds, also subjectivity and the narrative capacities of human beings can be semiotically examined. Giving the missing masses of the objects a voice is accordingly to articulate how subjectivity is dependent on and constituted by these voices.

Now, what are the consequences of Ricoeur's critique of semiotics for Latour's proposal to semiotically trace the constitution of subjectivity? Already in Latour's use of semiotics to study the coming into being of scientific facts, by staging the character of the dissenter, he does not only introduce a semiotic methodology but also a narrative structure to the story about science that Latour intends to tell, thereby actively inscribing a teleology of disbelief. With the introduction of this character, simultaneously a pre-understanding of what science can tell us is introduced. Accordingly, the dissenter cannot be understood as the position of a value-free observer suddenly introduced when applying semiotics but also reveals Latour's own, implicit pre-understanding of which story about science should be told. Similar to Ricoeur's critique of Greimas, we can thus say that Latour's story of science presupposes a hermeneutic moment through which a specific narrative teleology is inscribed.

Still, Latour's treatment of scientific facts remains quite removed from the human subject's own narrative capacities; his analyses of the morality of technical artifacts penetrate intuitively much closer to the heart of subjectivity. By showing that the actions that we undertake are shaped by specific technological artifacts that often surpass us unconsciously and by showing how these artifacts can—and must—be recognized as crucial figures in what human subjectivity is, Latour introduces a way to show how they actively are of *our* concern. To paraphrase Latour: they are not matters of fact that we must passively relate to, but are matters of concern whose reality and working are and must be actively shaped (cf. Latour 2004). Only in this way does it become possible to recognize how subjectivity is shaped by them.

With Ricoeur, we hold that only inasmuch as an actor-network or trajectory "speaks" to us in terms we can understand, only to that extent are we challenged to further explore its semiotics scientifically, rather than superimposing our own preconceived notions on it. We explain more, as Ricoeur says, to understand better. And isn't this what Latour ultimately also aims for: that a manifold of overlooked actants can be given a *voice*, as participants in the "parliament of things," not simply governing our systems of knowledge and moralities of conduct, but taking part in an ongoing conversation? Without narrative understanding, therefore, a parliament of things will present nothing but noise to human ears, in this Ricoeur is right. However, without semiotic explanation, we will remain deaf to the nonhuman voices, which is what Latour shows. Such a Kantian pun is not only directed at Latour who likely sees in Ricoeur's dialectic between explanation and understanding still a vestige of the Modern Constitution. It is also an invitation to Ricoeur's philosophy, to try and follow Latour in what he calls his "counter-Copernican revolution."

For there is a moral to Latour's story, a teleological frame of reference, a grand narrative in which he inscribes his own project. This narrative reveals Modernity as a myth to be disenchanted, proving that we have never actually been Modern (Latour 1993b). Latour thereby structures his own story teleologically, makes use of a historical plot offering the reader a quest: to escape the narrow confines of a Western, human-centered, imperialist worldview. By breaking this spell of an assumed a priori distinction between human and nonhuman, Latour also challenges our concept of subjectivity. This is to say that for Latour, every way of substantializing "subjectivity" is to be avoided, as only specific "subjectivities" exist within the specific trajectories in which they participate. This last idea is guiding Latour's recent book *An Inquiry into Modes of Existence*, where he draws on semiotics to show from a non-anthropocentric point of view how we "*became* humans—thinking, speaking humans—*by dint of association* with the beings of technology, fiction, and reference. [We] became skillful, imaginative, capable of objective knowledge by grappling with these modes of existence" (2013, 372). By tracing how humans are transforming and being transformed in networks in which natural, fictive, and technological beings figure as well, Latour intends to develop a narrative that shows how the form of subjectivity—which is one among other possible subjectivities—particular to the Moderns can be understood as arising within a specific trajectory.

Again, with Ricoeur, we can say that this new way of conceiving of subjects and objects cannot be reduced to a supposed neutral position introduced by semiotics, as if we can simply watch the influence that things have on a semiotic screen unfolded in front of us. Instead, the narrative that is opened up when the missing masses are given a voice is historically situated, and one of growing concern. Blowing life into the objects—that is, turning them into matters of concern—that were initially muted by the Modern Constitution is therefore to blow new life into subjectivity. The narrative that makes the Modern Constitution tremble is accordingly a moment in which subjectivity starts to tremble as well, thereby opening up a space for renewing its own emplotment.

REFERENCES

Achterhuis, Hans, ed. 2001. *American Philosophy of Technology: The Empirical Turn*. Translated by Robert Crease. Bloomington: Indiana University Press.

Akrich, Madeleine, and Bruno Latour. 1992. "A Summary of a Convenient Vocabulary for the Semiotics of Human and Nonhuman Assemblies." In *Shaping Technology/Building Society: Studies in Socio-Technical Change*, edited by Wiebe E. Bijker and John Law, 259–264. Cambridge: MIT Press.

Beetz, Johannes. 2013. "Latour with Greimas: Actor-Network Theory and Semiotics." Accessed May 28, 2020. https://www.academia.edu/11233971/Latour_with_Greimas_-_Actor-Network_Theory_and_Semiotics.

Bloor, David. 1999. "Anti-Latour." *Studies in History and Philosophy of Science* 30, no. 1: 81–112.

Coeckelberg, Mark, and Wessel Reijers. 2016. "Narrative Technologies: A Philosophical Investigation of the Narrative Capacities of Technologies by Using Ricoeur's Narrative Theory." *Human Studies* 39, no. 3: 325–346.

Feenberg, Andrew. 2017. *Technosystem: The Social Life of Reason*. Cambridge: Harvard University Press.

Greimas, Algirdas Julien. 1983. *Structural Semiotics: An Attempt at a Method*. Translated by Daniele McDowell, Ronald Schleifer, and Alan Velie. Lincoln: University of Nebraska Press.

Greimas, Algirdas Julien, and Joseph Courtés. 1982. *Semiotics and Language: An Analytical Dictionary*. Translated by Larry Crist, Daniel Patte, James Lee, Edward McMahon II, Gary Philips, and Michael Rengstorf. Bloomington: Indiana University Press.

Greimas, Algirdas Julien, and Paul Ricoeur. 1989. "On Narrativity." *New Literary History* 20, no. 3: 551–562.

Høstaker, Roar. 2005. "Latour: Semiotics and Science Studies." *Science Studies* 18, no. 2: 2–25.

Kaplan, David M. 2003. *Ricoeur's Critical Theory*. New York: State University of New York Press.

Kaplan, David M. 2006. "Paul Ricoeur and the Philosophy of Technology." *Journal of French Philosophy* 16, no. 1/2: 42–56.

Krarup, Troels Magelund, and Anders Blok. 2011. "Unfolding the Social: Quasi-Actants, Virtual Theory, and the New Empiricism of Bruno Latour." *The Sociological Review* 59, no. 1: 42–63.

Latour, Bruno. 1987. *Science in Action: How to Follow Scientists and Engineers Through Society*. Cambridge: Harvard University Press.

Latour, Bruno. 1988. "A Relativistic Account of Einstein's Relativity." *Social Studies of Science* 18, no. 1: 3–44.

Latour, Bruno. 1990. "On Actor-Network Theory: A Few Clarifications Plus More than a Few Complications." *Soziale Welt* 47, no. 4: 1–14.

Latour, Bruno. 1992. "Where Are the Missing Masses? The Sociology of a Few Mundane Artifacts." In *Shaping Technology/Building Society: Studies in*

Sociotechnical Change, edited by Wiebe E. Bijker and John Law, 225–258. Cambridge: The MIT Press.
Latour, Bruno. 1993a. "Pasteur on Lactic Acid Yeast: A Partial Semiotic Analysis." *Configurations* 1, no. 1: 129–146.
Latour, Bruno. 1993b. *We Have Never Been Modern*. Translated by Catherine Porter. Cambridge: Harvard University Press.
Latour, Bruno. 1994. "On Technical Mediation: Philosophy, Sociology, Genealogy." *Common Knowledge* 3, no. 2: 29–64.
Latour, Bruno. 1999. *Pandora's Hope: Essays on the Reality of Science Studies*. Cambridge: Harvard University Press.
Latour, Bruno. 2000. "The Berlin Key or How to Do Words with Things." In *Matter, Morality, and Modern Culture*, edited by Paul Graves-Brown, 10–22. London: Routledge.
Latour, Bruno. 2005. *Reassembling the Social: An Introduction to Actor-Network-Theory*. Oxford: Oxford University Press.
Latour, Bruno. 2008. *What is the Style of Matters of Concern? Two Lectures in Empirical Philosophy*. Assen: Van Gorcum.
Latour, Bruno. 2013. *An Inquiry into Modes of Existence: An Anthropology of the Moderns*. Translated by Catherine Porter. Cambridge: Harvard University Press.
Latour, Bruno, and Jim Johnson. 1988. "Mixing Humans with Non-Humans? Sociology of a Few Mundane Artefacts." *Social Problems* 35: 298–310.
McGee, Kyle. 2014. *Bruno Latour: The Normativity of Networks*. London: Routledge.
Ricoeur, Paul. 1984a. *Time and Narrative: Volume 1*. Translated by Kathleen McLaughlin and David Pellauer. Chicago: The University of Chicago Press.
Ricoeur, Paul. 1984b. *Time and Narrative: Volume 2*. Translated by Kathleen McLaughlin and David Pellauer. Chicago: The University of Chicago Press.
Ricoeur, Paul. 1989. "Greimas's Narrative Grammar." In *Paris School Semiotics Volume I: Theory*, edited by Paul Perron and Frank Collins, 3–32. Amsterdam: John Benjamins Publishing Company.
Ricoeur, Paul. 1990. "Between Hermeneutics and Semiotics: In Homage to Algirdas J. Greimas." *International Journal for the Semiotics of Law* 3, no. 8: 115–132.
Ricoeur, Paul. 2016. *Hermeneutics and the Human Sciences: Essays on Language, Action and Interpretation*. Edited by John B. Thompson. Cambridge: Cambridge University Press.

Chapter 4

Ricoeur's Critical Theory of Technology

David M. Kaplan

Paul Ricoeur had very little original to say about technology, other than to echo the critique of instrumental rationality that was commonplace among European philosophers in the aftermath of World War II. In his few scattered remarks on the subject, he contrasts the calculating rationality of science and technology with more humane conceptions of experience. Like other critics of the day, Ricoeur warned about treating thought and action as neutral instruments, like devices or tools. Such an understanding of humanity is, at best, incomplete, at worst, a hidden danger that eventually leads to an impoverished life of cold efficiency and mindless conformity. Ostensibly, as societies become increasingly technocratic, we become less free.

This mid-century critique of instrumental reason is shared by Frankfurt School Western Marxists such as Horkheimer, Adorno, Marcuse, and Habermas, who claimed that reason itself has become equated with formal techniques geared only toward efficient manipulation. As a result, our lives are emptied of their meaning and purpose as our social interactions are increasingly amoral and bureaucratic. The Frankfurt Marxists agree with Weber that "rationalization" was an autonomous cultural process that organizes social life into objects of administration and control. They analyze the totalitarian effects rationalization and the historic and utopian alternatives that would free us from its grip. Ricoeur agrees with both the diagnosis and the cure to instrumental reason, which is not necessarily to his credit.

The chapter briefly examines the main strands of the Frankfurt Marxist philosophy of technology and how Ricoeur generally agrees it. Next, it presents Feenberg's criticism of this direction of thought. Finally, it discusses how Ricoeur's work could add crucial dimensions missing from early and recent critical theory. Specifically, his conceptions of ideology and utopia, and political fragility could add much-needed nuance to a twenty-first-century critical theory of technology.

1. FRANKFURT SCHOOL CRITICAL THEORY OF TECHNOLOGY

Frankfurt School philosophy of technology is transcendental in the Kantian sense. It argues that a particularly perverse form of rationality underlies the social order concealing the oppressive character of both welfare state capitalism and state socialism. This instrumental rationality is said to drive societies, like an independent force that follows its own imperatives—and we conform to it, not the other way around. This critique of technology was crystallized in the 1940s through the 1960s, in the aftermath of the Holocaust and the atomic bomb, and continued through at least the 1980s. While science and technology were often celebrated for their immense successes in medicine, agriculture, and space travel, critics noted that technical progress was also creating a dehumanized, mechanized world that demanded conformity at the expense of our individuality and freedom (Ellul 1964; Illich 1971; Mumford 1967).

There are two main forms of this critique of technology. It is said to be either too empty and value-neutral (where technical means are independent of ends); or it is said to be value-laden (where technical means are connected to ends) by privileging efficiency and control above all else.

Marcuse argues for the latter that technological rationality contains substantive ends that treat people and nature as neutral objects to be controlled. Echoing Heidegger, he speaks of a "technological *apriori*" that projects objects as potential instrumentality, easily co-opted by economic and political power (Heidegger 1982). In his 1941 article, "Some Social Implications of Modern Technology," Marcuse (1982) argues that technological rationality undermines "individual rationality" (or autonomy) by taking efficiency to be the sole standard of judgment. Industrialized societies employ it to convince people to accept things like mass production and bureaucracy that benefits capital but diminishes the quality of life. Marcuse argued that traditional appeals to autonomy appear quaint, conformism seem rational, and protest irrational (Marcuse 1982).

The result is a "one-dimensional" society that erodes the capacity for critical thinking—and a corresponding one-dimensional personality that willingly conforms to a society that limits freedom, imposes false needs, and stifles creativity (Marcuse 1964).

Marcuse did, however, express the hope that one day we will develop technologies that would alleviate suffering and promote happiness. Developing those technologies would require a complete political reversal—a radical break from capitalist modes of production and a new science and technology that would be instruments of liberation. The new science and technology would lead to new forms of cooperation, new modes of production, and new kinds of communities. It would serve our true interests and help satisfy genuine human needs (Marcuse 1962, 203–247).

By contrast, Habermas argues that technologies are neutral with respect to ends. He agreed with Marcuse that technology is a form of instrumental, calculative reason, but disagreed that it is destined to subsume all forms of thought and action. For example, in the 1970s, he attempted to find a middle ground between the realms of rational, efficient technical control and the free, communicative social interaction. However, we do not have to choose one or the other: either objective, value-free, and quantitative reason, or interpretive, normative, and qualitative reason. Instead, we need to "translate technical progress into practical consciousness" by rethinking the relationship between technology and the social world. We should make technology serve democratic ends and not the other way around (Habermas 1971).

In the 1980s, Habermas argued for a two-tiered concept of social rationalization. Bad, one-sided rationalization, based on science and technology, operates in self-regulating economic (money) and administrative systems (power). Good, social rationalization, based on communication, operates wherever people are free to speak and act together. The problem that the advanced industrialized societies face is that economic and political systems have broken free and have colonized the social world, subjecting people to market imperatives, economic crises, and state oppression (Habermas 1987).

In the 1990s, Habermas continued to frame critical social theory in terms of two-tiers but argued that *legal* communication mediates technical systems and the social lifeworld. The law has two faces: a normative one that reflects participants' own moral understandings of themselves and societies, and an empirical one that is subject to economic and institutional mechanisms of power. The role of the law "is to tame the capitalist economic system, to restructure it socially and ecologically in such a way that the deployment of administrative power can be brought under control" (Habermas 1996, 410). Now, public debate about legal reform is the key to decolonizing system by the lifeworld.

Habermas's 2000s works on genetic enhancements, embryo research, and preimplantation genetic diagnosis continue his argument that anything in the realm of technical expertise should be subject to public deliberation and the law. Unregulated genetic interventions threaten to end the liberal ideal of autonomy. Now, biotechnology needs taming (Habermas 2003).

2. RICOEUR'S PHILOSOPHY OF TECHNOLOGY

Ricoeur's take on technology sometimes resembles Marcuse, other times Habermas. Sometimes he treats technology as tool usage, other times as a value-neutral reason, and still other times as an ideology of dominant group interests.

For example, in *Freedom and Nature*, Ricoeur briefly discusses tools and instruments in his analysis of the role of the body in voluntary actions. He distinguishes between the way we "use" our bodies from the way we use tools. On the one hand, the tool is incorporated into action and extends it as an "organic mediator." One's attention is focused on the pragma (what is to be done by me), not on "the indivisible pair of organ and tool seen as an extension of the organ" (Ricoeur 1966, 213). On the other hand, the tool is entirely in the world, functioning like a natural force subject to the laws of physics. "The physical, industrial character of the relation of tool to work absorbs the organic character of man to the tool" (Ricoeur 1966, 213). Like Marx, we become appendages to machines.

The will-organ-tool-work relation is ambiguous because it can be read in two directions: from the will (and point of view of phenomenology) or the work (and the point of view of empirical science). The tool is "the point in which these two interpretations meet" (Ricoeur 1966, 214). A phenomenology of tool use reaches its limits in the objective characteristics of artifacts, which can only be *explained* in the manner of a thing, not *understood* in the manner of interpretive experience. On Ricoeur's account, technology is a neutral means, understood empirically, that extends but does not modify the will. Explanation and understood are dialectically related in his qualified use of the term.

Elsewhere, Ricoeur argues that neutral techniques of reason, not instruments, are the essence of technology. In *History and Truth*, for example, he writes: "The technical world of material tools and their extension into machines is not the whole of man's instrumental world. Knowledge is also a tool or instrument. . . . Knowledge becomes stratified, deposits of knowledge accumulate like tools and the worlds which result from them" (Ricoeur 1965, 83). The quest for knowledge is like a "technical pursuit" that uses the acquired knowledge of the past "as a tool or instrument" to carry history forward (Ricoeur 1965, 83).

For this reason, he says we can speak of "technological revolutions" that uniformly advance societies regardless of their particular cultures and traditions. The resulting "world-wide technics" helps developing nations "approach a certain cosmopolitanism" (Ricoeur 1965, 272). Technology, on this reading, is the neutral instrument of historical progress. "It would be absurd to condemn machines, technocracy, administrative apparatus, social security, etc. Technical procedures and, in general, all 'technicity,' have the innocence of the instrument" (Ricoeur 1965, 107).

But elsewhere in *History and Truth*, Ricoeur warns about the dehumanizing effects of scientific-technological progress. For every advancement, it brings about, it "at the same time constitutes a sort of subtle destruction . . . of the ethical and mythical nucleus of mankind" (Ricoeur 1965, 276). The kind of world civilization we are creating is "mediocre civilization" that is "wearing away" at the cultural resources of the "truly great civilizations of the world" by creating a uniform, standardized culture. Ricoeur notes that "everywhere throughout the world, one finds the same bad movies, the same slot machines, the same plastic or aluminum atrocities, the same twisting of language by propaganda, etc." (Ricoeur 1965, 276). The triumph of such a conformist consumer culture, where everything is identical and everyone anonymous, would "represent the lowest degree of creative culture" and a danger "at least equal and perhaps more likely than that of atomic destruction" (Ricoeur 1965, 278). Ricoeur here echoes Marcuse.

3. FEENBERG AND TWENTY-FIRST-CENTURY CRITICAL THEORY OF TECHNOLOGY

Feenberg astutely notes that both Marcuse and Habermas (and their generation of post-war critics of technological rationality) treat technology as a juggernaut rather than something that bends to the will. Instrumental, value-free theories and substantive, value-laden theories both share what he calls "take it or leave it" attitude. We can either accept or reject technology; we cannot change it—or refute the idea of efficiency. But Feenberg asks if that is really the case (Feenberg 2002, 8).

Instrumental theories treat technologies as neutral with respect to ends, with no valuative content of their own. They are simply instruments, indifferent to the ends they are employed to achieve, indifferent to culture and history, functional in any context, and always bound to a universal standard of efficiency. By contrast, substantive theories treat technology as value-laden but based the values of efficiency, standardization, and control. It is an autonomous cultural system that overrides all other competing values and restructures the entire social world as an object of control. Feenberg, however, argues that both theories misinterpret technology as a destiny beyond change or repair (Feenberg 2002, 9–37).

Consequently, proposals to reform technology policy assume all that can be done is to limit it or place boundaries around it. For example, if there is too much pollution, we should all reduce, reuse, and recycle. If reproductive technologies are too invasive, we should resist them. If genetically engineered foods are questionable, we should stop making and eating them. If communication technologies are outpacing our moral political frameworks, we should slow them down. According to both instrumentalist and substantive theories, the only way to respond to technology is to protect ourselves against it.

According to Feenberg, both theories ignore the possibility of simply changing whatever it is we do not like. Technologies are in fact products of our intentions and desires and we can design whatever purposes and values we like into our things. We can change them because they are contingent, not necessary. They are flexible, social objects open to interpretation that have technical properties incorporated into their very structure. The "double aspect theory" that he proposes explains that social meaning and functional rationality are intertwined as two inextricable aspects of same object (Feenberg 2002, 12–20).

Critical theory of technology aims to identify the social horizons in which devices are produced, to remove any illusion of necessity, and to expose the relativity of technical choices. The danger of the apparent neutrality of functional rationality is that it is often enlisted in support of a hegemony, which systematically decontextualizes the technical aspects of things in order to secure their power and authority. These aspects hide behind technical rules and procedures.

In place of Weberian rationalization, Feenberg proposes a "democratic rationalization" that would incorporate democratic values into industrial design. It is not just a dream: actual examples include union struggles over workplace health and safety, community struggles over development and pollution, environmental design and LEED Certification, and universal design for people with disabilities. The key to a just and happy society is to design our world to create more meaningful and livable environments. Feenberg combines the substantivism of Marcuse but with the liberal sympathies of Habermas (Feenberg 2002, 162–182).

4. RICOEUR'S CONTRIBUTION TO TWENTY-FIRST-CENTURY CRITICAL THEORY OF TECHNOLOGY

Whatever Ricoeur's contribution to philosophy of technology might be, it is not a substantive-democratic theory. His philosophical temperament is too cautious and tentative for anything that bold—his work is more like Habermas' than Marcuse's.

He's a self-described "post-Hegelian Kantian," which means that he attempts to both mediate and not-mediate, overcome limits and accept limits. Like Hegel, Ricoeur believes we can find reason, truth, and moral right through philosophical mediation; like Kant, he believes that human experience and philosophy are riddled with *aporia* that have only practical responses, not theoretical solutions. Ricoeur's "third term" mediates without reconciling: it is like an arc drawn from one philosopher to another, that both relates the work of another and leaves it alone (Ricoeur 1974, 398–420).

Olivier Abel calls this method of non-synthetic reconciliation Ricoeur's "ethics of method." For moral reasons, Ricoeur takes great pains to respect the differences among the philosophies he brings together (Abel 1993).

For example, in his rapprochement of hermeneutics and critical theory, Ricoeur claims not "to fuse a super-system which would encompass both," but instead to let each "recognize the other, not as a position which is foreign and purely hostile, but as one which raises in its own way a legitimate claim" (Ricoeur 1991, 271). Usually, he relates something humanistic to something objectifying. In *Freedom and Nature*, it is phenomenology and empirical science; in *Freud and Philosophy*, hermeneutics and psychological explanations; in *Conflict of Interpretation*, hermeneutics and structuralism; in *Oneself as Another*, the ethical aim with the moral law. Each time, Ricoeur preserves an objective, explanatory, transcendent element without entirely subsuming it to a broader, interpretive framework.

We see this non-synthetic mediation, or poetics, again and again throughout his career. In *Freedom and Nature*, it is a dream of transcendence and embodiment; in *Symbolism of Evil*, it is a surplus of meaning; in the *Rule of Metaphor*, it is a proposed world we could inhabit; in *Time and Narrative*, a discordant concordance; in *Oneself as Another*, the call of conscience in response to tragic action. His works themselves are examples of a post-Hegelian Kantian hermeneutic phenomenology.

Yet, while remarkable in its originality, Ricoeur's commitment to non-synthetic mediation does not lend itself very well to the kinds of creative applications some of his readers have undertaken. To treat his works as solutions or frameworks to be applied betrays his ethics of method. In particular, it does not support the argument that humans and their technologies are inextricably linked and co-constituting. For philosophers of technology who rely on mediation theory, they rightfully turn elsewhere for support; or they recruit Ricoeur into a project that betrays the *ethos* of his philosophy (Ihde 1990; Verbeek 2011). Even careful readers can miss the difference between his version of non-synthetic mediation and ordinary, Marxist mediation (Kaplan 2009).

Granted, sometimes he does what he says he does not, and proposes the very mediations he claims to eschew (Kaplan 2003). But if readers today want to follow his lead and approach issues in a Ricoeurean fashion, then we should recognize that he does not propose answers as much as creative and practical responses to what could be—as-if our proposal were true. This approach can be maddeningly tentative, but so is Ricoeur.

For example, both Habermas and Feenberg argue that technologies should be more democratic but importantly in different ways. Habermas claims that technical-scientific reason is *entirely different* from and should be subordinate to social-communicative reason. Feenberg argues that *technology itself* should be democratized by incorporating our values directly into industrial design. Now, which conception would Ricoeur endorse: Democratic regulation of technology or democratic rationalization? Gadamer used to complain that Ricoeur sided with Habermas in the hermeneutics-ideology critique debates of the 1970s, and he is right (Ricoeur 1991, 494).

There are, of course, important differences between Habermas and Ricoeur. For example, when Habermas speaks of ideology, he takes over the Western Marxist conception of it as false consciousness, legitimation of power, and social pathology that hides the truth and thwarts freedom. Habermas brings the linguistic turn to the critique of ideology with the idea that systematically distorted communication prevents us from truly understanding each other and acting together.

But when Ricoeur speaks of ideology, he is not only informed by Western Marxism but also by sociologist Karl Mannheim, anthropologist Clifford Geertz, and his own works on metaphor and narrative. His thesis is that ideology should be understood in contrast to utopia as two poles of a single "cultural imagination" (a concept borrowed from Cornelius Castoriadis). The cultural imagination is made up of ideas, stories, and images a society has about itself that organizes social life (Ricoeur 1986).

What distinguishes ideology and utopia from other interpretive schemas are the way they function. Ideology consolidates, integrates, and orders a society according to the interests of a dominant group; utopia calls a society into question, and seeks to shatter a social order for the sake of liberation. Both are ultimately about power: ideology legitimates it, utopia challenges it.

According to Ricoeur, they each have both positive and negatives senses. The positive, constructive function of ideology is to constitute and conserve social relations; its negative, destructive function is to resist the transformation of an order that has frozen social relations in such a way that sustains domination. Ricoeur claims that there is "no social integration without social subversion" (Ricoeur 1986, 16).

As for utopia, its positive function is to call a society into question from an imagined vantage point; the negative function is to provide flight from social reality. A utopia is a view from nowhere in terms of which we might examine—or flee from—our social reality. "Utopia is the mode in which we radically rethink the nature of family, consumption, government, religion, and so on. From 'nowhere' emerges the most formidable challenge to what-is" (Ricoeur 1991a, 184).

Ideology and utopia are imaginary and false perspectives that are "non-congruent" with social reality yet vital for members of a community. From a Marxist perspective, non-congruence with reality is either false-consciousness or naïve escapism. But from the perspective of the cultural imagination, they are simply imaginative interpretations of what a society is and what it could be. Ricoeur goes so far as to say that "we only take possession of the creative power of the imagination" through a relation to such figures of false consciousness as ideology and utopia (Ricoeur 1986, 311). In other words, social understanding relies on fictions and there is no way around it. There are no noninterpretive perspectives to view the world objectively.

Ricoeur's nuanced take on ideology and utopia has bearing on a critical theory of technology. For example, if neither is entirely good or bad, then the very idea of a critique of ideology needs to be qualified. Both Habermas and Feenberg treat instrumental rationality as ideology that should be subjected to communicative action, or transformed to serve, not limit, us. Both see functional rationality and technical decision-making as false universals. The true universal is some kind of broader linguistic-interpretive frame that contextualizes and relativizes the technical within the social.

From the perspective of Ricoeur's thought, technological rationality may be ideological in either a negative or positive sense. In the 1960s, he treated it in the manner of the day as the antithesis of human thought and action. But recently, as more nations retreat from a liberal global order and embrace ethno-nationalism, the open-minded now turn to science and technology to rescue objectivity and neutrality from the current close-minded climate. Like the positivists of the 1930s, progressives hold onto the hope that our political leaders will rely on uncontestable truths and objective facts, and reject conspiracy theories and pseudo-science.

In the United States, one's political identity is largely constituted by whether one accepts or rejects science and technology. You either believe in climate science or you do not; you heed the advice of public health experts in the name of safety, or you ignore it in the name of liberty. The entire U.S. political stage is divided not only along party lines but also by attitudes toward the authority of reason. The vast majority of Democrats want science and scientists to play an active role in policy debates, while the majority of Republicans say scientists should stick to science (Kennedy and Funk 2019).

After Obama was reelected in 2012, one governor warned his fellow Republicans that they "must stop being the stupid party" and "stop insulting the intelligence of the voters" (Sink 2013). Shortly after his plea for self-reflection fell flat, he stepped back in line with party and continued its caricatures of American liberalism as the party of big government paternalism that believes "32 oz. sodas are evil; red meat should be rationed . . . trans-fat must be stopped . . . wild weather is a new thing" (Jindal 2013). He realized there was no need to challenge a party orthodoxy that was already relying on fabrications and ignorance to maintain the status quo.

Meanwhile, the Democratic Party identifies itself as the party of climate change and medical science—more attuned with Silicon Valley than the Rust Belt. Their political candidates openly embrace brighter technological futures in ways that ring of post-war ideology, yet in the present context, it helps to consolidate the political identities of millions in a positive, constructive way. In other words, the same phenomenon functions as ideology albeit in opposite ways: one to conserve a social order, another to galvanize a community. The mask-wearers and mask-refusers not only display their group membership but their views on science and technology.

Or, perhaps we can say, in keeping with Ricoeur's work, that Democrats invoke science and technology as utopia (in the positive sense) to call into question the way Republicans invoke science and technology as ideology (in the negative sense). One way to raise critique within the hermeneutic circle is to contrast ideology to utopia, and utopia to ideology. As Ricoeur says, we "cannot get out of the circle of ideology and utopia, but the judgment of appropriateness may help us understand how the circle can become a spiral" (Ricoeur 1986, 314).

Another way Ricoeur's work differs from both old and new critical theory is his generally positive view of objectifying rationality. In most of his mediations he treats it sympathetically, even as the *critical* moment within a broader interpretive philosophy—the exact opposite of Habermas and Feenberg!

For example, in the 1970s, he distinguished his hermeneutic theory by the "positive and productive" character of *distanciation* that enables communication in and through distance in a way Gadamer supposedly fails to recognize. Distanciation allows for a critique of ideology to be incorporated, as an *objective and explanatory segment*, in the process of communication and self-understanding. Ricoeur explains that "distanciation, dialectically opposed to belonging, is the condition of possibility of the critique of ideology, not outside or against hermeneutics, but within hermeneutics" (Ricoeur 1991, 268).

Ricoeur wants to show that hermeneutics, properly conceived, is also critical and evaluative. Distanciation opens the possibility of achieving critical distance from oneself and one's tradition because the very medium of understanding itself is distanced from itself. We are never so beholden to our historical perspectives that we cannot detach and assess them in relation to methodological explanations, such as Marxist and Freudian theories. Ricoeur's is a critical hermeneutics thanks to the objective and explanatory moments that are never fully overcome. "Distanciation," he argues, "is the soul of every critical philosophy" (Ricoeur 1974a, 249).

In fact, Ricoeur treats objectifying explanations (as well Habermas's regulative ideal of unconstrained communication) as neutral, functional, and modular as the critical element that should be *preserved*, not the ideological elements that should be exposed. Ricoeur is far more science and technology-friendly than either Habermas or Feenberg.

He also tempers the democratic ambitions of critical theory with his notion of the "political paradox." Democracy, for Ricoeur, is always fraught. On the one hand, any political authority is legitimate if it comes from the rational consent of the governed; on the other hand, political practice is often so violent and coercive that it is something to which individuals, in principle, cannot consent. The political sphere is a fragile balance of authority and domination, reason, and tradition. No political regime can ever be entirely legitimate because they are all potentially too violent to be entirely just. The paradox of political power is ineliminable.

Elsewhere, Ricoeur calls it the "fragility of politics," which stems from fragility of political language, situated in a "vulnerable zone" between rational argumentation and rhetoric. It is vulnerable because ideology and utopia are permanent features of our social life, always present in politics, frustrating any attempt to "purify our language into a vehicle for transparent political representation." So, instead of pure democracy, Ricoeur advocates an impure democracy—one that is always tied to particular situations, specific practices, and ineliminable conflicts (Ricoeur 1987).

For example, there are conflicts over the idea of consensus itself: neither the rules of discourse, the subject matter, the participants, nor the conclusions are ever fully decidable. There are conflicts over the ends of government. Terms like "security," "freedom," and "justice" are necessary for public debate yet open to interpretation and subject to ideological appropriation. And, finally, there are conflicts over what counts as the good life. Ricoeur explains that "from such conflicts, such a plurality of ends, and such a fundamental ambivalence comes the fragility of political language" (Ricoeur 1987, 43).

The democratic politics of technology of Habermas and Feenberg appear naïve in comparison. Forget for the moment the practical difficulties of asking people to participate to technological decision-making: how it would work, who would make decisions, how people would be informed (Winner 1991; Sclove 1995). Ricoeur questions more than the feasibility of democratic politics but the limits and dangers of it. The ideal of civic participation in technological affairs would empower people to fundamentally transform the shape of our daily lives as workers, commuter, patients, and media users—everywhere our lives are suffused with man-made things. The potential for use and abuse of power would be magnified to unimagined dimensions: transportation, biotechnology, infrastructure—geo-engineering?! It's not immediately clear why it would be desirable to give responsibility for such things to anyone but those who are solely committed to the public good.

That said, even though political discourse is fragile, Ricoeur is confident that we can use political means to "complete the Enlightenment project" and "recover the unfulfilled promises of the past." He implores us:

> not to flee the field of political confrontation, but to enter it with a sense of measure that leads to great respect for the extreme fragility of the "good life" a

life for which "good" government serves as the most proximate figure open to us as political animals (Ricoeur 1987, 41).

5. CONCLUSION

A critical theory of technology in the manner of Ricoeur would build on Habermas and Feenberg: expose injustices encoded into things and expand political and legal deliberation into the realm of things. But it would do so cautiously, with eyes wide open to human fallibility and political fragility. Democracy is, after all, no panacea—even if (somehow) it was purged and purified of conflicting interpretations and ideology. Furthermore, Ricoeur maintains the need not only for Enlightenment ideals but also the ideals of our pre-modern heritage that extends back to antiquity. We need to take a longer look at our history and relearn lessons from the failures and successes of the past—and to learn from the experiences of others from different traditions. Above all, we should be mindful that political discourse is always vulnerable to rhetoric and distortions; and that ideology and utopia are permanent parts of the politics of technology—in both positive and negative senses. Ricoeur moves twenty-first-century critical theory of technology beyond politics, into poetics: imagining new ways to see and to be in the world of artifacts.

REFERENCES

Abel, Olivier. 1993. "Ricoeur's Ethics of Method." *Philosophy Today* 37, no. 1: 23–30.
Ellul, Jacques. 1964. *The Technological Society* (J. Wilkinson, trans). New York: Knopf.
Feenberg, Andrew. 2002. *Transforming Technology: A Critical Theory Revisited*, 2nd edition. New York: Oxford University Press.
Habermas, Jürgen. 1971. *Toward a Rational Society: Student Protest, Science, and Politics* (J. Shapiro, trans). Boston: Beacon Press.
Habermas, Jürgen. 1987. *Theory of Communicative Action, Vol. 2: System and Lifeworld* (T. McCarthy, trans). Boston: Beacon Press.
Habermas, Jürgen. 1996. *Between Facts and Norms* (W. Rehg, trans). Cambridge: MIT Press.
Habermas, Jürgen. 2003. *The Future of Human Nature* (H. Beister and W.Rehg, trans). New York: Polity Press.
Heidegger, Martin. 1982. *The Question Concerning Technology and Other Essays* (H. Lovett, trans). New York: Harper Perennial.
Ihde, Don. 1990. *Technology and the Lifeworld*. Bloomington: Indiana University Press.

Illich, Ivan. 1971. *Deschooling Society*. New York: HarperCollins.
Jindal, Bobby. 2013. "The GOP Needs Action, Not Navel Gazing." In *Politico*, June 18, 2013. https://www.politico.com/story/2013/06/bobby-jindal-opinion-gop-needs-action-092933#ixzz2WaFBxNKU. Accessed June 2020.
Kaplan, David M. 2003. *Ricoeur's Critical Theory*. Albany: SUNY Press.
Kaplan, David M. 2009. "Paul Ricoeur and the Philosophy of Technology." *Journal of French and Francophone Philosophy* 16, no. 1–2: 42–56.
Kaplan, David M. 2011. "Thing Hermeneutics." In *Gadamer and Ricoeur: Critical Horizons for Contemporary Hermeneutics*, edited by George H. Taylor and Francis J. Mootz. London: Continuum Press.
Kennedy, Brian, and Cary Funk. 2019. "Democrats and Republicans Differ Over Role and Value of Scientists in Policy Debates." *Pew Research Center*, August 9, 2019. https://www.pewresearch.org/fact-tank/2019/08/09/democrats-and-republicans-role-scientists-policy-debates/. Accessed June 2020.
Marcuse, Herbert. 1964. *One-Dimensional Man*. Boston: Beacon Press.
Marcuse, Herbert. 1982. "Some Social Implications of Modern Technology." In *The Essential Frankfurt School Reader*, edited by Andrew Arato and Eike Gebhardt. New York: Continuum.
Mumford, Lewis. 1967. *The Myth of the Machine*. New York: Harcourt Brace Jovanovich.
Ricoeur, Paul. 1965. *History and Truth* (C. Kelbley, trans). Evanston: Northwestern University Press.
Ricoeur, Paul. 1966. *Freedom and Nature: The Voluntary and the Involuntary* (E. Kohak, trans). Evanston: Northwestern University Press.
Ricoeur, Paul. 1974a. *Political and Social Essays*. Edited by David Stewart and Joseph Bien. Athens: Ohio University Press.
Ricoeur, Paul. 1974b. *Conflict of Interpretations: Essays in Hermeneutics* (W. Domingo, trans). Evanston: Northwestern University Press.
Ricoeur, Paul. 1986. *Lectures on Ideology and Utopia*. Edited by George H. Taylor. New York: Columbia University Press.
Ricoeur, Paul. 1987. "The Fragility of Political Language." *Philosophy Today* 31, no. 1: 35–44.
Ricoeur, Paul. 1991a. "World of the Text, World of the Reader." In *The Ricoeur Reader*, edited by Mario J. Valdés. Toronto: University of Toronto Press.
Ricoeur, Paul. 1991b. "Hermeneutics and the Critique of Ideology." In *From Text to Action: Essays in Hermeneutics, II*, edited by Kathleen Blamey and John B. Thompson. Evanston: Northwestern University Press.
Sclove, Richard. 1995. *Democracy and Technology*. New York: Guilford Press.
Sink, Justin. 2013. "Bobby Jindal: GOP Needs to 'Stop Being the Stupid Party.'" *The Hill*, January 25, 2013. https://thehill.com/in-the-news/279243-jindal-republicans-must-stop-being-the-stupid-party. Accessed June 2020.
Verbeek, Peter-Paul. 2011. *Moralizing Technology: Understanding and Designing the Morality of Things*. Chicago: University of Chicago Press.
Winner, Langdon. 1991. "Artifact/Ideas and Political Culture." *Whole Earth Review*, no. 73: 18–24.

Chapter 5

Free the Text!
A Texture-Turn in Philosophy of Technology
Bruno Gransche

Paul Ricoeur is famous for his hermeneutics and oeuvre on understanding language, discourse, and literature. He understands the human condition as inherently intertwined with stories, and the human-world relation as essentially narratively mediated. His focus lies on language and writing—he barely mentions technology. Nonetheless, his insights can help to better understand today's socio-technical ensembles and our relationship with technology.

The mechanism enabling a transposition of Ricoeur's findings from language to technology is to interpret both as two notions of their common denominator: *texture*. The basic etymological meaning of texture is *weaving*, meaning a technical process of configuring separate parts into a whole that is then characterized by the properties of those parts *plus* the properties of the weave. This article makes use of this notion of texture, which is a powerful thinking tool both metaphorically and literally and a great integrator of philosophy, narratology, and technology. To interpret language and technology by their textuality allows us to investigate their *similarity*, which means both sameness *and* differentness. Philosophical discussions on the topic are predominantly focused on the differentness: philosophy of language and philosophy of technology are quite different and language is—as in Ricoeur's oeuvre and generally in hermeneutics—by far more prominent. The *linguistic turn* affected many philosophical investigations and especially philosophy of technology fought some *turn-wars* to overcome this turn and counter it with a *material* and *empirical turn* (Franssen et al. 2016), *return to things (themselves)* (Verbeek 2005; Wiltse 2020), or *object-oriented ontology* (Harman 2013), and so on.

On the other hand, the sameness side of language and technology was far less considered, even though it holds the power to settle some of the conflicts of the turn-wars. It shows that text is not merely a language phenomenon and that considering certain realms as text-like is not an incursion of linguistic schemata into nonlinguistic realms, but a poietic-hermeneutic perspective that addresses phenomena as being made, that understands *what cloth they are cut from*—which also brings in a material focus. This article proposes a texture-oriented interpretation of language and technology with a focus on the sameness side—without implying the difference side to be minor or neglectable. This will pave the way to read Ricoeur as a hermeneutics of texture rather than of language, and thus to apply his findings not only to linguistic texts but to *techne*-texts as well.[1] This texture-hermeneutics can then be linked to other philosophical hermeneutics and to the concept of technology as a medium with a focus on recent and "classic" German philosophers; the link is to think of media as textured-texturing spaces of possibilities where modal tracks and trails are to be traced, read, and understood. This chapter will propose a set of selected tools based on this link and finish by questioning the notion of "understanding machines" to exemplarily demonstrate the potential of those tools.

1. NEW TOOLS FOR THE SAME OLD MISSION

Three main sources can help to support this similarity operation under the common denominator *texture* and it is part of the intention of this article to show their suitability and to propose a more extensive use thereof in the current international philosophy of technology.

Firstly, there are traces in Ricoeur's writing that suggest the similarity between language and technology: Ricoeur's thoughts on language, the realization of language (discourse), text, writing, and reading can be understood as rather technical indeed; he grasps the idea of the configuration of making sense as a *techne* and authors as "artisans who work with words" (Ricoeur 2012, 45): "Discourse thereby becomes the object of a *praxis* and a *techne*" (Ricoeur 2016, 98). Such remarks as well as his prominent use of *poiesis* suggest links to technology in his philosophy and offer glimpses of the sameness side.

Secondly, there are some "dusty" nineteenth- and twentieth-century German philosophers—for example, Hegel, Kapp, Dilthey, and Cassirer—that offer helpful philosophical tools even and especially for today's being-with-technology. They are without a doubt philosophical heavyweights, but somehow considered outdated in the international philosophy of technology debate—not much unlike Ricoeur—until now. Even though their work has several shortcomings or—from a current point of view—questionable suitability, they have great potential for a somehow eclectic fresh use of some of their insights: for example, G. W. F. Hegel's teleological development of the world spirit or his notion of the absolute spirit need not necessarily be affirmed to use his helpful notion of the objective spirit. Or, even though Ernst Kapp's idea of technology as organ projection and way of self-understanding cannot be affirmed today (see Gransche 2020a), he merits the rightful and momentous inclusion of technology as part of the objective spirit.

Thirdly, apart from the career of mediation theory in English-speaking philosophy of technology, recent German philosophy—especially of technology—introduced a fundamental *medial turn*[2] that is very well grounded in German philosophical tradition and connected to other (especially French) philosophical traditions. Additionally, central works of this German medial turn—represented particularly by Christoph Hubig[3]—offer a selected, commented, and updated link to the first source. Unfortunately, they have not been translated into English and are therefore hardly considered outside the German-speaking world, a fact I seek to change with this article as well as in others (e.g., Gransche 2020b).

How could those sources support a similarity-focused analysis of language and technology and the interpretation of Ricoeur as texture hermeneutics, thus using its potential in philosophy of technology?

First, the concept of texture is understood in its earlier meaning as a weave and therefore as a (human-)made configuration of raw material.[4] In this context, what is textured is (man-)made, fabricated, technically put together. All texturing is responsible for the possibility of future texturing because the latter happens under the condition of the former. Therefore, the resulting texturing condition must be considered by the individual as given, as restricting and enabling. Second, this notion of the realm of textured entities can be linked to Hegel's definition of the objective spirit, as man-*made* entities that are *given* to the individual in the form of necessity. Classically, language is probably the most prominent part of the objective spirit, but Ernst Kapp and Ernst Cassirer were right to include technology as well. So, not just being *textured* is a common denominator of language and technology, but being part of the objective spirit, or sphere of culture, which means that they both are also *texturing*; they are characterized by a *made-yet-given* structure. This suggests that Ricoeur's insights on language can be transposed to technology especially where he focuses on this trait as made-yet-given. Third, this focus connects language with technology, society with the individual, and Ricoeur's hermeneutics with today's philosophy of technology.[5] Within the latter, it connects English-speaking mediation theory and postphenomenological tendencies with the German philosophy that considers technology as a medium. Fourth, Ricoeur's thoughts about understanding text and "oneself in front of the text" (Ricoeur 2016, 106) can help achieve a better understanding of (digital) technology and oneself in front of—or rather intertwined with—that technology.

To respond to a philosophical duty as coined by Cassirer, Ricoeur's work on the conditions of the possibility of language needs to be complemented by a respective focus on technology: "If philosophy," Cassirer writes (2012, 18), "wants to remain loyal to its mission, . . . it must also enquire into the 'conditions of the possibility' of technological efficacy and technological formation, just as it enquires into the 'conditions of the possibility' of theoretical knowledge, language and art." This chapter, as part of the joint efforts of the research group *Ricoeur and Technology*, accepts that mission by proposing a respective focus on the conditions of the possibility of texture, including both language *and* technology.

2. TEXTURE AND TECHNE—THE FREED TEXT

"What structural analysis discloses as the texture of the text, is the very medium within which we can understand ourselves" (Ricoeur 2016, 105). The *texture of the text* is for Ricoeur the medium of understanding; this notion will be subsequently unfolded and linked to (especially German) philosophy of technology as a medium. The phrase *texture of the text* seems to be pleonastic. Both terms share the same root, the Latin verb *texere*, which means to weave, fit together, construct, and fabricate and its past participle *textus*—something woven, cloth—which refers to words being woven into a written page just like threads or yarn into cloth. *Texture of the text* refers to the weave of the cloth or the structure of a construction; one is the composed result—*text*—and the other is the way the previous elements were combined—*texture*.[6] Both the Latin *texere* and the Greek *techne* go back to the same Proto-Indo-European root *teks*, which means to weave or fabricate. The notion of *weaving* is an apt tool to understand texture, techne, and text, which will be utilized in this article. Both, the texture of text and the texture of technology, are only pleonasms prima facie that are worth being examined further.

One type of text is *narrative text*, which is a paradigmatic and the most common notion of text. Clifford Geetz understands "culture as an assemblage of texts" (Geertz 2006, 448) and suggested "an extension of the notion of a text beyond written material" (Geertz 2006, 448). Interestingly, he grounds this extension on Ricoeur, who found earlier: "This notion of text—thus freed from the notion of scripture or writing—is of considerable interest" (Ricoeur 1970, 25). So the operation to free, in turn, Ricoeur's insights on text from its obvious language context and to apply it to other texts, for instance, in a technological context, seems to find at least Ricoeur's interest, if not approval. The narrative function of weaving (*emplotment*) transforms "the irrational contingency into a controlled, meaningful, intelligible contingency" (Ricoeur

1986, 14, my translation)[7]—narrative texturing makes sense of otherwise erratic heterogeneous elements. The process of this transformation appears in Ricoeur's works as a configuration or *mise en intrigue* (Ricoeur 1983, 102), a concept based on Aristotle's *mythos* (Ricoeur 2012, 31–51). The German translation of *mise en intrigue* (*Einfädeln* einer Intrige) means to *thread* or to *interweave* events into a tale. The weaving metaphor combines the *mise en intrigue* with two more recent philosophical perspectives on technology, namely technology as *texture* and as *medium* (Gransche 2015, 157–178) as well as to the phenomenological philosophy of history of Wilhelm Schapp for whom man is anthropologically the *being that is interwoven in stories* (Schapp 2012). The topos of a sacrificing hero, for instance, is a technique of making sense of death, just as stories about the afterlife, rebirth, and so on help coping with bereavement. The real threat in life is not death, but meaninglessness.

The ancient metaphor of a poet weaving stories like an artisan weaves cloth leads to the mystical *techne* of the Goddess Athena.[8] Her *techne* meant not only to *weave* material but also signs and symbols as well as actions and social relations. Athena grants and teaches the weaving of (a) material elements like thread or wood into cloth or buildings, of (b) intelligible elements like characters, notes, or numbers into books, music sheets, or calculations, and of (c) action elements like declaring war or marrying into routines and customs.[9] All three are interconnected: intelligible and social *technai* have material bases. Material and intelligible *technai* affect or aim at other people and their relations. Material and social *technai* are only *technai* in the first place because they are also intelligible (repeatable, teachable, etc.). This threefold techne-concept can be linked to Ricoeur's work. He focuses on the intelligible level—characters, words, sentences, writing—yet he thinks of language as a formable material and links the material and symbolic level with the artisan metaphor, so for him "the author is the artisan of a work of language" (Ricoeur 2016, 100). Through discourse, which is the realization of language, the social level comes into play: "So discourse not only has a world, but it has an other, another person, an interlocutor to whom it is addressed" (Ricoeur 2016, 95–96).

In the field of digital technologies, those three techne-levels are firstly the hardware or the combination of chips, cables, hard drives, batteries, and so on. Second, on the symbolic level, there is the basic binary code of 0 and 1—which makes them *digital* technologies—and with this symbolic 0-1-threads, all sorts of intelligible "cloth" are formed, stored, and transmitted in the form of texts, pictures, videos and audios, but also other information or routines, algorithms, procedures, scripts, and so on. Thirdly, on the social level, the digital technologies actualize effects like connecting people, networking (social networks), communication (email), inclusion/exclusion of groups, connecting customers or employees with companies (advertising, data economy), voters with candidates, future partners, dates, wives or husbands, and so on. Thus, the three techne-levels are three texture-levels that concern both language and technology. A rather recent development on the material level with extensive consequences for the action and social relation level is the far-reaching pervasion of digital technology of ever more parts of our life. The hardware spreads out from the factories or labs into our cities, cars, homes, and bodies thus extending the reach of digital systems to allegedly everywhere—hence terms like ubiquitous computing or smart everything. So, the material sphere potentially affected by digital technologies is growing, and it outgrew the desktop device by far.

3. GIVEN TRACKS AND TRAILS MADE

This article conceptualizes both language *and* technology, realizations of language *and* of technology, as objectivations of life, that is, entities of the objective spirit. However, this does not mean that language and technology are the same, though they are of the same kind in terms of *texture*, their *textured texturing* function, and their status as *objectifications of life*.

One aspect that can help to place today's technosphere in relation to Ricoeur's work on language—building primarily on Hegel, Kapp, and Cassirer—is that both language and technology are part of the *objective spirit*. Ricoeur understands the "objectifications of life," the entire "social and cultural worlds" as "appearances of the Hegelian objective spirit" (Ricoeur 2016, 12). These *made-yet-given* worlds are media that connect the individual with its present as well as past—and future[10]—society, the society to itself, and the individual to itself. Technology is a basic power of culture (Cassirer 2012, 22) and technologies are an essential part of those worlds (Kapp 2018, 23). Communication between individuals and collectives—especially "communication in and through distance" (Ricoeur 2016, 93)—is paramount for the formation of cultural entities. Ricoeur sees text as "the paradigm of distanciation in communication" (Ricoeur 2016, 93) and so are the communication *technai*. Technologies, and digital systems in particular, create distance and mediate the interpretation and influence of life across that distance.

To recapitulate: Text is for Ricoeur a structured work, a realization of discourse. Discourse is the realization of language. The work of discourse is a projection of the world. Discourse and the work of discourse mediate understanding of life by objectifications of life. "Taken together, these features constitute the criteria of textuality" (Ricoeur 2016, 94). To transpose this definition of language-textuality to techne-textuality, the analogy could look like this (table 5.1).

Table 5.1   Criteria of Textuality

I	II	III
Language	*Discourse*; Realization of language	*Writing*; language-text; Work of discourse
Technology	*Artifact*; realization of technology	*Technical system*; techne-text; work of artifacts

Artifact is used, here, in the sense of manifestation of technology (literally what is "technically made" *arte-facere*). This notion of artifact is not limited to material things: artifacts as realizations of technology include all three spheres of techne: material, social, and intelligible. For example, the stadium wave (Mexican wave), smoke signals, data biases in surveys, and so on are artifacts as well. As mentioned, social and intelligible techne always have material bases (audience bodies, smoke particles, hard drives or ring binders, etc.), yet the artifact goes beyond that. It is further important to note that level I shows mere abstractions and that level II entities are never isolated from a configured (use or symbolic) context—discourse is "indeed an idealism of lingual life" (Ricoeur 2016, 56);[11] we perceive neither as such but understand I through II and II through III.

Yet, all three levels are interconnected: for instance, what can be measured depends on the theory of information or worldview as well as on the possibility to build adequate sensors and the measuring skill, the craft of using the sensor in a way that gives reliable results according to the theory, and so on. Technology determines which technical realizations are possible, yet the objectifications of technology (artifacts and their ensembles) determine what insights about the world can be gained (using laboratories and instruments or Baconian "methods"); those insights, in turn, are exteriorized in technologies. Which technical systems can be configured depends on the parts and threads that are available and their availability again depends on basic insights and structures like thermodynamics or availability of energy. Just as what can be written as text depends on what thoughts, topoi, schemata, and words are available, and their availability again depends on what a language provides as grammar or in terms of potential for structuring words, meaning, tenses, cases, and so on. Technology textures the possibility space of the doable, makeable, and scientifically knowable[12] just like language textures the one of the tellable and thinkable. Both structure the realm of what can possibly be *textus*. Language and technology both fundamentally condition the human-world relation, as Cassirer aptly emphasized: "All mental handling of reality is bound to this double act of 'grasping'—'comprehending' reality in linguistic-theoretical thought [language] and 'gripping onto it' through the medium of efficacy [technology]. This is true for both mental and technological forming" (Cassirer 2012, 24).

How exactly does it help to remember the objective spirit in this context? Its *made-yet-given* characteristic is indeed a medial thought: technology as a medium is man-made, yet it conditions the possibilities of individuals in the form of necessity ("Freiheit als vorhandene Notwendigkeit"[13]). With digital technology, an increasing decoupling of means from specific ends appears: Digital devices (e.g., smartphones and computer) no longer represent specific means (e.g., a calculator, a flashlight, a camera, and a telephone), but are a space of possibilities for many means, they are a medium. As a space of possibilities, the medium determines what can be considered as possible means and possible ends. It is the "medium of efficacy" (Cassirer 2012, 24), but also a hermeneutic medium of comprehending, a pragmatic medium of deciding and acting, and a texture medium of weaving potential. It is as such a textured possibility space that allows for certain insights, interpretations, manifestations, and work-configurations and withdraws others.

The concept of *texture* and the *weaving* of fabric may serve to illustrate this notion of the *medium* as a textured space of possibilities that is applied here to both language and technology: In a cloth, threads are woven together to form a texture. This cloth represents the medium, which is open for further specific formations (it textures), but already has a certain structure (is textured), by which the possible formations it offers are also limited. Thus, shirts or trousers can be made from woven fabric, which would not be possible without the cloth-medium. However, no sword or spear can be made from woven fabric, as this lies outside the structured space of possibility of this medium. In the systems theory vocabulary, the medium could be described as a loosely coupled system, as a repository of possible couplings (woven fabric), which can be shaped by fixed coupling (sewing). Analogous to the universalization of means, the mediality of modern high technology is characterized by an increasingly open and adaptive structure of organizing, up to the vast indeterminacy of modern computerized technology.

The made-yet-given structure can be traced to the notion of *tracks for* and *trails of*, which can be used analogously to the notion of *medium* (Hubig 2006, 148–155). Technology, as a medium, offers certain *tracks for* events or actions like railroad tracks offer a structured possibility space for trains. Only events or actions *for* which *tracks* are offered by a certain medium can be chosen. Each medium delivers its own modal "structure" that makes them a *texture:* Their being made corresponds to being textured by *trails of*, their givenness corresponds to being texturing via *tracks for*. Media offer specific modal textures or potentials; if you want to actualize ends or use means for which one medium holds no tracks, you need to either want less or change or switch the medium. Accordingly, there are "natural" media,[14] like water and air, and there are technical media like cloth or program code, and any technically retextured natural media like supercritical CO_2[15] or vineyards can be considered technical media as well. The modal *tracks for* can be altered on a meta-medial level (man-made), but they are binding on an inner-medial level (given).

Media exhibit—in addition to *tracks for*—*trails of* such alteration efforts or actualization instances. A natural jungle (seen as a medium) offers tracks for passing through, each of them being potential pathways. Each actual pass-through leaves trails of such passing that can be identified as tracks for others—for example, a beaten path—and so on. Depending on the ability to identify the actual structure of the possibility space and depending on the capabilities to actualize a certain potential, concrete actualizations (of passing through) then alter the modal structure.[16] Tracks for actualizations are used and leave trails of those actualizations, those *trails of* then become *tracks for* future actualizations, and so on, which is why all texturing is responsible for future texturing. Media transformation by human actualization and technology makes the respective media an objectification of life that is man-made (trails of humans), yet in its modal conditioning, it becomes law-like, granting and refusing specific options at a time and no other (tracks for individuals).

This circle of media transformation is similar to Ricoeur's circle of understanding and indeed *tracks for* and *trails of* acting are similarly both textured and texturing as *tracks of* and *trails for* understanding. This comes as no surprise, given Cassirer's reminder that the media for comprehending and gripping onto reality are intertwined and that we act oriented by understanding and understand actively. To perceive life as *prefigured*—linking Ricoeur and Hubig—means to perceive the tracks for understanding and acting, to see it as specifically receptive to meaning and alterations. *Configuring* works (while always *refiguring* previous works), either of language or of technology, means "actually passing through chosen tracks" thus transforming the structure of the medium (of understanding and efficacy). The (re-)textured medium exhibits altered *prefigurations* for future *refigurations* and *configurations*. Refiguring works is to interpret the trails of their configuration while at the same time shaping different tracks for understanding and acting.

Language not only represents but also builds, and technology not only builds but also interprets the "'form' of the world," which "whether in thought or action, whether in language or in effective activity, is not simply received and accepted by the human being; rather, it must be 'built' by him" (Cassirer 2012, 24). To build—or retexture—the form of the world in action and in effective activity is a primary aim of technology. Cassirer stresses—and Ricoeur builds on him as well[17]—that the form of the world is also built in thought and in language. Ricoeur emphasizes the creative innovative *building* aspect of language, or linguistic text as configured works of language, and draws on Aristotle's understanding of *poiesis*. Understanding the world (including works) is not simply receiving *the* meaning of a phenomenon but it is a poietic procedure that makes *a* meaning of a phenomenon as an entity that has been made: "The *poiesis* of emplotment is a making that, also, bears on what is made" (Ricoeur 2012, 245 FN31). Or: Poiesis is a texturing of what is textured. What is *made* in the world includes per definition the realm of the objective spirit and, therefore, technology; including all technical media, infrastructures, works, and so on. Technology and language are both parts of the objective spirit and mediate how we can "grasp" the world or build the form of the world, which makes technology as well—not only language and its works—a hermeneutic task. Being media of meaning and action makes both fundamental conditions of being-in-the-world:

> Through fiction and poetry, new possibilities of being-in-the-world are opened up within everyday reality. Fiction and poetry intend being, not under the modality of being-given, but under the modality of power-to-be. Everyday reality is thereby metamorphised by what could be called the imaginative variations which literature carries out on the real. (Ricoeur 2016, 104)

What Ricoeur states for literature holds for technical works as well: Through technology, new possibilities of being-in-the-world are opened up. Technologies intend being under the modality of power-to-be, which once again recalls Cassirer—"Technology does not initially ask what is but what can be" (Cassirer 2012, 44–45). The imaginative variations which technology carries out on the real generate a metamorphosis of everyday reality. Cognitive schemata are also cultural entities, and our individual schema-set—as a version of shared cultural schemata—conditions our interpretations of world and self. These schemata can be altered by the being-given but also by the imaginative variations of literature and technology. The variations of the form of the world are but the imagined space of possibilities, which are the perceived *tracks for* in a medium. Which variation is chosen as an action purpose and actualized (or attempted to actualize) is open to the actor, yet no option can be chosen deliberately that is not seen as a variation before. What does not (seem to) have the power-to-be cannot be pursued—even Don Quixote did not pick his fight with the sun or gravity. So, the modal judgments of what is considered to be possible (erroneously or accurately so) condition what can be pursued; those judgments, in turn, are conditioned by world- and self-understandings; those understandings are shaped by imaginary variations, that is, by language and technology. Thus, these variations dictate actions that reform the (actual and modal) form of the world. "Technological work, however, never binds itself to this pure facticity, to the given face of objects; rather it obeys the law of a pure anticipation, a prospective view that foresees the future, leading up to a new future" (Cassirer 2012, 44–45).

The works of technology manifest the possibilities of its components and their combinatory potential (parts *and* weave),[18] where each part or system reveals modal structures of the world and new possibilities of being-in-the-world. Ricoeur's insights about language can be applied to technology: The work of artifacts—that is, technical ensembles/techne-text—are the *projections of a world* or a form of the world, just as the work of discourse—that is, language-text—is "the projection of a world" (Ricoeur 2016, 94). Furthermore, artifacts and the work of artifacts mediate self-understanding, just as discourse and the work of discourse do. Finally, taken together, these features—technology, artifacts, and technical systems—constitute the criteria of technicity or *techne-textuality*—which is an expression of the same *textus*-realm as *language-textuality*. Note that the differentness of techne-textuality and language-textuality is an analytical contrast. There is a sameness perspective to it as well: Due to the three textuality levels—material, symbolic, and social—and especially the symbolic dimension, any technology just as any action is symbolically mediated. "If, in fact, human action can be narrated, it is because it is always already articulated by signs, rules, and norms. It is always already symbolically mediated" (Ricoeur 2012, 57). So any techne-textuality is always already structured on a symbolic level and not only by a material one. Every sign has a material basis that manifests it—the metal of a road sign, a waving hand, the paper of a Euro-bill, and so on—even though the meaning of the sign does not necessarily depend on a specific material. Every sign, a gesture, the character A, and so on, also has a techne-basis because it is brought about technically: waving or flag semaphore must be learned and trained, letters are written with a pen, chisel, or printed by a press, printer, and so on. So, the work of discourse and the work of artifacts have both a techne-textuality as well as a language-textuality.

The technology-language similarity in focus here—that is, both being textured by life and texturing life—allows to grasp technology as another part of the hermeneutic situation, it can be seen as a "hermeneutical horizon" (Gadamer 2011, 397) and objectified "experience of the world" (Gadamer 2011, 436).[19] Text as the structured work of discourse and artifacts both are projections of the world, of life as it was and is, and—this is special—as it *can be*. Specifically, linguistic and technical texts connect people with what can be. Both are poietically made in the transcendental condition of the objective spirit (structured by tracks for) and both actualize manifestations and shape the sedimentation of this realm (as trails of). Both mediate understanding in specific ways: language, discourse, and the works of discourse primarily in terms of tellability and intelligibility; technology, artifacts, and the works of artifacts primarily in terms of doability and efficacy.[20]

4. TRACKING SKILLS TO TRACE INTERPRETATIONAL MACHINES

Understanding technology and technological interventions or transformations of our lifeworld requires techno-hermeneutical capabilities. It is necessary to try and unfold the foldings of the cloth and to retrace the threads as far as possible; doing so means (causally) explaining and (poieticly) understanding technical works. Unfortunately, the threads in our vastly interconnected digital technosphere are dissolving (universalization of means, indeterminacy of computerized technology), and the *trails of* get lost[21] so the *tracks for* are ever less identifiable. This trend could be countered by explicitly making the trails visible again or—and this is what this article calls for—by developing better *tracker skills*. Of course, the former helps the latter, so, surely, a combination of both would be best.

Ihde speaks of a hermeneutical human-technology-world relation—among other relations—in which technical systems represent the world on an interface and users relate to the world only via the technical representation thereof, thus being required to interpret the displayed information and try to understand what form of world is projected by it. There are at least two other relations to be considered beyond Ihde. First, people relate not only to technology as an other (Ihde's alterity relation) but to other people. Second, one could argue that learning technologies "read" people. More precisely: They trace their data trails with digital trackers and "interpret" them in a technogenic way. This would suggest that "interpretational machines" (Romele 2020, 83) are already emerging today. We understand the world through technology, but technology "understands" (a) people through their digitizable activity, their data-trails, technical action, and interaction history, as well as (b) the world as detectable through sensors. These technically aggregated profiles that can be seen as a sort of technogenic "interpretation" of a human/user are partly fed into other connected systems in the technosphere and restructure other technical "interpretations" that then might be used for fully automated decisions.[22] Partly, they are presented to human "readers" that then interpret the individual by this technical representation. We are thus led to understand others through technology. These "reversed hermeneutic" relations raise tricky questions: How is technical, highly-automated, AI-learned "understanding" to be grasped in contrast to human understanding? How does technology mediate—distort/transform—the interpretation of people by people?

To investigate the contrast between machine "understanding" and human understanding, the following—strongly simplified—schema[23] (as understood by Gethmann) can be used to show that understanding must be distinguished from explanation (mainly as defined by Dilthey), and that understanding of life is to be distinguished from an understanding of the objectifications of life (mainly on the basis of the works of Ricoeur and Hubig).

First of all, the objects of "understanding" can be classified into three sorts of events: occurrences, behaviors, and actions. Those events can be respectively "understood" by way of reduction from effects to causes, expression to experience, and means to ends. These operations "explain" occurrences, "understand" (in a first sense I) behavior, and "understand" (in a second sense II) actions (see table 5.2). Explanation reconstructs events as the effects of a cause, that is, as causal occurrences. "An event is explained when it is 'covered' by a law and when its antecedents are legitimately called its causes. The key idea is that of regularity" (Ricoeur 2012, 113). Explanation, therefore, is far away from understanding life, because our experience has a "prenarrative quality" (Ricoeur 2012, 74), narration deals with events, and one narrative characteristic of eventfulness is the singularity of events, their non-iterativity and unpredictability.[24] Explanation is not part of hermeneutics but of the (natural) sciences (according to Dilthey's distinction). Technology is, first and foremost, explained and not understood. Interpretation—as *understanding I*—reconstructs events as expressions of experience, that is, as behavioral phenomena. Interpretation—as *understanding II*—reconstructs events as the intentionally brought about goal, that is, as phenomena of action.[25] Explanation, understanding I and II each have a realm they rightfully preside over, yet confusions of these realms are rather common. Technologies do not have genuine preferences, no desired ends or goals, they have no normative autonomy,[26] they cannot *want* something, prefer A over B, change their mind, and so on. They can simulate these capacities, but autonomy is not the simulation of autonomy. Understanding II is performed by concluding from means to ends. Without normative autonomy, neither can be grasped and understanding II cannot be performed.

Understanding I is performed by concluding from expression to experience. Understanding the processes of a technical system as expressions of an experience (instead of explaining it) would be a case of animism or mentalism and formulations such as artificial *intelligence*, *learning* algorithms, *autonomous* agents, and so on drive this kind of misunderstanding. Understanding "is the domain of all who are actively involved in human affairs, and differs

Table 5.2 Minimal Ontology of Understanding

			Occurrence (cause-effect)	Explanation
Event	Not brought about			
	Brought about	Not intended	Behavior (experience-expression)	Understanding I
		Intended	Action (means-end)	Understanding II

from explanation by participating in life, which is possible only on the basis of life" (Dilthey 1989, 439). We belong to the life that we interpret—we are part of the reality that we construct as an object that we objectify. We are able to interpret the world, life, and others because they are receptive to meaning *to us*. In short, Gadamer and Ricoeur claim: "Life interprets itself. Life itself has a hermeneutical structure" (Gadamer 2011, 221), or "it is thus possible to say of life what Hegel said of spirit: at this point, life grasps life" (Ricoeur 2016, 12–13). Even though technical systems might increasingly get involved in (certain) human affairs, they do not have expressions and experience in the Dilthey sense—they do not live nor participate in life and, therefore, have no "lived experience" (*Erlebnis*: Dilthey 1997, 223)—which is at the basis of this understanding I concept. Therefore, technology cannot genuinely understand in this sense either. At best, it explains or contributes to explanations of human affairs. According to this minimal ontology of understanding, technology is not capable to understand at all and any mention of "interpretational machines" or attempts to attribute interpretational agency to technology must be read metaphorically. The denial of interpretational *agency*, however, does not imply a denial of interpretational *relevance*.

5. MIRROR MIRROR IN THE CLOUD . . . SELF-UNDERSTANDING VIA TECHNOLOGY

One case of interpretational relevance is the technical mediation of self-understanding: "Perhaps it is at this level that the mediation effected by the text can be best understood. [...] we understand ourselves only by the long detour of the signs of humanity deposited in cultural works" (Ricoeur 2016, 105). With Cassirer's mission of philosophy in mind and in light of the discussions earlier, we can read Ricoeur as a statement on text beyond writing. He was referring to *language-text*, but it holds for *techne-text* as well: text mediates self-understanding. "All self-knowledge is mediated through signs and works" (Ricoeur 2016, 12). We need to take into account the long detour of the objective spirit if we want to understand ourselves. Therefore, self-understanding must necessarily be an understanding of "oneself in front of technology." We can read Ricoeur as a texture hermeneutic including both techne- and language-text:

> Henceforth, to understand is to understand oneself in front of the text. It is not a question of imposing upon the text our finite capacity of understanding, but of exposing ourselves to the text and receiving from it an enlarged self, which would be the proposed existence corresponding in the most suitable way to the world proposed. So understanding is quite different from a constitution of which

the subject would possess the key. In this respect, it would be more correct to say that the *self* is constituted by the "matter" of the text. (Ricoeur 2016, 106)

Such a Cassirer mission-driven, techne-oriented reading of Ricoeur yields the following insights: the self is constituted by the "matter" of the text—or—the self is constituted by the cloth of the technical systems. Furthermore, understanding is a question of exposing ourselves to technical systems and thus receiving from—or rather through—it an enlarged self as proposed existence in a proposed world. To understand the process of this exposure and enlargement is one task for which better techno-hermeneutical tracking skills are needed. Part of said tracking skills would be to account for previous (historic) re- and configurations of the idea of understanding oneself by analyzing technology.[27] Text mediates self-understanding, yet less in the Kappian notion that we can learn about ourselves by analyzing the texture of technical works and interpret it as a projection of our own texture. It should rather be seen as a way to understand oneself *in front of the text* or—today more accurately—*in the middle of* and intertwined with technology. As man "individuates himself in producing individual works" (Ricoeur 2016, 100), the mediation of this relation structures both relata at the same time, it is "strictly correlative" (Ricoeur 2016, 100).

Such self-understanding requires specific hermeneutic tools and concepts that can grasp the turn from language-text to an extended notion of text including technology. The challenge is to unfold (*auslegen*) or to deconstruct the construction, the texture of the weave to analytically separate the respective threads. If those threads are not properly distinguished, then, for example, technological biases, data artifacts, or third-party interests and manipulations might be taken for techne-text expressions of oneself. Thus, the hoped-for (at least by Cassirer and Kapp) potential "self-revelation, self-knowledge, and self-consciousness" might just be projections of technogenic texture or thoughts of others onto the self. This would not be to understand oneself in the middle of technology by the detour of signs of humanity, but rather to understand oneself as prefigured by technologies, which were configured by others. This would not mean "by the detour of signs of humanity," but by the detour of being use and profit for others, of being traded in a data economy, by their live-bidding, measuring, and sorting algorithms. It is a matter of techno-hermeneutical tracking skills which determines what kind of understanding can be made out of the proposed worlds and existences and how a life can be oriented accordingly.

6. TECHNOLOGY PREFIGURES LIFE—
INTERPRETATIONAL RELEVANCE WITHOUT AGENCY

What could it mean then that technology "reads" or "interprets" people via machine learning, if it cannot be read literally? Are there interpretational machines? Technology in general and specifically digital interconnected, background related learning IT-systems are media of life today. We can perceive, know of, decide between options that are modally structured by this techne-texture. Before we go out and feel the weather, we check the weather app to see how hot or cold it is. The same holds for our contact with others. Before we consider, for example, a candidate for a job, we check the formal data of the education system and we check the data available from Google, Facebook, News, Xing, Research Gate, and so on. These data configurations must be refigured, which as a result prefigures further investigations or an actual meeting in person.[28] Thus, how we understand each other, "from what cloth the other is cut," is increasingly prefigured by technical configurations that are based on technical refigurations (processing data-profiles) of technical configurations (data aggregations).

Was it the threshold of a sensor, the granularity of a sorting algorithm, the digitized phenomenon, the aggregation or rounding specifics, and so on that resulted in person A having this or that position in an output ranking of creditworthiness or date-matchings (Gransche 2020b, 72–73)? All this influences the systems' "perception" of humans, the world, and other technology. In this (metaphorical) sense, technologies "understand" people and the world always as *put in* by other technology. The technically aggregated data-sets or profiles can be seen as a sort of technogenic "interpretation" whose human conditioning contribution has been lost in the background relations of the technosphere. The loss of *tracks for* and *trails of* means that the single contributions, the plethora of causes that result in an effect, are no longer discernible.

It seems more accurate—keeping Dilthey in mind—to say that technology "explains" people, the world, and other technology rather than "understands" them. Yet, the detachment of effects from causes due to the loss of tracks/trails distances systems' data processing even from the form of explanation. This does not, in turn, approximate it to an understanding I or II, though. The systems' output can be used as other systems' input again as in automated decision-making, which is, in fact, a rule-based sorting mechanism. Algorithms are not much more than—yet complicated versions of—such sorting rules like "if-then-else"-chains. Whether directly presented to human deciders or executing predetermined decisions (rules), the systemic configurations (e.g., data profiles) are understood, interpreted, and made sense of by humans. For now: technologies cannot (even) explain anything, because they only detect correlation and not causality—people interpret output correlations as causation; technologies cannot understand (I) anything, because they have no lived-experience; technologies cannot understand (II) anything, because they have no normative position to a desirable event (ends). Strictly speaking, all that technologies do—even the most advanced ensembles—is "just" sorting and correlating data; everything else is human refiguration of its output.

7. CONCLUSION

Others, the world, technology, and we ourselves are phenomenologically given to us predominantly and increasingly mediated by technology. Technology does not understand people, but it conditions how people understand others, the world, and technology. Understanding, trying to make sense, understanding differently, trying to create a different sense, is a fundamental aspect of being-in-the-world. Hitherto, people were understood only by people via and according to configurations of the objective spirit such as stories, schemata, and so on. With learning technologies, people are "understood" by technologies—or, to be precise: their understanding and their being understood by others is technically prefigured. How to understand this extensive technical prefiguration and how to understand oneself intertwined in techne-textures is a challenge today; a challenge, however, that texture-hermeneutics including the tools described here can help to meet. Those tools are:

- The notion of *texture* as the weaving and as the common denominator of language and technology, and media as *textured texturing entities*;
- The *objective spirit* as the realm of texture entities with the made-yet-given characteristic—including technology;
- The *threefold texture levels*—material, intellectual, social—and their application to both technology and language;
- The notion of *technology as a medium* with a made-yet-given modal structure understood as *textures of tracks and trails*;
- *Medial track reading* in a texture hermeneutics and understanding as tracking or tracing—as in "Sherlock-Holmes-abductions";
- A clear distinction between *occurrences, behavior, and action*
- A clear distinction between *explanation and understanding* and between a life-based understanding (I) and an understanding (II) as based on objectifications of life.

NOTES

1. Their sameness invites the application of Ricoeur's thoughts about language to technology—their differentness puts limits to this endeavor. Both need to be considered, yet because the philosophical works on the differentness of language and technology are legion compared to the few on their sameness (e.g., Cassirer), the perspective of the differentness is not extensively discussed in this chapter.

2. See for the *medial turn* in the German discussion of philosophy of technology of the past two decades, for instance (Hubig 2006; Margreiter 1999; Münker 2009) and for an overview: Gransche (2015, 141–177).

3. Hubig (2006).

4. This being-made does not necessarily refer to human agency, for example, spiderwebs, beehives, or beaver dams are made as well. Yet, in contrast, sand waves or lotus leaves could rather be considered not-brought-about structures. If considered textures, then with the loss of the "being-made" aspect of the use of the word proposed here—for example, "made by wind and weather" being a metaphor then. The main interest of this chapter lies in human-made textures.

5. See for the latter connection, for instance, initial works especially of the *Ricoeur & Technology Research Group*: (Coeckelbergh and Reijers 2016; Gransche 2017; Kaplan 2006; Reijers and Coeckelbergh 2016; Romele 2020).

6. The interpretation of Ricoeur's texture and text as weave and cloth has even some biographical correlation: "My greatgrandfather was an artisan who wove cloth" (Ricoeur et al. 1998, 5).

7. The German extended version of a talk Ricoeur gave in Tübingen on May 15, 1985 reads: "transformiert die narrative Operation die irrationale Kontingenz in eine geregelte, bedeutsame, intelligible Kontingenz" (Ricoeur 1986, 14).

8. For instance, Hard (2020, 154–159).

9. See Hubig (2006, 37–41).

10. It is a shortcoming of Ricoeur that although he considers the past and has a brilliant way of analyzing the connection between the past and present, he fails to address the future and our relation to it. What connects vanished with present worlds connects to future possible worlds as well. Literature—or more broadly schemata—is a residue of our past worldviews but it shapes our interpretation of today's world and is a condition of potential future interpretations.

11. This "idealism" reveals an idealistic tendency of Ricoeur, as Romele correctly noted: "On the one hand, Ricoeur externalizes and materializes language and interpretational processes, in general, more than all other ontological hermeneutists. On the other hand, however, his understanding of the materialized language remains paradoxically idealistic" (Romele 2020, 6).

12. Technology has been described as applied science (e.g., by Mario Bunge), yet the opposite is also true: science is applied technology.

13. Hegel is rather difficult to translate. The original is: "Form der Realität als einer von ihm hervorzubringenden und hervorgebrachten Welt, in welcher die Freiheit als vorhandene Notwendigkeit ist, -objektiver Geist" (Hegel 1986, 32).

14. Given the global impact of human agency in the Anthropocene, "natural" media par excellence do no longer exist, just as the idea of an untouched nature.

15. Other supercritical fluids occur naturally. The supercritical phase is achieved by raising both the pressure and temperature above a substance-specific critical point, which can be achieved technically or in special environmental conditions like in a volcanic environment.

16. For instance, the potential routes are different for an elephant, a human, or a horse, but if an elephant once actually walked a certain route, it then becomes passable for a human, and if enough humans created a trail while passing, it then might become passable for horses, and so on—that would be *modal media transformation via actualization*. Or humans could cut through the wood with machetes or heavy

machinery, not leaving trails of their passing-through as a side product, but intentionally altering the tracks for future passability—that would be a *modal media transformation via technology*.

17. Ricoeur speaks of "symbolic mediations of action, in a sense of the word 'symbol' that Cassirer made classic" (Ricoeur 2012, 54).

18. Configured systems are actualizations of the *inner mediality* of technology (Hubig 2006, 149–159), they are concrete tokens of the makeable and technizable (incl., e.g., maximum climbability or drag, etc.). Analogously, each story is a specific actualization of the *inner mediality* of language, a token of the tellable.

19. The discussion of technology as hermeneutical horizon cannot be further discussed here for reasons of brevity.

20. This formulation focuses on differentness of the language-technology similarity. Yet, sameness aspects could be put forward as well, hence the addendum "primarily." Of course, language does things, and has effects and the power to produce successful results in speech acts or social technai such as insulting and contracting. And surely, technology tells things and renders nature intelligible, for example, through lab and experimentation devices and so forth.

21. Hubig speaks of the "virtualization of technology and the loss of trails" ("Die Virtualisierung der Technik und der Verlust der Spuren") (Hubig 2006, 183–192).

22. Such as not inviting and applicant to a job interview and not even presenting the aplication as an option to recruiter in the case of AI use in human ressource procecees.

23. This schema was introduced in Gransche and Gethmann (2018).

24. See for a discussion of a full set of criteria of eventfulness: Gransche (2015, 257–262).

25. Ricoeur clearly must be positioned in the understanding II or action-means-ends-related concept of understanding: "Understanding, in the second case, is grasping the operation that unifies into one whole and complete action the miscellany constituted by the circumstances, ends and means, initiatives and interactions, the reversals of fortune, and all the unintended consequences issuing from human action" (Ricoeur 2012, x). And: "Understanding is always more than simple empathy" (Ricoeur 2012, 97).

26. For the multilevel autonomy approach of *operative, startegic, and normative autonomy*, see: Gransche et al. (2014), Gransche (2020b), and Hubig (2014, 130–144).

27. See especially Kapp (2018).

28. Of course each nondigital encounter with a person—each social experience—potentially changes worldviews and conception of man and influences the subsequent interpretation of similar phenomena. Yet, the digital configurations are much more manipulable than real-world impressions, meaning the weaving process is receptive to interferences such as deliberate manipulation (e.g., webutation), alterations due to untraceable technospheric causes, and so on. This raises serious power questions, granting those in strategic positions within the technosphere the power to determine the social understanding of people.

REFERENCES

Cassirer, Ernst. 2012. "Form and Technology." In *Ernst Cassirer on Form and Technology: Contemporary Readings*, edited by Aud S. Hoel and Ingvild Folkvord, 15–54. New York: Palgrave Macmillan.

Coeckelbergh, Mark, and Wessel Reijers. 2016. "Narrative Technologies: A Philosophical Investigation of the Narrative Capacities of Technologies by Using Ricoeur's Narrative Theory." *Human Studies* 39(3): 325–346. doi:10.1007/s10746-016-9383-7.

Dilthey, Wilhelm, ed. 1989. *Introduction to the Human Sciences. Selected Works I*. Princeton, NJ: Princeton University Press.

———. 1997. *Poetry and Experience*. Edited by Rudolf A. Makkreel and Frithjof Rodi. Selected Works V. Princeton, NJ: Princeton University Press.

Franssen, Maarten, Pieter E. Vermaas, Peter Kroes, and Anthonie W.M. Meijers, eds. 2016. *Philosophy of Technology After the Empirical Turn*. Philosophy of Engineering and Technology 23. Cham: Springer International Publishing.

Gadamer, Hans-Georg. 2011. *Truth and Method*. 2., rev. ed., Reprint. Continuum Impacts. London: Continuum.

Geertz, Clifford. 2006. *The Interpretation of Cultures: Selected Essays*. [Nachdr.]. New York: Basic Books. Accessed May 26, 2020.

Gransche, Bruno. 2015. *Vorausschauendes Denken: Philosophie Und Zukunftsforschung Jenseits Von Statistik Und Kalkül*. 1st ed. Edition panta rei. Bielefeld: Transcript.

———. 2017. "The Art of Staging Simulations: Mise-En-Scène, Social Impact, and Simulation Literacy." In *The Science and Art of Simulation I: Exploring—Understanding—Knowing*, edited by Michael M. Resch, Andreas Kaminski, and Petra Gehring, 33–50. Cham: Springer International Publishing.

———. 2020a. "Datenschatten Und Die Gravitation Fast Richtiger Vorhersagen." In *Datafizierung Und Big Data: Ethische, Anthropologische Und Wissenschaftstheoretische Perspektiven*, edited by Klaus Wiegerling, Michael Nerurkar, and Christian Wadephul, 129–150. Wiesbaden: Springer Fachmedien Wiesbaden.

———. 2020b. "Handling Things That Handle Us: Things Get to Know Who We Are and Tie Us down to Who We Were." In *Relating to Things: Design, Technology and the Artificial*, edited by Heather Wiltse, tbt. London: Bloomsbury Visual Arts.

Gransche, Bruno, and Carl F. Gethmann. 2018. "Digitalisate Zwischen Erklären Und Verstehen: Chancen Und Herausforderungen Durch Big Data Für Die Kultur- Und Sozialwissenschaften - Eine Wissenschaftstheoretische Desillusionierung." Accessed February 28, 2019. http://www.abida.de/sites/default/files/ABIDA%20Gutachten%20Digitalisate.pdf.

Gransche, Bruno, Erduana Shala, Christoph Hubig, et al. 2014. *Wandel Von Autonomie Und Kontrolle Durch Neue Mensch-Technik-Interaktionen: Grundsatzfragen Autonomieorientierter Mensch-Technik-Verhältnisse*. Stuttgart: Fraunhofer Verlag.

Hard, Robin. 2020. *The Routledge Handbook of Greek Mythology*. 8th ed. Contemporary Issues. Milton: Routledge. Accessed May 27, 2020.

Harman, Graham. 2013. *Bell and Whistles: More Speculative Realism*. Alresford: Zero Books.
Hegel, Georg W. F. 1986. *Enzyklopädie der philosophischen Wissenschaften im Grundrisse: 1830. Dritter Teil. Die Philosophie des Geistes. Mit den mündlichen Zusätzen*. 9. Auflage. Werke in 20 Bänden 10. Frankfurt am Main: Suhrkamp. Accessed April 4, 2019.
Hubig, Christoph. 2006. *Die Kunst Des Möglichen I: Grundlinien Einer Dialektischen Philosophie Der Technik; Technikphilosophie Als Reflexion Der Medialität*. 2 vols. 1. Bielefeld: Transcript.
———. 2014. *Die Kunst Des Möglichen III. Macht Der Technik*. Berlin/Heidelberg: Transcript.
Kaplan, David M. 2006. "Paul Ricoeur and the Philosophy of Technology." *JFFP* 16(1/2): 42–56. doi:10.5195/JFFP.2006.182.
Kapp, Ernst. 2018. *Elements of a Philosophy of Technology: On the Evolutionary History of Culture*. Edited by Jeffrey W. Kirkwood and Leif Weatherby. Posthumanities 47. Translated by Lauren K. Wolfe. Minneapolis: University of Minnesota Press.
Margreiter, Reinhard. 1999. "Realität Und Medialität. Zur Philosophie Des *Medial Turn*." *Medien Journal. Zeitschrift für Kommunikationskultur* 23(1): 9–18.
Münker, Stefan. 2009. *Philosophie nach dem »Medial Turn«: Beiträge zur Theorie der Mediengesellschaft*. MedienAnalysen 4. Bielefeld: Transcript Verlag.
Reijers, Wessel, and Mark Coeckelbergh. 2016. "The Blockchain as a Narrative Technology: Investigating the Social Ontology and Normative Configurations of Cryptocurrencies." *Philosophical Technology* 46(2): 39. doi:10.1007/s13347-016-0239-x.
Ricoeur, Paul. 1970. *Freud and philosophy: An essay on interpretation*. 4. print. Terry Lectures 314. New Haven, CT: Yale University Press.
———. 1983. *Temps et récit: Tome 1*. Points essais 227. Paris: Éditions du Seuil. Accessed July 12, 2017.
———. 1986. *Zufall Und Vernunft in Der Geschichte*. Tübingen: Gehrke. Erweiterte Fassung des Vortrags "Kontingenz und Rationalität in der Erzählung," Tübingen. Accessed May 15, 1985.
———. 2012. *Time and Narrative: Volume I*. With the Assistance of K. McLaughlin and D. Pellauer. Chicago, IL: University of Chicago Press. Accessed July 12, 2017.
———. 2016. *Hermeneutics and the Human Sciences: Essays on Language, Action and Interpretation*. Cambridge Philosophy Classics Edition. Cambridge: Cambridge University Press. Accessed April 2, 2019.
Ricoeur, Paul, François Azouvi, and Marc B. de Launay. 1998. *Critique and Conviction: Conversations with François Azouvi and Marc De Launay*. European Perspectives. New York: Columbia University Press.
Romele, Alberto. 2020. *Digital Hermeneutics: Philosophical Investigations in New Media and Technologies*. Routledge Studies in Contemporary Philosophy 127. New York: Routledge. Accessed November 14, 2019.
Schapp, Wilhelm. 2012. *In Geschichten Verstrickt: Zum Sein Von Mensch Und Ding*. 5th ed. Frankfurt am Main: Klostermann.

Verbeek, Peter-Paul. 2005. *What Things Do: Philosophical Reflections on Technology, Agency, and Design*. 1. Paperback Print. University Park, PA: Pennsylvania State University Press.

Wiltse, Heather, ed. 2020. *Relating to Things: Design, Technology and the Artificial*. London: Bloomsbury.

Part II

RICOEUR'S ETHICS OF TECHNOLOGY

Chapter 6

Narrative Self-Exposure on Social Media

From Ricoeur to Arendt in the Digital Age

Annemie Halsema

Self-documentation has become increasingly popular in the digital age. Facebook, Twitter, blogs, and vlogs are important means of expressing oneself. We publish selfies and photos of family and friends on Facebook and Instagram, and keep in touch with our intimates by means of message services such as WhatsApp and Snapchat. Our sense of self therefore has more and more become entangled with digital technology (Koosel 2015). In the literature on digital technologies, the theme of the self and of identity formation has been important from the start. In fields such as media studies and sociolinguistics, the opportunities internet offers to create novel identities—distinct from one's "real-life" personality—have been explored (Turkle 1995) as well as the social pressures underlying identity performance such as our need for social inclusion (Koosel 2015).

My question in this chapter is what the use of digital technologies means for the notion of the self. To what extent does self-documentation in the digital age give way to novel forms of the construction of personal identity? In philosophy, the hermeneutical notion of the self comes closest to the self articulated on social media. In twentieth-century hermeneutics, the self is not considered as internal to the mind or the body, but it is expressed, that is, articulated and mediated by language. It is not metaphysical—although it does have an ontological ground—but it is the self of human beings that show who they are in a public environment. Hannah Arendt was one of the first to connect who somebody is to this person's life story and to articulating oneself in speech and action (Arendt 1958). Referring to Arendt, Paul Ricoeur in the nineties of the past century developed the concept of narrative identity to describe hermeneutical self-identity.[1] In the past decades, his work has become one of the most important philosophical sources for this notion.[2]

I investigate in this chapter how contemporary forms of self-documentation on social media,[3] which can be considered as a form of digital identity,[4] relate to Paul Ricoeur's notion of narrative identity. Narrative and digital notions of the self are forms of discursive identity, that is, a view on identity in which the self is not considered in an essentialist manner as stable and unified, but rather as plural, constantly negotiated, fluid, and enacted through the participants' discourse (Page 2012, 17). I will claim that the narrative and digital notion of the self are continuous and that it is important to consider their continuity in order to understand the identity constituting aspects of digital identity. The two notions are dissimilar as well, however, and should be taken as such in order to appreciate the specificity of digital identity construction. In short, I argue that the difference is that narrative identity implies expression of the self, whereas digital identity aims at self-exposure.

In the first section, I explore the notion of digital identity and the relation to narrative identity that scholars in the fields of sociolinguistics and media studies assume (Turkle 1995; Ryan 2004; Page 2012; Koosel 2015). I also consider the extent to which Ricoeur's notion of narrative identity is considered as not sufficient to address identity formation (De Lange 2010; Frissen et al. 2015). In the second section, the relevant characteristics of narrative identity that Ricoeur describes in *Time and Narrative* I (1984), III (1988), and in *Oneself as Another* (1992) will be discussed. In the third section, I specify the aspects of narrative identity that become questionable in the case of identity construction on social media on the basis of the named critiques. Ruth Page's empirical sociolinguistic study (2012) about identity construction in blogs, Facebook posts, and tweets will be an important source in this part. In the fourth and last section, I will suggest that identity construction on social media is not merely a form of self-expression, but rather a form of what Cavarero (2000) and Butler (2005) have called "exposure" of the self. Self-exposure is continuous with narrative identity but in the end will show to imply loss of self-ownership rather than constitution of the self.

1. DIGITAL IDENTITY FORMATION AND THE RELATION TO NARRATIVES

The phenomenon of digital identity is referred to with many different terms such as online identity, online personality, digiSelf, virtual identity, avatar, and online persona (Koosel 2015, 77). Digital identities can be described as human to human, or more specifically online identities interacting with other online identities in a virtual environment. Most scholars consider interactivity as the most important innovative facet of digital identity, compared to earlier forms of identity construction (Page 2012; Koosel 2015). Other novelties mentioned are the opportunities to create *personae*, in which you take control and become the author of the story of your life. Sherry Turkle in her renown study on identity in the age of internet considers computer games as laboratories for the construction of identity (Turkle 1995, 184) and considers them as parallel identities, that either help to escape reality or provide the relative freedom to virtually be who you want to be and explore other possibilities (Turkle 1995, 186–209). Whereas Turkle presented online identity as an alternative for identity in real life, others recognize that offline and online behavior overlap and that identity constitution is carried out both offline and online, or even that distinguishing between off- and online identity is no longer relevant (see Page 2012, 17–18). In this chapter, I argue that in order to specify the typical features of digital identity formation, it is important to distinguish it from offline identity construction. However, in order to understand how digital identity can be constitutive for one's personal identity, we need to take into account the relation between online and offline identity construction.

Computers and the internet facilitate specific styles of presenting the self (Koosel 2015, 78). Computer-mediated communication gives way to different sorts of stories: in some cases, entire stories are uploaded to the web as electronic files, or narrative interviews are recorded and published online as podcasts. Blogs sometimes reconfigure a genre such as diary writing; discussion forums are novel forms of conversational storytelling. Wiki software allows new forms of storytelling, such as single stories told by multiple authors (Page 2012, 186–187). This broad variety of storytelling in itself does not imply that the notion of narrative is too traditional to be applicable in the case of identity construction on social media. Most scholars in the field of narrative and new media claim that the notion of narrative also applies to the digital age (Ryan 2004; Page 2012; Koosel 2015). As Marie-Laure Ryan explains, narratives in narratology from the start have been understood as transcending various media (Ryan 2004, 1). In Ryan's definition, narrative encapsulates a wide range of variations of plots and styles of articulation.[5] Instead of defining narrative as (articulated) text, she argues for a notion of narrative that includes the sensemaking of experiences and events, and the capacity of relating events to each other causally. These causal relations do not necessarily need to be explicit in the text, but one should be able to infer them from the text (Ryan 2004, 11). The notion of narrative hence still applies when it concerns digital identity, because pictures, photos, and videos can be considered as an extension of linguistic narratives. The notion of narrative extends beyond the linguistic story and includes visual self-expressions (Ryan 2004; Page 2012).

Ricoeur's conception of narrative identity, although offering an excellent starting point for considering digital (and ludic[6]) identity construction, is considered by some scholars as faulty in particular respects. In the first place, according to Frissen et al. (2015, 34), Ricoeur almost exclusively pays attention to the art of the novel: "Because of his focus on works belonging to serious high culture, he seems to be blind to the often more frivolous ways identity construction takes place in everyday gossip and life stories, and in popular fictional accounts, such as movies, soaps, comics, and narrative computer games, among others." . Indeed, Ricoeur draws on Aristotle's notion of emplotment and narrative theories such as Greimas's (Ricoeur 1992, 140–168). He uses examples from literary classics, such as Virginia Woolf's and Robert Musil's novels, in order to illustrate his claims (Ricoeur 1992, 148–149). Frissen et al. therefore claim that Ricoeur's seminal model is high culture literature and that he does not pay attention to frivolous, playful utterances such as gossip and telling anecdotes. Digital identity narratives are considered by other scholars in the field as having more in common with spoken conversation than with written communication and literature genres.[7] A second problem with Ricoeur's theory that is mentioned by Frissen et al. is that his focus on mostly classical novels

> also results in a greater emphasis on elements of form that are connected with these kinds of novels, such as monomediality, linearity, and closure. The kinds of narratives we come across . . . in popular culture often have a different form; they are, for example, multimedial, interactive, connected, and open-ended. (Frissen et al. 2015, 34)

A third objection that may influence the relevance of Ricoeur's notion of narrative identity for thinking about identity construction on social media is that it concerns temporality mainly (i.e., our life story connects past events to the present and anticipates a future, and constitutes a certain permanence in time), whereas in digital identity construction place, spatiality is relevant as well. According to De Lange, "Narrative theory lacks a framework for understanding the role of spatiality, place and mobility in identity construction" (De Lange 2010, 235).[8] The narrative identity framework cannot account for factors such as spatial situatedness and mobility that have become more important in digital identity construction. To what extent are these points of critique feasible? In the next section, I first briefly introduce Ricoeur's notion of narrative identity and then evaluate the mentioned critiques.

2. NARRATIVE IDENTITY: EXPRESSION THAT CONSTITUTES THE SELF

For Ricoeur, the narrative account of the self fulfills more than one purpose. Narrative identity first of all is the continuity, the permanence in time that we assign to persons or communities, and that connects human time to cosmic time. Narratives humanize time (Ricoeur 1984, 3): by means of narratives we make events "our own," we give sense to them and put them in a meaningful order. The permanence in time Ricoeur ascribes to narrative identity is not absolute or fixed, because narratives about our lives may change. Instead, narrative identity mediates between the two poles of self-sameness (*idem*) and selfhood (*ipse*). The *idem*-pole can be understood in terms of a person's character and the *ipse*-pole is connected by Ricoeur to the minimal self-constancy that keeping a promise entails. Keeping a promise implies the will to "hold firm," even if I were to change my opinion or inclination (Ricoeur 1992, 124). One's narrative identity hence can be seen as neither completely coinciding with one's character, nor with keeping a promise, but it is situated somewhere in between its two poles. It includes stability, self-constancy as well as mutability, change. Because narrative identity includes permanence in time, the notion for Ricoeur also fulfills the function of forming an answer to the questions: "Who did this? Who is the agent, the author?" (Ricoeur 1988, 246). Narrative identity as such is a condition for ethical subjectivity: in order to be able to act ethically, the self needs to be able to give an account of itself. Last but not least, the concept of narrative identity fulfills the philosophical role of mediating between two major considerations of the self in the philosophical tradition, namely identity as an illusion (in the Humean and Nietzschean accounts of the self) and identity as self-certainty (in the Cartesian and Lockean accounts) (Ricoeur 1988, 246; 1992, 115–130). The hermeneutical notion of identity combines the features of these two philosophical ideas: it is a construction that at the same time offers the self a certain amount of assurance of itself.

Ricoeur uses the notion of "attestation" in order to draw the outlines of the ontology of his notion of the self (Ricoeur 1992, 299–302). I find this ontological ground of narrative identity relevant to discuss, because it implies that by speaking out, testifying, someone does not only utter herself about something, but at the same time assures herself and is confirmed of her existence by others. Ricoeur suggests a hermeneutics of the self in which the self displays the credence and trust of *existing* as a self that characterizes attestation, in which this self is mediated by language (Ricoeur 1992, 302). In claiming "It's me here"—the Levinasian expression (*me voici*) that Ricoeur adopts in order to express what attestation implies—the person uttering this not only testifies to something but also testifies to herself. Attestation is thus always also self-attestation (Ricoeur 1992, 22–23). In attestation, the self confirms its capability to speak, to act, and to take on projects and tasks. The fact that attestation is mediated by language has the consequence that self-affirmation is also intrinsically connected with the relationship to the other person. The trust in the self of which attestation accounts does not come about because of an inner self-assurance but rather is closely related to the trust of the other person in the self (which is the case in promising as well).

Whereas Ricoeur develops the ontology of his narrative conception of selfhood by means of "attestation," in the debate over narrative identity, his view is considered as ontological in another respect, that is, because it includes that the self is *constituted* by its life story. This claim is considered as "strong narrativism" (Hutto 2016).[9] Even though Ricoeur does hold this view, as we will soon see, this does not imply that persons can be conflated with the characters of stories completely. Instead a person shares the condition of dynamic identity that is peculiar to the story recounted (Ricoeur 1992, 147). What does this imply?

The temporal dimension of permanence in time is crucial for Ricoeur's notion of narrative identity. This implies that in the life story, events in the past are related to the present and to anticipations of one's future. In telling one's life story, the connectedness of one's personal life is constituted. The narrative places the past events into a particular order. Ricoeur refers to the Aristotelian notion of "emplotment" to designate this order of events (Ricoeur 1992, 141–143).[10] In *Time and Narrative* Ricoeur explains the function of the plot. He elaborates upon its mediating function in discussing mimesis2, configuration. Whereas mimesis1, prefiguration, considers the field of human action as structured in such a way that it can be captured in a narrative (i.e., actions have beginnings and endings, we can answer questions about them such as what, who, why, how?), mimesis2 concerns the actual narrating about these actions, putting them into a narrative order. It is here that plot shows its mediating function. In the first place, the plot mediates between the individual events or incidents and the story as a whole: it draws a meaningful story from a diversity of events or incidents, or transforms the events or incidents into a story (Ricoeur 1984, 65). Emplotment hence "is the operation that draws a configuration out of a simple succession" (Ibid.). The plot furthermore brings together heterogeneous elements, such as agents, goals, interactions, circumstances, and unexpected results (Ibid.). And third, the plot synthesizes heterogeneous elements into a temporal whole. What is important in this respect is that the elements are connected to each other. As Ricoeur emphasizes, Aristotle does not disapprove of episodes, but he does condemn disconnected episodes. A plot in which the episodes follow one another in no probable or inevitable sequence is an episodic plot for Aristotle, and as such is either improbable (when the events are summed up one after the other) or probable (when the events are causally related to each other) (Ricoeur 1984, 41). Ricoeur follows Aristotle in considering the plot as a configurational act, in which actions are grasped together, or to put it differently, in which "the unity of one temporal whole" is drawn from "the manifold of events" (Ricoeur 1984, 66). This will prove to be important in considering digital identity, because especially in interactive blogs, the unity of the plot will be in danger, as we will see in the next section.

In *Oneself as Another*, Ricoeur redefines the Aristotelian notion of emplotment as "the synthesis of the heterogeneous" (Ricoeur 1992, 141). This means in the first place that the plot synthesizes multiple events or incidents into a unified story. Second, the components of the action (intentions, causes, chance occurrences) are related in a sequence by the plot; third, the plot mediates between pure succession and temporal unity. Ricoeur specifies the configuration that characterizes emplotment with the Aristotelian notions of concordance and discordance. Concordance refers to the principle of order "that presides over the arrangement of facts" (Ibid.). It makes the plot of the story of one's life "an ordered transformation from an initial situation to a terminal situation" (Ibid.). Discordances are the reversals of fortune that disturb the unity of the plot; they threaten its identity. Ricoeur understands the narrative composition as a specific dialectic between the two, that is, as discordant concordance, which is similar to "the synthesis of the heterogeneous" (Ibid.). The person hence for Ricoeur is characterized by the synthesis of the heterogeneous elements of her life story.

A person is the character in a story and "characters . . . are themselves plots" (Ricoeur 1992, 143). Identity therefore is not something that exists *before* the person's life story. Instead, the narrative constructs the identity of the character. This suggests that even though the two overlap, for Ricoeur, there remains a distinction between the person telling the life story (the author or narrator) and the character in the story—which will prove to be important for understanding the relation between online and offline identity. Ricoeur claims that the character draws singularity "from the unity of a life considered as a temporal totality which is itself singular and distinguished from all others" (Ricoeur 1992, 147). It is in narratives about (parts of) their lives that persons express who they are. One's life is perhaps not exhausted by this narrative, but we need the narrative in order to articulate who we are.

Ricoeur works out this unity of a life in terms of practices that are governed by a life project (Ricoeur 1992, 158). He understands narrative identity in terms of "unity," which will become a problem when relating narrative to digital identity, but he conceives of this unity in an open way. He claims in the first place that the author or the narrator, and the character in narrative identity do not coincide. Especially the author does not overlap with a person's personal identity: Ricoeur contends that we are co-authors of our life stories rather than authors (Ricoeur 1992, 160). In the second place, there is no narrative closure in life stories, because they do not have a clear beginning or end. A strict "narrative unity of life" does not exist for this reason. Related to this problem is the difficulty that we can weave more than one plot and may recount several stories of our life (Ricoeur 1992, 161). Next, our life stories are caught up in the histories of others. My life story is part of my parents life story, of my partner's, of the ones of my friends and colleagues. It is "entangled in histories" (Ricoeur 1992, 161). For these reasons, speaking of "the narrative unity of life" is problematic, and identifying literary fiction and life is as well. Ricoeur suggests to understand "the narrative unity of a life" instead in terms of an unstable mixture of "fabulation and actual experience" (Ricoeur 1992, 162). We need fiction in order to organize life retrospectively, and we need to take the narrative as provisional and be prepared to revise the plot of our lives.

Narrative identity for Ricoeur thus is a testimony of oneself, co-created by other people, in which persons recognize themselves. It forms a temporal unity that has a coherent configuration, in which narrated events are situated in the plot. Through identification with the narration, the narrator gains an identity that is open to change—you can tell different stories about yourself in the course of a life time—but that functions as the constancy with which a person's identifies at the same time. Earlier we have seen that some scholars claim that Ricoeur's narration identity for different reasons is not feasible for understanding digital forms of self construction. In the next section, I discuss these and other points of critique.

3. SELF-NARRATION ON SOCIAL MEDIA

The problems mentioned by Frissen et al. (2015) are that Ricoeur had in mind classical literary high culture examples when conceptualizing narrative identity and that the style of writing on internet is not in line with that. His notion of narrative identity according to these scholars furthermore implies a monomedial and linear model of self-construction with a closed ending that does not apply to identity construction on the internet. The problem De Lange (2010) notices furthermore is that the notion of narrative identity merely considers temporality in identity construction, whereas spatiality place seems more relevant on social media.

The claim that the notion of plot does not apply as much to digital identity construction can easily be refuted, because we have seen that for Ricoeur, the plot of the narrative does not have a closed ending: there is no narrative closure in life stories because they do not have a clear beginning or end, and because our life stories are caught up in the histories of others who contribute to our life story. This counterargument and the one of De Lange is refuted by Ruth Page in her sociolinguistic study of identity construction in blogs, Facebook posts, and tweets (Page 2012). In contrast to De Lange's critique that spatiality should be accounted for in computer-mediated communication, Page claims that social media stories reaffirm temporality as the core property required to identity narrative (Page 2012, 187). Page signals that there is "rarely a unified, dichotomous contrast between the stories told in old and new forms of technology" (Page 2012, 186). Instead, classical narratives and social media narratives to a large extent overlap. For instance, linearity in the sense of the temporal organization of events still applies to social media stories. The main difference with classical storytelling is the resources used to construct the chronological frame of reference, she claims. In the case of social media stories, chronology is not drawn from the story content alone, but also from other sources, such as dates in blogs, time stamps in the template. The most important shift between classical storytelling and social media stories is that recency has become a governing characteristic of the latter (Page 2012, 189). On Facebook, adverbs such as "today," "tonight," "tomorrow" are frequently used, and conventional narrative adverbs such as "then" are underused. Temporal recency as a form of narrative linearity, Page concludes, "is well suited to the moment-to-moment updating that has become increasingly characteristic of social media practices" (Page 2012, 191). Page herewith refutes De Lange's objection to Ricoeur's notion of narrative identity: time remains an important reference on social media and is not replaced by place. What does change, however, is the importance of temporal recency.

In discussing temporality, Page and De Lange refer to the time references mentioned by the users in their accounts on social media. In the case of De Lange, however, also another notion of temporality is at issue. The narrative organizes time; it integrates one's past and future. Ricoeur develops this idea in terms of "permanence in time," and considers *idem* and *ipse* as the two forms of permanence in time (Ricoeur 1992, 113–118). Also when time is not explicitly mentioned in the life story or post, the narrative therefore constitutes permanence in time, as we have seen, and integrates events, intentions, causes, and chance occurrences into a coherent whole.[11] Yet, also in this respect, social media stories do not differ so much from classical narratives. Page's study shows that social media stories are not necessarily open-ended, and still contain a narrative sequence. The extent to which a story is open-ended depends on the social media format: anecdotes in a blog may contain stories in their entirety, but sometimes stories are told episodically across units, or sites are linked to other sites (tweets linked to photos, Facebook, blogposts). In this case, the reader of the tweet can gather the tweeter's story from other sources. Even in the case a person frequently updates her Facebook status, by means of daily episodic brief accounts of experiences or events, we can still consider this in terms of narrativity, because all these seemingly unconnected events refer to a named person's life experiences (Page 2012, 188). The unifying frame is the person's life, which in this case finds an articulation on the internet.

Even though Ricoeur does not explicitly consider storytelling on social media, it hence is not its temporal recency and fragmentation or its "episodic linearity" (Page 2012, 194) that requests a reformulation of his theory of narrative identity. His theory offers ample opportunities to account for episodic storytelling and moment-to-moment updating of one's experiences and events in life.[12] Nevertheless, the use of social media for self-documentation also leads to gradual changes. Even though temporal sequence remains a core property of narrative, because of social media communication "teleologically focused narrative progression appears less like a norm and more like a subtype of narrative," as Page infers (2012, 194). The more we use social media to report about ourselves and the more influence these media obtain as a means of self-expression, the less temporal sequence will be relevant.

On the one hand, it is important to distinguish digital from narrative identity formation (in order to understand the specificities of digital identity construction and the shifts it means for considering the self) but, on the other hand, to understand that the two are continuous. Without the notion of offline narrative identity, the coherence of the self and the fact that one's life story is constitutive for the self cannot be understood. We just saw that, on the basis of her empirical study, Page argues that the coherence of one's life story as articulated on the internet is dependent upon the connection between online and offline identity. As mentioned before, the opportunity to create an identity online, a fake-identity, for some scholars such as Turkle, was considered as a space for a person's growth (Turkle 1997, 262–263). Other scholars claim that the internet facilitates "disembodied" identities: physical markers such as race, gender, age, dress, and other visual information can be concealed; aural or oral markers such as language, accent, and tone of voice as well as non-verbal cues such as gestures, facial expressions and emotional information are obscured by mediated communication (Koosel 2015, 26). Page alludes to this variety of possibilities on the internet as well and concludes that indeed new identities can emerge, but in the end "the patterns of interaction are often shaped by familiar indexes of transportable identity, like gender." I support her conclusion that "even while technological features found in the textual and generic context of social media change, some contextual factors, like the construction of social identity, have stayed convincingly familiar" (Page 2012, 206).

Even though computer-mediated interaction does offer the opportunity to create new discourse identities, the factors that defined identity offline also remain to shape identity online. A person's identity, according to Ricoeur, is shaped by the narrative about her life. This narrative is not a singular story articulated by its narrator once in a life time, but consists of the stories of others about this person as well and only ends when the person stops being remembered by others. Yet the character of a life story draws her singularity from the unity of a life considered as a temporal totality, which is singular and distinguished from all others (Ricoeur 1992, 147). In most forms of storytelling on social media, multiple tellers contribute to the story. As mentioned before, compared to earlier forms of identity, interactivity is considered as one of the novel aspects of digital identity formation. In itself this is not an argument to consider Ricoeur's notion of narrative identity obsolete. Yet, we might ask to what extent multiple tellership still adds up to a coherent self-identity? Is it not precisely the constituting aspect of narrative identity that gets lost on social media because of its interactive character?

Page distinguishes between different forms of multiple tellership, depending on control of the text by the co-tellers. In discussion fora, for example, all contributors can post equally, but there may be a site administrator who moderates. In blogs, most of the time the blogger retains the right to moderate the comments on her webpage and the author of a comment to the blog therefore cannot control the publication of her contribution. In other words, multiple tellership on social media varies from one teller who maintains control of the text, to all tellers contributing equally (in wiki's). Instead of there being a binary opposition between a single and multiple tellerships, Page concludes that there is a spectrum of possibilities that move from single to co-tellership and multiple narrators (Page 2012, 199–200). Because Ricoeur's "narrative identity" does not limit the self-narrative to one person, it keeps an opening to identity construction on social media. Whereas the main aim of narrative identity is to "gather" the life events of a person and constitute a relative unity, however, digital identity seems to serve another purpose, namely interaction with others. This implies that digital identity increases the interactive aspects that the notion of narrative identity already contains. Besides the growing importance that is attached to recency and the declining relevance of temporal sequency, it is this interactive aspect of personal identity that in the case of digital identity gains significance as facet of personal identity.

The analysis of Ruth Page's empirical investigation of identity construction in blogs, Facebook posts, and tweets, in this section, has brought us to the conclusion that narrative and digital identity should be understood as different, but not as completely distinct. Digital identity in several respects differs from narrative identity. The most important differences are that digital identity centralizes temporal recency instead of narrative progression and that it includes multiple (and unknown) tellers instead of one or a few persons. In order to understand digital identity as a form of identity construction, however, the relationship between the identity constituting aspects of narrative identity and digital identity should be maintained. First, this concerns the relationship between the expressions of the self ("the posts") and the self (offline). Second, this concerns the idea that we articulate who we are by means of expressing ourselves and that our identity is constituted by the connectedness of our life story. In the final section, I will draw the consequences of this relationship.

4. FROM SELF-EXPRESSION TO SELF-EXPOSURE

For Ricoeur, self-understanding is an interpretation that in the narrative finds a privileged form of mediation (Ricoeur 1992, 114). Self-understanding is not instantaneous but necessitates the detour of signs, symbols, and narratives. It is the mediation by "symbols"—meanings "incorporated into action and decipherable from it by other actors in the social interplay" (Ricoeur 1984, 57)—that makes that human action can be narrated. Taken as such, narrative identity does not so much pertain to a person's strictly individual sense of self, but rather is the symbolization of human action that has become readable for others. Although, as we saw, Ricoeur conceives of the character as drawing her individuality "from the unity of a life considered as a temporal totality which is itself singular and distinguished from all others" (Ricoeur 1992, 147), it is a symbolized singularity that she aims at. This implies that it is a singularity shared with others, in symbols readable for others. In expressing ourselves on social media, such as Facebook, blogs, and tweets, we document the events in our life while at the same time keeping in touch with others. As such, self-documentation on social media contributes to our narrative identity but brings to the fore other aspects as well: instead of self-gathering, fragmentation and instead of singularity of tellership, multiplicity. The posts on social media need to be considered as elements of one's life story and retrieve their integration from this narrative, but the mentioned facets of the declining relevance of temporal sequence and interactivity demand that the specificity of digital identity construction needs to be addressed as well. I suggest that instead of self-expression, which is critical in the case of narrative identity, digital identity construction can be described as a form of self-exposure.

"Exposure" is the notion Judith Butler emphasizes in developing her poststructuralist ethics (Butler 2005).[13] She considers the "I," that gives an account of itself and that we have called a narrative self, as implicated in the social conditions of its emergence, "in a social temporality that exceeds its own capacities for narration" (Butler 2005, 7-8). This self that is encapsulated in its relations to others is also referred to by her with the notion of "exposure." "Exposure" refers to uncovering, making known, disclosure to another. It points at singularity and vulnerability at the same time (Butler 2005, 31). Butler derives the notion from the Italian feminist philosopher Adriana Cavarero, who considers the self as fragile and unmasterable and completely given over to others (Cavarero 2000, 84). My suggestion in this chapter is that self-documentation on social media can be seen as such a form of self-exposure, that has both identifying and effacing effects. In social media, we do not express ourselves but expose ourselves to others. As such, "digital identity" is not completely similar to Ricoeur's notion of narrative identity, even though I have argued that it is continuous with it. It deserves to be studied as a distinct phenomenon.

What differentiates self-exposure from self-expression? In order to explain the difference, we will need to return to the source of narrative identity, Hannah Arendt. Adriana Cavarero in *Relating Narratives* (2000) rereads Arendt's hero of the life story. For Arendt "the hero" of a story is not the product of the author or the producer of the life story, because products reveal a person's whatness, and not his or her whoness. Stories reveal agents, also when the person did not write a single line himself, such as Socrates (Arendt 1958, 186). Arendt is more radical than Ricoeur, in claiming that "nobody is the author or producer of his own life story" (Arendt 1958, 184). For her, action and speech need to be distinguished from production. "Who" someone is can "only become tangible *ex post facto* through action and speech" (Arendt 1958, 186), while everything else we know of this person "including the work he may have produced or left behind, tells us only *what* he is or was" (Ibid.). Whereas Ricoeur considers the self as not the author but co-author of its life story and this story as the basis for a self's singularity, Arendt accentuates that "whoness" is revealed within the web of human relationships and that the disclosure of the who through speech always falls into an already existing web (Arendt 1998, 184). It is precisely this point that Cavarero emphasizes and radicalizes. She reads Arendt's "hero," who according to her is often accused of exhibitionist narcissism, as a fragile and unmasterable self that exposes itself in its uniqueness. It is a self of action and a narratable self, Cavarero concedes, that is completely given over to others. "In this total giving-over there is . . . no identity that reserves for itself protected space or private rooms of impenetrable refuge for self-contemplation" she claims (Cavarero 2000, 84). In recounting one's life story, intimate self-reflection becomes externalization. Telling one's own story therewith implies distancing oneself from oneself, doubling oneself, making oneself into an other (Ibid.).[14] Narrating the self in short implies an irremediable exposure to others (Cavarero 2000, 87).

Instead of understanding the narrative self in terms of its self-gathering, Cavarero lays claim on its exposure to others, that—in the words of Judith Butler—constitutes its singularity (Butler 2005, 33).[15] In Ricoeur's philosophical anthropology, the prevailing idea is that human beings are speaking, narrating agents. It is the active, capable human being that expresses itself and gathers itself in its narrative, which forms the center of his philosophy. Human beings are born into a world of language that precedes and envelops them (Ricoeur 1986, 27), and in narrating they individuate and become singular beings. In the digital age, and specifically on social media, action has turned into speech and visualization. Documenting about ourselves in words and visuals and therewith exposing ourselves to others has become our main activity. Self-documentation on social media therewith includes an amplification of self-reflection, which contributes to our sense of self, to our narrative identity. Perhaps at first sight it might seem a thickening of self-love but this exposure to others signifies the loss of self-ownership as much. Instead of expressing ourselves about who we are, we expose ourselves to others. In showing what we do with whom and where in visuals and words, we lose ourselves.

5. CONCLUSION

In this chapter, I have investigated whether contemporary forms of self-documentation on interactive social media, such as discussion forums, blogs, and social network sites, can be understood in terms of Ricoeur's notion of narrative identity. Is Ricoeur's theory too much focused upon the classical novel, with its linear plot and closure of the story, to be of use for digital identity notions, as is claimed by some scholars (De Lange 2010; Frissen et al. 2015)? I have argued that for Ricoeur narrative identity needs to be understood in terms of the self attesting something and thereby also confirming herself as capable human being. Self-attestation is what happens on social media as well: self-documentation on social media can thus be understood as self-affirmation, as testifying to one's existence. On the basis of Ruth Page's empirical sociolinguistic study (2012), I have argued furthermore that this happens in the form of a narrative, and that it is not necessary to reformulate Ricoeur's theory because of being focused too much upon the novel, linearity and closure. This does not imply that narrative and digital identity are completely similar, however. Even though we need the notion of narrative identity in order to understand how identity construction by means of narratives takes place (in terms of understanding one's life as a relative and open unity to which others contribute), digital identity formation has its own characteristics. The main differences between narrative and digital identity are that temporal sequence is less relevant and that variable forms of tellership appear (from single, to co-tellerschip to multiple narrators). Again, this does not imply a complete break with narrative identity, instead digital identity reinforces aspects that are already part of the theory of narrative identity. In order to conceptualize these differences, I have suggested to understand digital identity formation in terms of self-exposure instead of self-expression. Exposure is the concept that Adriana Cavarero (2000) in her reading of Hannah Arendt uses in order to explain that the self is fragile and unmasterable. In her interpretation, narrating the self implies an irremediable exposure to others. This reading of Arendt, in contrast to Ricoeur's, accentuates the self's being completely given over to others. On social media we do not only "gather" ourselves by means of narratives, but also expose ourselves to others, and in this process we risk to lose ourself.

NOTES

1. See Ricoeur (1988, 1991a, 1991b, 1992). See for the reference to Arendt: Ricoeur (1992, 58–60), and also Ricoeur (2005, 250, 253).

2. Apart from Ricoeur, influential philosophical theories of narrative identity have been developed by Schechtman (1996, 2014) and Dennett (1991).

3. I use the term "social media" to refer to applications on the internet that promote social interaction between participants, such as discussion forums, blogs, social network sites, and video sharing.

4. "Digital identity" encompasses the various ways in which individuals express who they are on the internet. Stacey Koosel explains digital identity as follows: it "may exist as a small blurb of text under the subheading of "Biography" or "About Me" (often used in social media platforms). Other times in less structured environment, or web-platforms with more flexible and less stringent design in how to communicate, such as commenting or blogging, a digital identity can be created and maintained by a single or series of small stories." Also "snippets of identity information shared on social media platforms such as 'status updates' and 'comments' on Facebook can be seen as small but significant parts of identity construction in social media" (Koosel 2015, 73).

5. The brief version of her definition of narrative is: "Narrative is . . . a mental representation of causally connected states and events that captures a segment in the history of a world and its members" (Ryan 2004, 337).

6. Ricoeur's theory of narrative identity construction is used by Frissen et al. (2015) and by De Lange (2010) in the context of what they call the "ludification of personal and cultural identity" (Frissen et al. 2015, 10). Following Huizinga, they speak of Homo ludens 2.0, pertaining to the anthropological condition of humans in the digital age.

7. The written form of communication used online has been referred to as "spoken written communication" (Kacandes 2001) or "secondary orality" (Ong 1982).

8. De Lange (2010) in his dissertation aims at a theory of playful identities that he contrasts to Ricoeur's narrative identity, claiming that the first offers a way out of narrative's "closed circularity and is open to its own revisions" (2019, 230), that it allows for mobility and change, and does not morally prescribe how to live (2010, 231), and that the theory of playful identities "does not play by the rules set by narrative but instead offers a new lens to look at identity mediation" (2010, 232). This critique does not do justice to Ricoeur's narrative identity, as I hope to show in this chapter. Narrative identity instead is never ending and always open to change, Ricoeur's little ethics is not prescriptive and the rules set by narrative are not fixed. Other points that De Lange mentions are the Western character and the simplification of what culture entails in Ricoeur's theory (2010, 234). And finally, he claims that the role of spatiality, place, and mobility in identity construction can be accounted for by considering mapping instead of temporality (De Lange 2010, 235). Whereas De Lange in his dissertation contrasted the theory of ludic identity to Ricoeur's narrative identity, but in many respects did not do justice to Ricoeur, in Frissen et al. (2015), he and other scholars claim that Ricoeur's theory needs to be expanded. The latter is more in line with my claim that narrative and digital identity need to be understood as continuous yet different.

9. Self-constitution is the central idea in Marya Schechtman's account of narrative identity (1996). In the debate over narrative identity, it is mainly her conception that is discussed—see Strawson (2004), Hutto (2007, 2016), Zahavi (2007, 2008), and Gallagher (2014).

10. For Aristotle in the *Poetics* the plot forms the structuring and integrating process that gives an identity to the story. It is not a static structure but a process that composes the story and that makes it into a complete story that, in the end, can only be finished by the reader (Ricoeur 1991a, 21).

11. Thanks to one of the reviewers for pointing out this difference.

12. An extra argument in this respect are the narrative capabilities and intentions of the readers of social media, who often make connections between different narrative items. The intention to constitute a story out of seemingly unconnected posts may be explained with the help of Ricoeur's notion of mimesis$_2$.

13. In the literature on identity construction on social media, Judith Butler's and Erwin Goffman's notions of performance are often used to explain digital identity (Goffman 1959; Butler 1999, 7–8). Therewith the social pressures underlying identity performance and our need for social inclusion are addressed. My point is more radical, namely that the self is not a singular construct that is distinct from social norms, but becomes a self within social conditions. This is in line with Butler's *Giving an Account of Oneself* (2005).

14. For Cavarero, the unity of narrative identity is nothing more, or less, than a promise and a desire. The narratable self she calls "an exposed uniqueness that awaits her narration" (Cavarero 2000, 86).

15. For Butler, Cavarero's exposure amounts to our fundamental dependency on the other, to our sociality, that is, the foundation for ethics.

REFERENCES

Arendt, Hannah. 1958. *The Human Condition*. Chicago: University of Chicago Press.
Butler, Judith. 1999. *Gender Trouble: Feminism and the Subversion of Identity*. New York and London: Routledge.
Butler, Judith. 2005. *Giving an Account of Oneself*. New York: Fordham University Press.
Cavarero, Adriana. 2000. *Relating Narratives: Storytelling and Selfhood*. London and New York: Routledge.
Dennett, Daniel. 1991. *Consciousness Explained*. Boston, MA: Little, Brown.
Frissen, Valerie, Sybille Lammes, Michiel de Lange, Jos de Mul, and Joost Raessens. 2015. "Homo Ludens 2.0: Play, Media, and Identity." In *Playful Identities: The Ludification of Digital Media Cultures*, edited by Valerie Frissen, Sybille Lammes, Michiel de Lange, Jos de Mul, and Joost Raessens, 9–50. Amsterdam: Amsterdam University Press.
Gallagher, Shaun. 2014. "Self and Narrative." In *The Routledge Companion to Hermeneutics*, edited by Jeff Malpas and Hans-Helmuth Gander, 403–412. London and New York: Routledge.
Goffman, Erving. 1959. *The Presentation of Self in Everyday Life*. New York: Doubleday Anchor Books.
Hutto, Dan, ed. 2007. *Narrative and Understanding Persons*. Cambridge: Cambridge University Press.
Hutto, Dan. 2016. "Narrative Self-Shaping: A Modest Proposal." *Phenomenology and the Cognitive Sciences* 15: 21–41.
Kacandes, Irene. 2001. *Talk Fiction: Literature and the Talk Explosion*. Lincoln: University of Nebraska Press.
Koosel, Stacey. 2015. *The Renegotiated Self: Social Media's Effects on Identity*. Talinn: Alfapress.
Lange, Michiel de. 2010. *Moving Circles: Mobile Media and Playful Identities*. PhD-dissertation, Erasmus University Rotterdam, https://www.bijt.org/wordpress/wp-content/uploads/2010/11/De_Lange-Moving_Circles_web.pdf.
Mul, Jos de. 2015. "The Game of Life: Narrative and Ludic Identity Formation in Computer Games." In *Representations of Internarrative Identity*, edited by Lori Way, 251–266. London and New York: Palgrave Macmillan.
Ong, Walter. 1982. *Orality and Literacy*. London and New York: Routledge.
Page, Ruth E., eds. 2012. *Stories and Social Media: Identities and Interaction*. New York and London: Routledge.
Page, Ruth E., and Bronwen Thomas, eds. 2011. *New Narratives: Stories and Storytelling in the Digital Age*. Lincoln: University of Nebraska Press.
Ricoeur, Paul. 1975. "Phenomenology and Hermeneutics." *Noûs* 9: 85–102.
Ricoeur, Paul. 1984. *Time and Narrative*, vol. 1. Chicago: Chicago University Press.
Ricoeur, Paul. 1986. *Fallible Man*. New York: Fordham University Press.
Ricoeur, Paul. 1988. *Time and Narrative*, vol. 3. Chicago: Chicago University Press.
Ricoeur, Paul. 1991a. "Life in Quest of Narrative." In *On Paul Ricoeur: Narrative and Interpretation*, edited by David Wood, 20–33. London and New York: Routledge.

Ricoeur, Paul. 1991b. "Narrative Identity." In *On Paul Ricoeur: Narrative and Interpretation*, edited by David Wood, 188–199. London and New York: Routledge.
Ricoeur, Paul. 1992. *Oneself as Another*. Chicago: Chicago University Press.
Ricoeur, Paul. 2005. *The Course of Recognition*. Cambridge: Harvard University Press.
Ryan, Marie-Laure, ed. 2004. *Narrative Across Media: The Languages of Storytelling*. Lincoln: University of Nebraska Press.
Schechtman, Marya. 1996. *The Constitution of Selves*. New York: Cornell University Press.
Schechtman, Marya. 2014. *Staying Alive: Personal Identity, Practical Concerns, and the Unity of a Life*. Oxford: Oxford University Press.
Strawson, Galen. 2004. "Against Narrativity." *Ratio* XVII: 428–452.
Turkle, Sherry. 1995. *Life on the Screen: Identity in the Age of Internet*. New York: Touchstone.
Zahavi, Dan. 2007. "Self and Other: the Limits of Narrative Understanding." In *Narrative and Understanding Persons*, edited by Dan Hutto, 179–201. Cambridge: Cambridge University Press.
Zahavi, Dan. 2008. *Subjectivity and Selfhood: Investigating the First-Person Perspective*. Cambridge: The MIT Press.

Chapter 7

Digital Hermeneutics

Will the Real Quantified Self Please Stand Up?

Noel Fitzpatrick

Throughout the vast work of Paul Ricoeur, there is no one place where he deals directly with questions concerning contemporary technologies (Digital or otherwise). Rather, it could be argued, he deals with aligned questions of language, identity, memory, abstraction, and inscription. These aligned questions have now become kernel to the development of contemporary philosophy of technology.[1] It could also be claimed that Ricoeur's understanding of the text as a process of human mediation in the world can be extended to technology as a form of being-in-the-world; writing and data here understood as forms of inscription of the self in the world. At the core of these debates is the question of data itself, questions not only about the structure data but also the use and repurposing of data.

A distinction of vocabulary is necessary between data and information—data will be understood as the empirical trace, and, as Floridi (2008) argues, information is the first level of interpretation, which is produced through the elaboration (description) of data. The abstraction of the individual as a form of data (or statistical body)[2] has ethical implications and these ethical consequences of a primitive, indeed reductionist, person are now coming to the foreground. The predicative behavior modeling based on data aggregations is having direct ethical impacts on individuals and their choices. The question of bias within the construction of data sets has become a point of much debate in computer science recently.[3] There is an urgent need, therefore, to develop ethical frameworks to address these issues. Such ethical frameworks will need to be both macro and micro. There is need for a personalized ethics, or to be precise, an individual ethics of practice, as well as ethical-legal instruments.

This chapter will argue that ethics of technology could be framed as a form of "micro ethics" or, as Ricoeur sets out in *Oneself as Another* (1992), as a form of "little ethics." This chapter will demonstrate how the work of Paul Ricoeur is not only relevant to developing an ethical framework for digital technologies but also gives a background for posing ethical questions within the philosophical underpinnings of a hermeneutic phenomenology. This allows us to update his thinking in relation to something akin to what I have termed "digital hermeneutics."[4] Digital hermeneutics includes the understanding that digital technologies function as interpretative processes, which are language-based, but also the wider implication of hermeneutics as a move toward action in the world which has a profoundly ethical dimension. Digital hermeneutics in this context is focusing on the process of self-understanding mediated through digital technologies; it is an expansion of the reflexive nature of Ricoeur's philosophical anthropology, an extension of the detour through language to a digitally mediated self-understanding. The notion of mediation or inscription being mobilized is underpinned by the concept of hypomnesis, the exteriorization of memory in the trace, the digital trace acting as a placeholder for memory, or, according to Stiegler (1998), a form of tertiary retention.

The ethical questions raised by the construction of the digital self, as a self which comes to understand itself through its mediation in digital objects, both as a narrative identity through digital technologies and as a digital hermeneutic self, need to take into account the specificity of digital technologies. The speed of development of digital hermeneutic devices has surpassed the speed of cultural integration and therefore legislation. One such example is a form of digital distanciation of the self from the data self which exists as traces left across the internet that are bought and sold by data brokers. There is a distanciation between the individual (self) who created the traces and the ability to comprehend these traces. Ricoeur's distinction between the ascribing to the "What" and the imputable to the "Who" is very useful in this case, where both the "what" of identity and the "who" of identity are necessary, these are dialectical and non-exclusive categories. The narrative identity acts a mediator between "the who" and "the what" of personal identity.

This chapter will focus on the two ethical challenges that digital technologies pose, first, the self as a form of ethical abstraction and second, the relation between data self as the narrative self and narrative identity. Both these aspects are present within one particular mode of the self which has come to the fore more recently and comes under the headings of the "shadow self," "the quantified self" or "statistical double." The shadow self or digital twin refers to the grouping of data present on the web to represent an individual, for example, the identity that Google represents of an individual through their use of Google technologies, Google Search, Google Maps, YouTube,

Google Drive, and Gmail. The notion of the Quantified self, in addition to the shadow self, has specific information including biometric information, for example, temperature, heart rate, sleep rate, and movement. As the name indicates, the "quantified self" refers to the quantification of the individual as a set of empirical (biological) data about the person. There has been a recent extension of the biological information captured from simple tracking devices to contract tracing as this biometric tracing can also include virologic information about whether a person has had a particular viral infection such as COVID-19. In this chapter, all three terms will be used under the umbrella term "data self."

In order to reach the ethical questions proper to the "quantified self" or "data self," it is necessary that the notion of the self and the concept of quantification are explored. Once we have established the difficulty of the quantification of the self, we will be able to pose an ethical framework of "little ethics" of the quantified self. The first section will explore the notion of data and quantification of the self. The second section will move on to expound on the distanciation between the ascribing of particularities of the "what," and prescribing the accountability of the "who." The third section will outline how Ricoeur's "little ethics" could be expanded into digital quantified self or data self.

1. THE DATA SELF

This section will explore some of difficulties related to the concept of the self in terms of the relationship between data and the individual person. It will also explore the terms of "quantified self" and "data self" by developing the analysis of measurement and identity. The personal identity will be constructed through identity as sameness of character (*idem*-identity) and identity as sameness of self-constancy (*ipse*-identity). The data self will act as form of narrative self which mediates between both forms of sameness. The quantification of the person through the use of empirical data about individuals is nothing new. Foucault (1976) demonstrated that the governance of individuals through empirical means can be understood as a form of biopower or biopolitics. For example, biopolitics can encompass, on the one hand, the use of abstract census data to calculate and control populations, and on the other, the use human body measurement, anthropometry techniques, to categorize peoples. However, the amount of empirical data about individuals has exploded exponentially with the development of the internet, digital technologies, GPS, and, more recently, smart devices. The sheer volume of data produced by individuals is becoming increasingly problematic in terms of storage (data warehousing). Some of the data is intentional production of data through email, SMS messages, Instagram, TikTok, and YouTube. Other data, however, is produced non-intentionally: GPS position, search engine queries, website traffic analytics, natural language processing through smart speakers, online video platforms, and so on. This section will explore how these new forms of empirical data gatherings are posing questions about how they influence not only the person's understanding of themselves but also how they impact how the other's understanding of the individual. To use Ricoeur's language: how the oneself (objectified self) is understood as the other; the objectification of the individual person as data and its interpretation through data analysis is impacting on the very construction of (narrative) self.

The use of the noun "quantified self" has become widespread. For example, in the book *The Quantified Self,* Lupton (2016) makes a distinction between the "Quantified self in Precarity" and other forms of tracking of the person; one aspect of the Quantified self is self-monitoring, a form of surveillance of the self:

> Several terms in addition to self-tracking are used to describe the practices by which people may seek to monitor their everyday lives, bodies and behaviours; such terms are lifelogging, personal informatics, personal analytics and the quantified self. (Luton 2016, 14)

On the one hand, the Quantified Self refers to the monitoring of the self by the individual themselves but, on the other hand, it refers to quantification of the self by other means of monitoring including without the person's knowledge or consent. However, in addition to the semantic ambiguity of both senses of quantified self as passive activity and as active intentional activity, there is a fundamental ambiguity about the notion of the quantified self as quantification or measurement of the self. The quantification of the person or the self is sometimes referred to as the datafication of the self, or the datafied self, and this is captured by the phrase "we are data"; our identities and hence our futures are being pre-determined algorithmically. As Cheney-Lippold points out:

> We are likely made a thousand times over in the course of just one day. Who we are is composed of an almost innumerable collection of interpretive layers, of hundreds of different companies and agencies identifying us in thousands of competing ways. At this very moment, Google may algorithmically think I'm male, whereas digital advertising company Quantcast could say I'm female, and web-analytic firm Alexa might be unsure. (Cheney-Lippold 2017, 11)

The identification of our embodied individuality with a set of vectors in the numeric string of the algorithm, or ones and zeros, is highly problematic.

What is the relationship, if any, between my identity as captured by my online activity and myself? There is a tension between the empirical notion of the quantification of the individual and the identity building or profiling that takes place by companies such as Google. For our purposes, the notion of quantification or datafication includes all forms of numerical data about an individual: from heart rate, to number of steps to number of times a word searched on a search engine. This empirical data is turned into data points that become vectors which are used to create the profile made by the internet companies. Therefore, the fundamental philosophical question is: can the self be measured, is there a way to have metrics of the self, can the self be quantified? As Ricoeur points out the primitive concept of the person will hold the prominence of the physical sameness as the defining attributes of the person: it is the same thing that weighs sixty kilograms that has this or that thought (Ricoeur 1992). The philosophical anthropological question of what is means to be human is updated by asking the question in relation to a specific form of technology, a specific form of mediation in the world through processes which capture the activity of being human through measurability, statistics, calculation, and ultimately, through algorithms, machine learning, and neural networks.

The different meanings of the self and quantification will be primarily foregrounded as a movement toward a hermeneutics of the self, to borrow the term from Paul Ricoeur, which he developed in *Oneself as Another*, published in 1992.[5] This will be demonstrated as a Digital Hermeneutics of the self. Hermeneutics is first understood here as way of being in the world as *homo interpretans*,[6] or hermeneutic self, and second, as form of activity which is reflexive, the traces and crumbs of data as "signs, symbols, texts" call upon us to interpret. The well-known distinction between *idem* and *ipse* identity is at the core of the thesis proposed in *Oneself as Another*. It is the articulation between sameness as identical *idem*, something that is always the same which never changes and sameness as *ipse* identity across and through change. The *idem*-identity is that which is the *same* in the person while the *ipse*-identity is the one which is the selfhood. This should not be misconstrued as forms of opposition but rather as dialectic or dialogical process between sameness and selfhood. To quote Ricoeur:

> The equivocity of the term "identical" will be at the center of our reflections on personal identity and narrative identity and related to a primary trait of the self, namely its temporality. Identity in the sense of *idem* unfolds an entire hierarchy of significations. . . . In this hierarchy, permanence in time constitutes the highest order, to which will be opposed that which differs, in the sense of changing or variable. Our thesis throughout will be that identity in the sense of *ipse* implies no assertion concerning some unchanging core of the

personality. And this will be true, even when selfhood adds its own peculiar modalities of identity, as will be seen in the analysis of promising. (Ricoeur 1992, 2)

This distinction between the forms identity as sameness and selfhood will enable us to revisit the notion of the quantified self where the type of selfhood presupposed by the notion of the quantified self can be called into question. According to Ricoeur, the *idem*-identity and the *ipse*-identity are both part of the construction of the narrative identity. The analysis that Ricoeur gives to the notion of identity as sameness will be key to our understanding of how the quantification of the self can take place as a form of self-abstraction. The forms of empirical data of the body are now in real time and from multiple sources and are highly dynamic; the sameness of the physical person is revealed through the dynamic nature of entropic biological processes. It is the transparency of the bodily processes that is enabled by digital biometric devices which are being incorporated into the narrative identity of the individual. In Ricoeur's terms, this can be seen as part of the ascription: the empirical data generated through digital datafication of the person/self gives, on the one hand, generic attribution and on the other, elements that be conceived of as character. The ascription of the bodily process happens now in real-time monitoring which can be made public (most tracking devices enable the sharing of data not just to like-minded sports enthusiasts but also to insurance companies). The incorporation of this data monitoring of physical and psychic ascription into self-description and into identity is something new: its inclusion in the narrative identity. To put it simply, the sameness of physical and psychical ascription is presumed through the datafication of the self: it is the apparent sameness of the self, revealed through technology. The *idem*-identity of the sameness of character is a presupposition in how the data is interpreted as meta data, for example, in profiling individuals. The ascribing or describing through the empirical data become part of the descriptive self, part of the story of the narrative self, of self-description. The technologies that enable self-description have proliferated through technologies which I choose (or not) to portray myself. There is an overlap between the capturing of physical bodily processes and the portrayal of the self to the other through digital platforms. Hence, two elements of the Ricoeur's analysis of description and narration are closed intertwined the technologies such as Facebook and TikTok. Indeed, the prescriptive, or ethical, aspect of narrative self of what I am calling "technologies of the self" will be explored in the next section. However, questions of quantification of the self should not be confused with simplified understandings of the "technologies of the self" to borrow Michel Foucault's expression. For Foucault, the "technologies of the self" are technologies of the self, which permit individuals to effect by their own means or with the help of others a certain number of operations on their own bodies and souls, thoughts, conduct, and way of being, so as to transform I themselves in order to attain a certain state of happiness, purity, wisdom, perfection, or immortality. (Foucault 1983, 18)

For Foucault, the "technologies of the self" are not technical or digital devices, but ways through which one operates on her own body and soul. Foucault analyzes these "technologies of the self" of the western Greek and Christian traditions. What is of interest is that the new technologies, namely digital technologies, have enabled new forms of "technologies of the self," new forms of control of the self and new forms of mediation in the world, where the becoming the world is mediated through forms of the digital technologies whether they be social media, search engines, or biometric devices.

The question of mediation is central for Ricoeur from the outset of his philosophical anthropology; indeed he states that we mediate ourselves through the "signs, symbols and texts" of man. The new signs and symbols are today the signs and symbols of digital technologies, where mediation takes place through all forms of inscriptions in the world with and through digital technologies. The reductionist datafication of the person as a quantified self poses fundamental questions in relation to reason, measurability, and calculation. It is now necessary to explore how this tension between the primitive notion of the notion of the self can be coupled with other aspects of the self. In the next section, this chapter will explore which primitive notion of the self is being foregrounded.

2. WILL THE REAL QUANTIFIED PERSON PLEASE STAND UP?

This section will explore how the quantification of the self is part of the sameness (*idem*-identity) and the selfhood (*ipse*-identity) of narrative identity and also identifies the risks that are involved in the reduction of the identity and the self to quantification and calculation. In addition, this section will explore the relation between quantification, not simply as measurement but within the wider philosophic problematic "trace" or "inscription." Within digital technologies, our digital identity is constructed through traces left on the internet. These have become our modes of inscription in the world, exteriorized traces which can be standardized and optimized. The prominence of the datafication of the personal identity in the traces left on the internet is one form of identity. When one speaks of the quantification of the self, prominence is given to the self as something which is quantifiable, measurable. With the development of smart devices, the datatification of the self contains the biometrics of bodily activity but also includes all activity on the web that make our statistical other or our Internet profile. In order for the shadow self or statistical double to have monetary value, the emphasis is given to temporal sameness of the *idem*-identity which permits the ascription of objective features, quantified features. Therefore, the "What" of ascription necessarily dominates the quantified self, the statistical double and the shadow self. However, the question of "who" cannot be contained within something in general, as a general description (ascription), it has to have singularized dimension: "Keeping one's word expresses a self-constancy which cannot be inscribed, as character was, with the dimension of something in general but solely within the dimension of *who?*" (Ricoeur 1992, 123).

Within Paul Ricoeur's analysis of philosophies of the subject, it is important to point out that we are not speaking of a philosophy of the subject nor of some version of the Cartesian cogito, a subject "I": we are speaking of a hermeneutic of the self, a *homo interpretans*. The first characteristic of the hermeneutics of self is reflexive, a reflective self, a self-designating self. The second characteristic is enabled by the distinction between *idem*-identity and *ipse*-identity: the question of "who" is speaking in the quantified self and the "what" of the person. As Ricoeur asserts, the question "who?" introduces "all the assertions relating to the problematic of the self" (Ricoeur 1992, 16).

Here the question of who is speaking and what is the person enables the analysis of what to include in the quantification of the person as a statistical body. The statistical body is the description of the "what" of the person at the most primitive level; we could also include, as Ricoeur does, physical and mental states, moods humors, and so on. This is the lower limit of the narrative identity, the bodily activity; however, within the use or repurposing of this biometric data, it becomes part of the narrative identity and is used to develop the multiply features (vectors) of my online profile (shadow self). The statistical body becomes part of the self-designation of the person; it becomes part of the narrative self. To put it simply, the statistical body refers to the sameness in the world of the person as a "thing."

Sameness of *idem*-identity is the characteristic of permanence over time: the character. The quantification of the person as statistical body emphasizes the thingness of the person which is part of what being a person is. However, the fundamental misconstruction is to reduce the "self" to the thing. The presupposition of sameness over time is extended into the construction of the profile created through dynamic information. However, this very profile itself is not fixed and changes also over time. My profile will also change and be reconstructed over and over again through my mediation in and through digital technologies. In addition, if we follow Ricoeur's reasoning, the very notion of the quantified self needs to be called into question. The quantification of the thingness of the person does not include the notion of the self. The ascription is that same for everyone, hence the distinction with description which describes the qualities of the individual character, and with prescription which encapsulates the moral and the ethical of "who." As Ricoeur points out, it is the power of self-designation that makes the person not merely unique type of thing but a self. The concept of selfhood lies beyond the sameness of the body. However, it is worth pointing out that both are necessary for narrative identity: the sameness is part of the account, the story that I tell about myself. To help clarify how this could be the case, that is, that there is a self beyond the sameness of statistical body, a self beyond the quantification of the self, we need to firstly determine what is being referred to one when speaks of the narrative self or narrative identity. The narrative identity can be defined as follows:

> Our own existence cannot be separated from the account we can give of ourselves. It is in telling our own stories that we give ourselves an identity. We recognize ourselves in the stories we tell about ourselves. It makes little difference whether these stories are true or false, fiction as well as verifiable history provides us with an identity. (Ricoeur 1985, 213)

The distinction between fiction and history is at the core of this quotation, whether something is true or false, whether something is fiction or verifiable history. This raises questions about the ontological status of fiction, which we do not have the time to develop here but which I have done elsewhere (Fitzpatrick 2017). Nonetheless, this quotation highlights that narrative identify is something that is constructed over time and not a fixed identity. The second aspect in the quotation is that we cannot separate our identity from the stories we tell about ourselves. The narrative identity is beyond ascription of singularities to the person thing and incorporates a description of the person as selfhood. The dynamic nature of narrative identity, one could argue, is paralleled by the dynamic nature of the data gathered in real time about the individual, their movements, as well as likes and dislikes. The quantified self is a real-time collection of ever-changing data flows but at the same time, it is from this dynamic data stream that enclosure of identity takes place. It is this ability to make static categories from dynamic data that has monetary value. The algorithmically processed data enables the metadata of categorization of the traces into a meaningful actionable clusters—gender, age, sexual orientation, religious beliefs, and political persuasion. This clustering enables the target marketing, to drive the desire of consumption through the libidinal economy (Lyotard 1974). Importantly, the self-designation of story is denied, and there is no reflexive element to how this identity is constructed:

> All the while, these algorithmic interpretations are rarely known to us. They make who we are from data we likely have no idea is being used, a denial of explicit reflexivity that suppresses the "social" component of our identity's social constructionism. (Chiney-Lepplod 2017, 32)

The reflexive components of the narrative identity are denied through a denied access to the elements of the data which I am not aware of and that are used make up the "who" of the data self. These elements of the description made by me but which I cannot control are part of the narrative about me, which amounts to denial of self-reflectivity. Due to the fact that I have no idea how my online profile is constructed and how it is changing, it remains a black box. While Ricoeur does not mention technologies of storytelling, he does refer to language and the characteristics of interlocution in his analysis of narrative identity. In *Time and Narrative* (1984), the examples of storytelling are either those of Greek mythology or autobiography (in the case of Saint Augustin) or the modern novel (Virginia Wolf, Kafka, and Marcel Proust), there is little or no mention of other ways of telling one's own story. It is here that the extraction of value from the quantification of the self through the generation of data is beyond the analysis of Ricoeur. In Ricoeur's work, the hermeneutic of the technologies of the self are limited; there is no mention of other forms of technologies for the development of the narrative identity. The development of the quantification of the self (or the data self) has only really come to prominence in the past ten years: Apple, Google, Amazon, and Facebook have become dominant since Ricoeur's death in May 2005. However, we can see possible glimpses of what his analysis might have been through parallels with the problematic of inscription.

It is in his publication *Memory, History, Forgetting* (2006) that Ricoeur addresses, although indirectly, the question of means or ways of storytelling as forms of inscription: the question of writing as a *pharmakon*. It is the analysis of the difference between memory and history that the concept of inscription is grounded, in order to move beyond my memories in my mind, the souvenirs need to be inscribed outside of the mind. To simplify this, we could think of the inscription in the world in multiple forms of drawings, paintings, graphics, writing, sculptures or monuments. In order for there to be collective, generalized souvenirs inscription is required: writing is necessary. The fundamental difference between memory and history is raised by the problematic of inscription. The question of inscription becomes the manner through which Ricoeur will establish the external markings of writing as a *pharmakon*. In his commentary on the *Phaedreus* for Ricoeur, it is the significance of inscription in generality which is at stake (Ricoeur 2004, 124). Inscription in the *Phaedreus* does not simply refer to writing but to all forms of inscription, including painting, as *graphein*. This is, of course, a reference to Derrida's seminal text on the *Plato's Pharmacy* where Derrida gives a very detailed reading of the *Phaedreus* and shifts from his grammatology to his project of deconstruction. However, it is only much more recently, with the work of Stiegler, that a fully developed understanding of question of exteriorization and inscription as forms of *pharmakon* comes to prevalence. Stiegler extends the analysis of pharmacology to all forms of technology and develops the concept of tertiary retention which acts as a mode of inscription, the forms of inscription as analyzed in terms of exosomatic evolution and exosomatic organs. The pharmacology of the means of storytelling is not developed in the work of Paul Ricoeur, except in relation to history where he poses the question in relation to History: "what fascinates me, as it does Jacques Derrida, is the insurmountable ambiguity attached to the *pharmakon* that the god offers the king. My question: must we not ask whether the writing of history, too, is remedy and poison" (Ricoeur 2004, 141).

It is worthwhile extending the question of the historization of the self as statistical data as a form of pharmacology; to do so it is necessary to look to a symptomology of the quantification of the self, that is, the poverty of the quantification of the person as a thing among other things. In itself, this poses ethical questions in relation to the treatment of the statistical body as something from which monetary value can be extracted. Writing, taken here in the broadest sense of data generation, acts as a form of quantification of the self, and it could be argued acts as a remedy and as a poison. A symptomology would point toward the symptoms of the toxic, poisonous aspects of the quantification of the self (Vignola 2018).

However, the use of quantification of the person through tracking devices such as biometric devices falls under the classification that we are putting forward as *idem*-identity as something which is giving characteristics and qualities to the sameness or permanence of *idem*-identity. The ethical dimension, therefore, comes to the fore when we move from the sameness of "what" is the biological person or the person to the question of "who." The fundamental distinction between sameness and self or selfhood is, who is keeping their word throughout the changes of life? Who is keeping their promise? It is here that the question of ethics comes to fore in Ricoeur's analysis and it is to this "little ethics" that we will turn to in the next section of this chapter.

3. LITTLE ETHICS (AS OPPOSED TO MICRO-ETHICS)

This section will outline how questions of ethics and the problematic of the data self could benefit from an exploration of Ricoeur's "little ethics" (Ricoeur 1992, 290). As discussed in the previous section, there is a dialogical nature of the narrative identity, which mediates between the *idem*-identity and the *ipse*-identity in the construction of the narrative self. The dialectical nature of the movement from sameness of identity to otherness of selfhood leads us the final aspect of the triad, that of prescription. The "who" is the same as the person that makes the promise, the "who" that is responsible. In Ricoeur's analysis, the "who is speaking" becomes the "who is keeping their word." As Ricoeur outlines, the "who" is a self-reflexive, a "who" of interlocution and it is perhaps here the kernel of the issue of prescription and the data self. In the data self, the movement from description to narration can take where self-reflexivity is possible (where the data is known by the person emitting the data intentionally or not as the case maybe). Who is responsible (imputability) is normally a question of action, intentionality, and agency. However, who is responsible in relation to the data self cannot be resolved through the question of agency. In order to understand what is at stake for Ricoeur in relation to agency and prescription, it is necessary to set out briefly what Ricoeur means by "little ethics" which ironically has big impacts.

In the case for Ricoeur's "little ethics," it is "a return to morality through the sieve of ethics" (Ricoeur 1992, 170), for Ricoeur both morality and ethics benefit from the dialogue. It could be argued that the little ethical dimension of the quantified self or data self will enable the exploration of the lack of identification of the self with the statistical representation of sameness and this leads to what could be termed as ethical gap. Ricoeur's development of an ethical framework originates in questions of narrative identity and suffering. In addition, it is here that the ethical dimension of prescription is developed by Ricoeur, building on the analysis of description, narration, and prescription from the three volumes of *Time and Narrative* (1984, 1985, 1988). For Ricoeur, little ethics is developed as "aiming at the good life for and with others in just institutions." We do not have the time here to explore his little ethics fully but suffice it to say that there are three ethical values underpinning this ethical aim, first, the good life (self-worth, self-esteem), second, for and with others (solicitude), and third, in just institutions (participative justice). The ethical aim or *telos* is the good life – there is a well-known distinction between Aristotelian ethics, aiming toward the good life, and morality as the Kantian deontological obligation, which is the moral indication of right and wrong. The good life is with other people, hence the movement toward the other as something different to the sameness but a form of identity that includes the other.

The interest here is to see how this ethical value system of self, other, and justice could be related to questions of quantification of the self, or the statistical body that were developed in the first and second section. However, as with the area of medical data, there is a dissymmetry between the individual giving up their data and the position of the clinician who uses the data. In the case of large hugely profitable companies handing over data about the self becomes an obligation in order to access the services offered. There are aligned questions of opting in or opting out; in most cases, opting in becomes the obligation (e.g., self-tracking and car insurance). In relation to Quantified Self data, there are a number of issues of self-worth and self-esteem which can be raised, first, at a primitive level there is an embedded expectation, for example, that the data itself is of value to my own self-esteem in terms of a normalization of behavior and statistics, that is, that somehow the statistics gathered by myself can be seen in relation to the "ideal form" or what Aristotle might call expert practitioner or standards of excellence of practice: "The standard of excellence are rules of comparison applied to different accomplishments, in relation to ideals of perfection shared by a given community of practitioners and internalized by the masters and virtuosi of the practice considered" (Ricoeur 1992, 176).

Second, the role of comparison and excellence is built into the gesture of collecting the data and giving consent for the data to be used which enables self-esteem is part of the very process of comparison itself. The question of comparison is allowed by the very structuration of data as it is collected and visualized to the individual user of the tracking device or tracking app. The data is presented as comparative in relation to oneself and others, for example, indicating whether you have slept well, enough hours of deep sleep for your age group, and so on. As Ricoeur states, "self-interpretation becomes self-esteem" (Ricoeur 1992, 179). However, there is an internal good toward which the action of tracking or gathering the data is oriented, the healthy life. The healthy life and the good life here overlap—the good life in the case of the quantified self is the healthy life, presented as an intrinsic good in itself. The healthy life is the aim toward which the action of self-tracking is geared and one which is unquestioned. To use the terminology of Stiegler, there is a curative, positive aspect to the pharmacology of technology, here the healthy life and a negative, toxic aspect to the pharmacology which could be the addictive nature of the technology or the toxicity of the repurposing of my data through monetization.

The second value (solicitude) of the Ricoeur's "little ethics" is in relation to the other. Outside the realm of self-esteem and self-worth as point of comparison of excellence, there is the ethical demand of the other. Solicitude is an extension of the value of friendship, here understood not simply in the manifestation of the inter-relation with the other but a value which could be understood beyond the relation with the other, for example, in terms of being a friend (kind) to oneself. It is important to note, according to Ricoeur, that self-esteem is not a value of myself but of the self or of oneself, and that friendship is also a value of oneself or the self. The ethical framework of friendship can therefore include things such as respect for the other and for the self and also a form of reciprocity of equality and equal distribution.

Within Ricoeur's little ethics, the second value of reciprocity and solicitude has direct implications for the use of data and the quantified self. First, one could think of the relationship of friendship in social media context where there is a foregrounding of an ethical relationship to the other as they present themselves on social media platforms. There have been some interesting studies carried out in relation to disinhibition in language usage and behavior; these tend to focus on issues related to flaming or trolling and insults (Fitzpatrick 2012). However, there is another ethical issue, which is more indirect, where the question of imputability is raised. This is the ethical question concerning my accountability in relation to data which I consent to being gathered and which is used to make decisions in relation to others. This is most obvious in relation to data that I intentionally create or give consent to be used. However, there is also the use and repurposing of the data which could be outside my control. For example, when my data indicates that I buy a certain newspaper and the fact of buying this newspaper is included, the neural network is designed for insurance premiums. I have intentionally consented to this data being collected by clicking on a tick box when I subscribed. However, my data leads to the exclusion of someone from a specific insurance product or the refusal on the ability to obtain a mortgage. The ability to profile my financial risk has led to the development of complex neural networks which can include the type of newspaper as vector in the risk calculation. A correlation has been made between the number of people who default on a load and number of people who read a certain newspaper; this correlation is both positive and negative. In this example there is no real authorship (agency) in the traditional sense of the decision that can be traced back to my activity.

This is perhaps where the dislocation or delocalization of data leads to impunity; however, the question remains: What is my ethical responsibility toward the other in the sharing of my data to large globalized neural networks? This is further highlighted by opting in or opting out to certain forms of surveillance—if I self-monitor my driving and demonstrate that I am safe driver, my insurance premium will decrease. However, if I do not self-monitor I am presumed to be at higher risk. More worryingly is the sharing of data which I am not aware of, the use of data collection across multiple platforms where the shadow self or data self is given profiles which are then bought and sold by data brokers where decisions are made in relation to my profile which impact on others without my consent or knowledge. The allocation of accountability has traditionally come from the concept of agency or imputed agency; however, as the data self which represents myself cannot have the same imputability the ethical gap appears. There is an ethical gap between ascription and description. Indeed, can the data self be held accountable for decisions or choices influenced by the clustering of the data? It would be naïve, therefore, to reduce the ethical question of the other to simple social media presence, and friendship in this sense is beyond the means–end relation of Facebook.

The third component of the "little ethics" is just institutions, the movement from self-esteem to solicitude encapsulates a movement from the self to the other, where in the value of friendship there is a notion of just or fairness. However, as Ricoeur points out that justice is toward the other, justice and friendship are not the same thing. The objective of equity of an institution is beyond the sharing of friendship. For Ricoeur, the justice is the institution as judicial system:

> The *just*, it seems to me, faces two directions: towards the *good*, with respect to which it marks the extension of interpersonal relationships to institutions; and towards the *legal*, the judicial system conferring upon the law the coherence and the right of constraint. (Ricoeur 1992, 197)

The just moves beyond the interpersonal, toward the legal. It is here that there have been famous examples of the misuse of statistical bodies or data self in relation to the judicial system, where algorithms have been developed to aid the sentencing of prisoners, only to discover that the biases which they were supposed to eradicate are embedded with the computational processes itself (O'Neil 2016, 28). It is perhaps this third component to little ethics that demands international legislation where individual data is considered to be individual's right to their data, the right to know what data is being collected and what purposes and also the right for that data to be erased (the right to be forgotten). The European legislation on GDPR has brought this debate into the public domain. In terms of the quantified self or data self, there is a presumption of arithmetic equality as just, it is not because we have measurability that we have equality, the presumption is that because the data is neutral that the measures are objective. Ricoeur argues that the equality provides to the self another who is an *each*. In this, the distributive character of "each" passes from the grammatical plane to the ethical plane (Ricoeur 1992, 202). It is here that the ethical questions of the data as composites of the self need to take into account the equality of distribution. The *each* in the ethical place poses questions of the self which is quantified, the each being accountable and being equal; however, this is not the case when the self is reduced to a thing to be measure among things.

4. CONCLUSION

The quantified self or data self acts as form of abstraction which does not possess the rights of the individual and yet the data sets are used to determine and impinge upon the rights of individuals within society. To conclude, the question of attestation which is at the center of Ricoeur's analysis throughout *Oneself as Another* raises the distinction between the "what" of the self and the "who." The distinction between the same data set and the same person is where the movement from ascription to description enables the shift toward the ethical consideration as prescription. This raises the need for an ethical framework which enables us to pose the question of immutability by extension of the ascription, description, and prescription. The ethics of the self and the data need to be determined by recognizing that the reductionist understanding of the person does not lead to a denial of ethical responsibility. The contemporary difficulty is that the technologies of the self that have been developed promote the self as something which can be quantified. A radical revising of the question of the self is needed.

NOTES

1. See, for example, Coeckelbergh (2020).
2. The statistical portrayal of the person can be dated back to the development of Life Assurance in 1840s. However, with the advent of machine learning and data science, the insurance industry has changed radically: "Now, with the evolution of data science and networked computers, insurance is facing fundamental change. With ever more information available—including the data from our genomes, the patterns of our sleep, exercise, and diet, and the proficiency of our driving– insurers will increasingly calculate risk for the individual and free themselves from the generalities of the larger pool" (O'Neil 2016, 140).
3. See Kelleher and Brendan (2018).
4. See Fitzpatrick (forthcoming). This notion of "digital hermeneutics" is distinct from that proposed by Romele (2019). Digital Hermeneutics is constructed after the work of Ricoeur but with the work of Stiegler, digital hermeneutics as the place from which thinking can begin through digital technologies, as negaunthropic or as a positive pharmacology.
5. See Ricoeur (1992).
6. See Michel (2019).

REFERENCES

Cheney-Lippold, John. 2017. *We Are Data: Algorithms and the Making of Our Digital Selves*. New York: New York University Press.

Coeckelbergh, Mark. 2020. *Introduction to Philosophy of Technology*. Oxford: Oxford University Press.
Fitzpatrick, Noel. 2016. "The Question of Fiction: Nonexistent Objects, a Possible World of Response from Paul Ricoeur." *Kairos: Journal of Philosophy and Science* 17: 137–153.
Fitzpatrick, Noel. 2019. "A Question Concerning Attention and Stiegler's Therapeutics." *Educational Philosophy and Theory* 52, no. 4: 348–360.
Fitzpatrick, Noel. forthcoming. "Mischievous Hermes: Digital Hermeneutics and Stiegler's Therapeutics." In *Aesthetics, Digital Studies, and Bernard Stiegler*, edited by Noel Fitzpatrick, Neill O'Dwyer, and Mick O'Hara. New York: Bloomsbury.
Fitzpatrick, Noel, and John Kelleher. 2018. "On the Exactitude of Big Data: La Betise and Artificial Intelligence." *La Deluziana* 7: 142–155.
Foucault, Michel. 1983. *Technologies of the Self*. London: Tavistock Publications.
Hildebrandt, Mireille. 2011. "Who Needs Stories If You Can Get the Data? ISPs in the Era of Big Number Crunching." *Philosophy and Technology* 24, no. 4: 371–390.
Johann, Michel. 2019. *Homo Interpretans: Towards a Transformation of Hermeneutics*. Lanham: Rowman & Littlefield.
Kelleher, John, and Brendan Tierney. 2018. *Data Science*. Cambridge, MA: The MIT Press.
Komesaroff, Paul A. 1995. *Troubled Bodies: Critical Perspectives on Postmodernism*. Durham: Duke University Press.
Lupton, Deborah. 2016. *The Quantified Self*. Oxford: Polity Press.
O'Neill, Cathy. 2016. *Weapons of Maths Destruction*. New York: Crown Books.
Ricoeur, Paul. 1985a. "History as a Narrative Practice." *Philosophy Today* 29, no. 3: 213–222.
Ricoeur, Paul. 1985b. *Time and Narrative*, vol. 2. Chicago: The University of Chicago Press.
Ricoeur, Paul. 1992. *Oneself as Another*. Chicago: The University of Chicago Press.
Ricoeur, Paul. 2006. *Memory, History, Forgetting*. Chicago: The University of Chicago Press.
Romele, Alberto. 2019. *Digital Hermeneutics*. New York and London: Routledge.
Stiegler, Bernard. 1998. *Technics and Time, 1: The Fault of Epimetheus*. Stanford: Stanford University Press.
Vignola, Paolo. 2017. "Symptomatology of Collective Knowledge and the Social to Come." *Parallax* 23, no. 2: 184–201.

Chapter 8

The Pedagogical Relation in a Technological Age

David Lewin

There is no doubt that modern technology has changed education, but these changes bring with them questions and challenges. How should educators respond to the widespread technological changes that herald the demise of the conventional school (Masschelein and Simons 2013)? Not only does school look rather nineteenth century, the figure of the teacher appears to be a quaint anachronism whose days are numbered. When students have access to the total(ized) knowledge of the internet, what kinds of educational authority are legitimate? Are we, indeed, witnessing what Postman, nearly half a century ago, called *The Disappearance of Childhood*?[1] More insidious yet are the widespread effects of a technological culture that render educational systems accountable exclusively to the reductive pressures of the bureaucratic iron cage (Dunne 1997). Education as the "production of learning outcomes" (Masschelein and Simons 2013, 18) appears as inescapable as it is alienating as it has become *mass* production. National or international school and university league tables have become a ubiquitous feature of our efforts to "improve" education, even if such improvements are not unequivocally "good" (Biesta 2011; Flint and Peim 2011). Most relevant to this chapter, however, is the conflation of education and learning, and the consequent erosion of the figure of the teacher who is able to exercise appropriate educational judgment (Biesta 2017).

Following Heidegger (1977), it seems to me that the worst excesses of our technological culture are expressed not only in the way things are set upon and organized as "standing reserve" but also how human beings are themselves subject to technological enframing.[2] From this perspective, we can see that the reduction of education to the efficient production of learning outcomes vividly illustrates Heidegger's analysis in the domain of contemporary educational relations (Lewin 2014a). Here lies a danger that the "pedagogical relation" is masked by a "technological relation," whereby educational subjectivity becomes rationalized and instrumentalized. The goal of education is dominated by a notion of efficient learning, a notion that almost entirely occludes the broader aims of education: to develop subjectivity. Enter Ricoeur who addresses many pertinent issues: the formation of subjectivity through recognition theory, narrative identity, and affirmation of subjectivity (and potentiality). I argue that these themes are best understood within the structure of the pedagogical relation, that is, the relation between an educator and a student.

This chapter principally concerns anthropology, the concept that most obviously connects the educational and the philosophical, and invites us to consider how technological thinking in the Heideggerian sense conditions the anthropological ideas of contemporary education. While education has long sought a "more explicit theory of human nature" (Peters 1983, 51) many educational theories have assumed autonomy as a key educational aim (Marples 1999). In the terms of Ricoeur's philosophy (1984–1988, 1992, 2005), the formation of subjectivity is defined as something like autonomy: by the capability to act in the world. In what follows, I offer an ontological account of the pedagogical relation between the teacher and the student.[3] I agree with Biesta as well as Masschelein and Simons (2013) that, although under threat from forms of technologization (where educational relations are reduced to measurable outcomes), the figure of the teacher is educationally essential. Moreover, the process of "becoming a subject" through education, what Biesta calls *subjectification* (Biesta 2006, 2011), is shaped by a pedagogical relation, which involves three elements: a student, a teacher, and a world (content). The breadth of this theoretical triangulation—between philosophy, education, and technology—demands a certain circumspection. I now turn to some contextualization of education studies as a discipline followed by a discussion of Ricoeur's thinking in the context of this discipline.

1. RICOEUR AND EDUCATION STUDIES

It should be noted that by comparison with other major twentieth-century continental philosophers (notably Foucault, Lyotard, Derrida, Levinas, Heidegger, and Arendt), Ricoeur is not regularly drawn upon in theoretical or philosophical discussions of education.[4] It is true that Ricoeur does not offer a systematic or lengthy discussion of education,[5] but education as a theoretical discipline is often poorly conceived and its disciplinary status underdeveloped.

I cannot explore the many reasons for this here, but certainly it seems related to the diverse contexts and traditions in which educational concepts have emerged.⁶ The application of Ricoeur to education is typically rather piecemeal: where his ideas are taken up, it tends to be without reference to significant theory of education (Moratella 2015) or only to particular ages (Farquhar 2012) or subject fields (Streib 1998). A more general and wide-ranging analysis of Ricoeur's relevance to education studies is harder to find though we see some indications here and there (Gallagher 1992; Kerland and Simard 2011), more particularly around general educational questions of pedagogical relation (Hoveid and Hoveid 2009), of understanding, and of hermeneutics (Leonardo 2003). My intention in this short chapter is not, of course, to develop such a wide-ranging analysis, but to focus attention on the educational significance of a central issue for Ricoeur: philosophical anthropology.

Before all else, education is a matter of anthropology: every intention to positively influence a person implies a normative anthropology.⁷ Educational influences more or less consciously assume answers to the following questions: What kinds of formative influence are desirable and legitimate? How do we justify formative influence? How are desirable influences achieved? Difficulties arise when we acknowledge that answers to these questions most often entail some more or less stable idea of what it means to be human in terms of, for instance, rational autonomy or moral agency (Peters 1966). This can be difficult because we don't all agree; such ideas about what education should be also risks denying personhood from those considered "ineducable." These issues require a mediation between an essentialist position that excludes many from being (or becoming) human and the postmodern negation of anthropology that gives little direction for educational practice. As a figure that stands for mediating between polarities, Ricoeur's contribution here could be considerable, especially because Ricoeur's anthropology is so central to his philosophy. For Ricoeur, personal identity is neither fully stable or self-transparent, nor incoherent or self-alienated: our self-relation is essentially one of active interpretation, rather than fully autonomous self-authoring. Thus, Ricoeur seems to have a desire for, as Anderson (1993) has neatly put it, "having it both ways," that is, both affirming and denying identity at the same time. In the realm of subjectivity that means that Ricoeur has been prepared to face off the attacks on the concept of subjectivity that claim it is either incoherent (*pace* Hume) or illusory (*pace* Nietzsche). Ricoeur seeks a post-critical, or *reconstructivist* (Romele 2014, 108), conception of narrative identity that does not avoid these attacks but absorbs them. It is, I would argue, the concept of narrative identity that offers educators something significant. As educators offer students ways to understand their world and their own selves, so students come to form their own narratives. Through encounters with history and culture, the child comes to terms with their own story. Furthermore, the recognition offered by the teacher is a critical component of the student's formation of narrative identity. Thus, the educator both presents and represents the other through which the child comes to form herself. In addition to the formation of self through narrative, educators also rely on certain assumptions or affirmations to undertake their work: that the student before them is educable and that education has an influence on an enduring, self-same subject. If there were nothing like the *subject*—an identity that endures over time—there would be no one to educate. With the subject in *position*—affirmed or posited rather than established as a foundation—the possibility and necessity for education arises.

In positing that the subject (in this case the student) is interpreted as a person capable of learning, the ipseity of the subject is affirmed. This is in contrast to an inert object like a stone, or an entity to be trained according to a limited horizon of outcomes (i.e., a wild horse being "broken"[8] or a dog being trained). This indicates that the positing of a particular kind of subjectivity is a condition for what is normally called education (rather than *training*), as well as being an outcome of education (Hoveid and Hoveid 2009). This projection of "educability" (German: *Bildsamkeit*; French: *plasticité*) takes on a particular hue in the context of the modern bureaucratic, technologized state: it is in this context that we consider how far pedagogical relations between educator and student—themselves constituted by the projection of an educable subject—then replaced by a technological relation between a producer and consumer of learning. In other words, how much do the present conditions (of the technological culture we live in) reduce pedagogy to a technical process: "the production of learning outcomes"? (Masschelein and Simons 2013, 18; see also Dunne 1997, Introduction).

2. TECHNOLOGICAL THINKING

By *technology* I mean something quite broad, "by no means anything technological" (Heidegger 1977, 4) in the sense of the devices that surround us in the present age. Rather, technology here is understood in terms of the *technological thinking* that constitutes those devices, and that underpins the technical processes just mentioned. Although a rather vague notion, allow me to offer the guiding approximation: technological thinking is a way of seeing things and people only in terms of their apparent utility. This way of seeing has three implications: first, through this focus on utility, we are encouraged to overlook reflection on the telos, or final purposes to which such utility may be put. Indeed, in the technological milieu, or device paradigm as Albert Borgmann (1984) called it, reflection of final purposes is (perhaps systematically) obfuscated, by, ironically, the achievement of user functionality (preliminary ends). Second, the capacity and agency of a person may not be measurable or even visible in the terms currently available. This brings me to the third point that renders the first two problematic: the technological mediation (or technological hermeneutic) appears absolute such that any other mediation becomes unthinkable. Like instrumentalism or efficiency in general, seeking to measure and to use are not, in and of themselves, problematic: they are essential for the continuation of life. Yet, as Heidegger and many others have long argued, if these ways of seeing (or *disclosing*) become all-encompassing or totalizing, then we risk losing touch with the world and ourselves (Heidegger 1977; Lewis 2001). So, this concept of technological thinking points to the reduction of the world and others to nothing more than functionality determined by users who are encouraged to avoid reflection on the final purposes to which their use is aimed. In the context of education, this can mean a failure to consider subjectification: what it means to become a subject. Rather learning outcomes can be determined by technical needs of society: for example, to service economic needs; to gain competitive advantage and so forth. Again, these are not unimportant considerations, but when they become totalizing, they encourage us to overlook the significant responsibilities of educators to recognize and support the development of the subject.

The concept of education derives etymologically from two terms, one emphasizing the forming or molding of the student (*educare*), the other emphasizing the idea of drawing out or bringing forth what is latent or innate within the student (*educere*) (Bass and Good 2004). To apply popular metaphors for the figure of the teacher, we see here the sculptor (*educare*) and gardener (*educere*) views of education, though neither metaphor is entirely adequate (Buber 1997; Veck 2013). Education suggests a mediation between forming, putting in, and releasing, drawing out, which should be kept in view since the reduction of education to the production of learning outcomes risks losing sight of an essential aspect of formation: that educational influence works upon something—a human with innate dispositions, tendencies and capacities. And that, therefore, "good" education must pay some attention to those tendencies and capacities in contrast to a more technical view of education, which sees a transaction take place between the educator and educated that ignores such questions of human relation and formation.

3. EDUCATION: BECOMING WHO WE ARE

Philosophical anthropology refers both to what it means to be human and to *become* human. Although the latter is a fundamentally educational concern, both are of central significance to Ricoeur's considerations of identity and capability (Ricoeur 2016). Yet the idea of *becoming* human introduces various questions: What (or "who") is the being before it has become human? What unifies the identity of the person who is changed, formed, or transformed by education: What is this ipseity that defines self-identity in the context of change? Is human identity ever something to be realized? What about those people who never realize the attributes of human identity that are settled on? These aporias reveal something of the inevitable instability of the concept of human nature. Nevertheless, it seems that no education can do entirely without such a concept. Speaking educationally, human beings are neither just preformed objects to be uncovered (*educere*), nor only matter to be formed (*educare*), but beings *in potentia*. That potential can only be realized by the interaction between a student and something other; hence, the self is not autonomous in the sense of being fully self-authoring. As potential, we are capable of being brought about, achieved, or formed, through education in the widest sense (von Humboldt 2015). Important aspects of human formation can be understood as our capacities to speak, to act and to narrate, capacities that are formed through the recognition of our agency (first by others and then our own recognition) as well as our own attestation of it (Ricoeur 2005; Hoveid and Hoveid 2009). But we must learn to narrate ourselves. The question of how human beings become what they are lies at the heart of the human sciences and has encouraged the development of what has been termed *human science pedagogy* (Friesen 2017b) which highlights the activities of the pedagogical relation that are essential to human formation.

This pedagogical relation is hermeneutical insofar as it takes different forms depending on how we interpret the concept of education as well as how we interpret the figures of the student and the teacher: whether we see education as the transmission of data from the active knowing agent to the passive ignorant patient, or as creating conditions for active growth, is a matter of interpretation. To describe education in one or another way is to inscribe certain assumptions about pedagogical relations that ought not to be naturalized, but to be made visible (and perhaps put into question). What it means to be a teacher or a student requires mutual interpretation or projection between two figures and the relation they enter into.[9]

The particular interpretation of the pedagogical relation in much contemporary educational discourse, at least at the level of general policy and practice, presumes to simply describe neutrally, reaching for language derived from a certain scientific view of the world that does not take account of what Luhmann and Schorr (1982) have helpfully described as the *technology deficit* fundamental to any educational process: the absence of a linear relationship between causes and effects in education. One could make a case for saying that it is the very absence of a causal relation in education that actually makes the education of *persons* possible, for otherwise we would not be engaged in education, but programming. Of course, this also suggests that education cannot be entirely controlled or predicted (Biesta 2011). From this point of view, it is the fact (or rather, interpretation) that we are not machines that can be programmed that allows for the possibility of (a human) education and a pedagogical relation.

4. THE PEDAGOGICAL RELATION

The notion of the *pedagogical relation* attempts to delineate the distinctive characteristics of the relation between persons whose relation is defined not (only) by kinship or friendship, but primarily and distinctively by education.[10] This definition of the relation is not meant to be exclusive: parents will inevitably be educators and these identities (parent; educator) often interact in complex ways. What, after all, is the role of the parent when reading a bedtime story, or the schoolteacher who acts *in loco parentis* during playtime? Although the pedagogical relation may be something of an abstraction, it is a useful one because it describes the conditions in which we try to influence others.[11]

This pedagogical relation is formed by the intention of the educator to influence the student in certain ways that improve the student's knowledge, skill, or capability in some respect. If this is to be truly relational, then the student must, in some sense, allow herself to be influenced by the educator. On what basis does the student allow herself to be influenced? Initially at least, children trust their parents or educators and so a relation of trust justifies influence.[12] Indeed, the simple act of listening is a form of basic trust, as is the decision to accept a promise of some future reward for doing something that is not immediately desirable (e.g., practicing piano). Both sides must engage with and sustain such a relation. An established tradition of German pedagogical theory, from Friedrich Schleiermacher, to Wilhelm Dilthey, to Hermann Nohl understands this relation to imply interdependency within the relation; there is a dyadic structure to the pedagogical relation, which means the relation constitutes the identity of both as distinct but also united in their distinction (Mollenhauer 1972; Friesen 2017b). This means that there is no educator or student before they come into a pedagogical relation for it is the relation itself, which constitutes the identities of both. For many educational theorists in this tradition, this dyadic structure also implies the rather unfashionable view that there is a fundamental asymmetry between educator and student, not because the educator knows more or can do more than the student (they may or may not be qualified in these senses), but because the educator is concerned to improve the relation of the student to some object (the "content"). What makes the relation educational is that the educator is not concerned with the life of the student in general, but with the student's knowledge or capacity: in other words, with the student's relation to some "content." The "relation to the relation" (see Kenklies 2020) that the educator takes up is enacted by way of creating conditions for growth, as John Dewey (1916) famously defined education, by, for instance, the processes of selection of "content" for optimal learning that the educator engages in. Figure 8.1 illustrates these relations in the form of the pedagogical triangle.

Having presented the pedagogical triangle in descriptive terms I now turn to the interpretive dimensions of these pedagogical relations for these relations are not simply "present-at-hand" but are themselves the product of a formative and interpretive process.[13] I have already noted that the educator must speculatively project (or perceive) the student as *educable*. Here we find a point that intersects strongly with Ricoeur's interests in the imputation and attestation of personhood: the educator sees the student as educable because they interpret the student as a person, not a thing. This "projection" of personhood is a kind of imputation that expects something of the student: that they are educable. Furthermore, this imputation of educability is not certain, but forms an extension of Ricoeur's hermeneutics of the self (Ricoeur 1992), what we might call the *hermeneutics of the pedagogical relation*. Educators employ this kind of interpretive "prejudice" constantly to guide their activities: they form learning objectives on the basis on the projected educability of the students; they enact the lessons *as if* the children can absorb lesson as intended. This remains speculative since, insofar as we understand that education is a human rather than a purely technical process, there is no secure knowledge that the child is indeed able to learn, and despite efforts to be inclusive, there will always be circumstances in which the educator's projections go awry. It is this interpretive prejudice that makes education possible

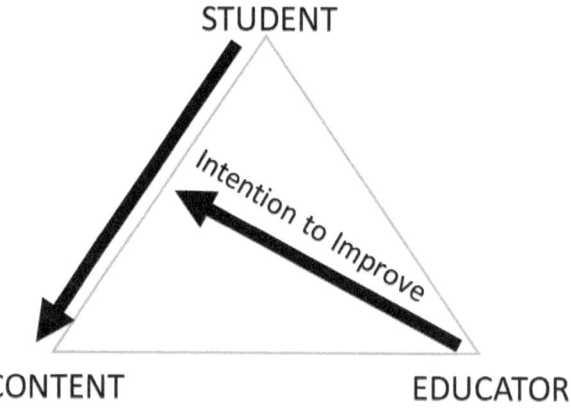

Figure 8.1 The Pedagogical Relation

while simultaneously giving rise to certain problems. One might say that all education entails a wager (Lewin 2014b).

Approaching the question of technology in terms of a basic educational relation indicates that technology is here interpreted as a way of seeing, what I have called a "technological hermeneutic" (Lewin 2014a, 34): that in the wake of Heidegger's analysis of technology, the manner in which things show up is shaped exclusively by the imperatives of modern technology. It should be acknowledged that the so-called empirical turn in the philosophical of technology (away from "classical" philosophy of technology)[14] has roundly criticized scholarship that remains focused on this rather pessimistic and "substantivist" if not "essentialist" (Feenberg 1999, Chapter 1) interpretation of our relation to technology (see Achterhuis 2001; Verbeek 2005) though one could argue that the empirical turn simply considers different things: raising empirical questions rather the conceptual issues, an approach in which the nature of technology must be assumed in order to proceed empirically. An adequate discussion of classical versus empirical approaches in the philosophy of technology is beyond my scope, but I raise it here to draw attention to a virtue of the classical approach: that it encourages us to consider how our interpretations of the world are just that, our interpretations. This interpretive condition is crucial to consider in education if we want to avoid the idea that the educational process is really only a matter of quasi technical enhancements of humans that might just as well (or even better) be undertaken by implants, smart drugs (Bostrom and Sandberg 2009), or technical exteriorizations of human beings, what Stiegler (2011) calls technical prostheses.

Yet and discussion of the technological hermeneutic does not provide criteria for deciding which technologies are good or bad since the devices themselves are consequent upon the technological revealing of things. We might argue therefore that this approach avoids the undecidable question of which technologies are educationally good or bad (e.g., should we employ tablet computers or mobile phones in classrooms, or ban their use?), even if we might wish to posit explicit policies regarding their educational use. So, we are encouraged to consider the question of technology in education by way of its broader influence on how human beings are interpreted (as educable) and how educational processes are "technologized" such that they seek to produce learning outcomes as effectively as possible. This concept of technologization is expressed in the idea that the relation between the teacher's input and the learner's outcomes, for example, can be interpreted as a simple ratio: a measurable proxy for educational efficiency. If this is established in explicit terms, then certain dimensions of education are at risk of being lost. The goods of education (as well as technology) are concealed by attempts to ameliorate consumers: the desire to satisfy the student, for instance, may very well be inimical to the experiences of disruption or alienation that many argue are central to education in the tradition of transformation or *paideia* (English 2012). Through technological enframing, education might be reduced in other ways such that it can seem justified, for instance, to pay teachers in proportion to their efficiency, whereby bonus payments for high levels of satisfaction are considered (Burgess 2018). The idea that we should minimize educational inputs and maximize outputs reflects the dominance of the concept of efficiency, which, as Jacques Ellul showed in the 1960s, may be characterized as the only value in a technological society, and one that obfuscates reflection of ends (Ellul 1973). Paulo Freire's critical account of the banking model of education illustrates this reduction of education to efficient transmission: "the more completely she fills the receptacles, the better a teacher she is. The more meekly the receptacles permit themselves to be filled, the better students they are" (Freire 2007, 72). From the perspective taken in this chapter, the banking model is a consequence of certain ways of seeing students, educators, and education itself, conditions, which I have characterized in terms of "technologization" and the technological hermeneutic.

Not only does this transformation of the pedagogical relation result in the power imbalance between educator and student that so exercises critical pedagogues like Freire, but it can also lead to the withdrawal of the figure of the teacher, which, in fact, has been widely identified within educational theory as a problematic feature of education in technological society. The process of learning is interpreted not in relational terms, but as a discrete function residing within the student only, and to be optimized through any number psychologically informed interventions that increasingly appear to not require the figure of a teacher. Biesta (2006, 2011, 2014) has developed one of the most sustained critiques of the notion that contemporary education is becoming dominated by a narrow notion of learning in which the figure of the educator is starting to look redundant. After all, learners are said to construct their own knowledge. In the Information Age, learners access all the information necessary for that construction without the intervention, interruption, or cost of a teacher-figure. Models of online learning are often presented as providing learning opportunities without the encumbrances, inefficiencies and questionable authority of the traditional pedagogical relation. Frictionless learning is celebrated without due consideration of what might here be lost.[15]

All of this makes something properly defined as the pedagogical relation itself virtually invisible. The technological organization of education as the efficient transmission from educator to student is an interpretation that has become naturalized and so appears just to be descriptive (see figure 8.2).

What is forgotten here is that pedagogical processes and relations entail interpretations of what it means to be a person, and what it means to become a person, interpretations of an anthropological nature. We return to thinking

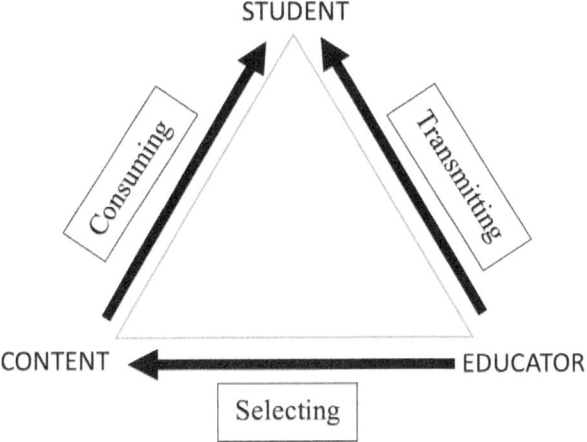

Figure 8.2 The Technological Structure of Education

about what kind of anthropology is implied in our projections of education, and so we turn to Ricoeur's concept of narrative identity.

5. RICOEUR AND THE EDUCATIONAL SUBJECT

> The self of self-knowledge is the fruit of an examined life . . . one purged, one clarified by the cathartic effects of the narratives, be they historical or fictional, conveyed by our culture. So self-constancy refers to a self instructed by the works of a culture that it has applied to itself. (Ricoeur 1988, 247)

Ricoeur's concept of narrative identity suggests something educationally significant: that the self is *instructed* (or formed) by narrativization. But, as this chapter has argued, that process of instruction entails the influence of another: the educator. Thus, students do not instruct themselves: they do so through the works presented to them by those who seek to influence them. Even where in the end, all education is self-education, there is still a role for another. Here we discover the main significance of Ricoeur for education: that students become themselves by learning to tell their own stories, through developing identity in relation to the many and varied narratives presented to them by others: specifically educators. To be sure, there will inevitably be the incidental influences of daily life that allow for various imaginative reconfigurations of the self, and surely the opportunities to engage imaginatively is what school ought to be able to offer (Masschelein and Simons 2013). Moreover, one might speculate that the concept of narrative identity and Ricoeur's (2005) considerations of recognition are directly illustrated in the context of the classroom where the child's capacity for action in the world may be significantly affected by the recognition conferred to them by the educator, and the corresponding self-recognition that ensues. Such self-recognition is essential to educational subjectification.

I have presented a hermeneutic account of education which rests upon the idea that what we know about ourselves and others is a matter of interpretation: of affirmation and (re)configuration. Ricoeur's hermeneutics of the self justifies the speculative move made by the self who, through attestation or imputation, sees the other as a *person*. Attestation or imputation are kinds of assurance or confidence, kinds of faith or belief in which persons are interpreted as agents capable of acting and suffering. It is a belief that the self has the capacity to act and to suffer, to do and undergo things that it can attest to itself (Ricoeur 1992, 21–22). It is on the evidential validity of attestation that Ricoeur can insist that persons are irreducibly different from things. This kind of distinction between persons and things is central to the account of pedagogical relations discussed earlier: it is an imputation made by every educator that the student is the kind of thing that is educable. To fully develop, this idea would take time, but for now, it is sufficient to recognize that, as a basic condition for thinking educationally at all, education involves speculative acts on the part of both the educator and the student. Moreover, Ricoeur's philosophy provides some account of how it is that the self-same "person" can exist over time, that is through (educational) change and development. Clearly this is an aspect of narrative identity: that time is understood through narrative, and that therefore our self-same existence through time is to be conceptualized through narrative identity. However we choose to define education, it is certainly something that entails change, and normally over extended periods of time (courses) and so a notion of self-sameness through time is also a basic condition for education.

But this chapter attempts to go further by thinking through this educational narrativization in relation to technology. It is worth keeping in mind that technology really concerns the means, that is, *how* a thing comes to be (or in the case of technological thinking, how a thing comes to appearance). There have always been techniques and technologies involved in educational formation, from the most basic tools of reading and writing, to the printing press, to new media (Postman 1994; Friesen 2017c). While pedagogy has also described the various techniques and technologies to realize educational aims, something else is at stake in contemporary technological culture: human beings are themselves subject to technological enframing to such an extent that education is interpreted as nothing other than a technique of enhancement. Enhancements could reasonably be defined as recognized features of improvement (hearing; eyesight; muscular power; memory), while leaving aside more complex dimensions of subjectification. As a form of what

Bostrum and Sandberg have called "conventional" enhancement, education is rather slow and inefficient by comparison with recent *unconventional* human enhancements such as "nootropic drugs, gene therapy, or neural implants" (Bostrom and Sandberg 2009, 312). Why not replace teaching with some form of implant? The danger of the technological hermeneutic applied to education is that we find ourselves without any criteria to distinguish conventional and unconventional enhancements. The dominance of the notion of efficient learning commands us to abandon the apparent niceties of the pedagogical relation and its reliance on complex human capacities such as pedagogical tact (Friesen and Osguthorpe 2018) in order to bring about the most effective enhancements. If learning outcomes can be produced through cheaper and easier online education, then how is it to be resisted?[16] Educational philosophers have also begun to wonder whether "knowledge insertion" is possible, desirable, or ethical and whether such a process will be tantamount to "cheating education" (Aldridge and Tillson 2018). Although not longstanding (see Ricoeur and Changeux 2002), Ricoeur's engagement with artificial intelligence demonstrates a commitment to understanding human identity as an interpretive process in light of technological change and suggests there may be further fruitful inquiries in this vein. So the question of the relation between education and anthropology in contemporary society opens further complex issues that I cannot develop more here. But what Ricoeur really offers us is a language with which to defend a richer conception of human identity in the face of technological reductions: narrative identity. We come to ourselves through a variety of longer routes: through recognition of parents and teachers, through the detours of the "text," and through the mediations of technologies—ancient (wax tablets) and modern (iPads).

6. CONCLUSION

Technological devices and processes are transforming education for the good. Moreover, the technological enframing of education is more totalizing and insidious, and requires, I argue, a philosophical response. This chapter has elaborated something of Ricoeur's potential to reframe the formation of narrative identity in education. But I have only been able to scratch the surface and have left many large questions unresolved. How exactly is self-understanding mediated by the figure of the parent, teacher, or other educators in our lives? Who initiates the processes of narration that allow narrative identity itself to form? If education is the intention to influence, where does this leave other informal influence upon children's lives and how are we to discern the outcomes of these fragile human processes. Where does any of this leave us when faced with the pressing issues of whether, or to what extent, we can allow modern technologies into the classroom? If education has always employed some kind of technical mediation, then what really defines the difference of our current technological milieu? Why is it that so many educators today look with suspicion at the creeping influence of modern technologies, and seek, in their stead, an educational environment free of any such influence?

I hope to have shown that education assumes certain basic conditions: the presence of the educable subject; the function of recognition in formation; and the formation of the self through narratives offered to the student. It is in the places of human formation that we find opportunities to catch sight of the limitations of technological revealing: for our (self) making seems to require the activities of interpretation that form the self through narrative. In recognizing this process, we may see beyond the one-dimensional structure of the technological hermeneutic.

NOTES

1. *Postman* (1994) was first published in 1982. The thesis of the book, that new media (particularly broadcast media like television) are eroding the distinctions between adults and children, seems only more relevant in an age of mobile devices and social media.

2. Heidegger defines "enframing" (*Gestell*) as follows: "Enframing means the gathering together of that setting-upon which sets upon man, i.e., challenges him forth, to reveal the real, in the mode of ordering, as standing-reserve. Enframing means that way of revealing which holds sway in the essence of modern technology and which is itself nothing technological" (Heidegger 1977, 20).

3. This relation is conceived ontologically rather than ontically: what makes a teacher is not age, professional status, or knowledge, but, as I argue later, a capacity and desire to influence someone's relation to some "content." The student is therefore defined in terms of their capacity to be influenced (*Bildsamkeit*).

4. For instance, a major recently published *International Handbook of Philosophy of Education* (Smeyers 2018) includes sections devoted to each of these figures (along with many others) while Ricoeur does not appear.

5. The Fonds Ricoeur has gathered papers and sections in Ricoeur which discuss education: http://www.fondsricoeur.fr/uploads/medias/doc/education-bibliographie-des-textes-de-paul-ricoeur-1.pdf. Accessed on September 1, 2020.

6. Education studies has only relatively recently been located within higher education and has remained primarily (and with some justification) framed by the needs of the profession of school teaching. In Germany, the discipline has a more established history through figures like Wilhelm von Humboldt, Hegel, Schleiermacher, Herbart, and Fichte. Simplifying, this tradition could be read as viewing the formation of a person as a fundamentally educational as well as philosophical question, making education intrinsically philosophical, and philosophy intrinsically educational.

Whereas in Germany philosophical discussion of education as formation (*Bildung*) is commonplace and exists not as a sub-field of "philosophy" but in its own right, things look rather different in French- and English-speaking nations (Westbury et al. 2015).

7. The tradition of humanistic education that reflects such a view is captured in Mollenhauer (2013).

8. Wittgenstein uses the concept of *Abrichtung*, which can be translated as training (Friesen 2017a).

9. The idea that the small child actively enters into a pedagogical relation may sound odd, since typically young children are born into a set of relations and processes that cannot be called chosen. However, on the whole children begin trusting their parents. Why this should be the case is a complex and interesting question.

10. Although hardly a new idea (Klafki 1970), the notion of the *pedagogical relation* as something worth systematic conceptual attention is not, at least in Anglo-American educational theory, widely recognized (Friesen 2017b; Friesen and Osguthorpe 2018).

11. It is possible for the pedagogical relation to be formed from a relation of self to self, as can be found in the German tradition of *Selbstbildung* (self-formation or self-education) (Schäfer 2005), but in general, the pedagogical relation is defined in terms of a relation between different people: an educator and a student. As Herman Nohl puts it "the unique (*eigene*) creative or generative relationship that binds educator and educand," see Friesen (2017b).

12. This does not mean, of course, that the trust cannot be abused or that the abuse of trust would be justified, but only that trust is often a necessary condition for a pedagogical relation for only then will the student allow themselves to be influenced.

13. From the perspective of Ricoeur's philosophy, where phenomenology and hermeneutics belong together, there is no simple description of these relations apart from the interpretive dimensions developed here, so this distinction is only an abstraction.

14. Classical philosophy of technology refers to the ontological or metaphysical approach of philosophers after Heidegger, such as Marcuse, Ellul, Arendt, and Borgmann (see Verbeek 2005, 4–9; Lewin 2011).

15. See Friesen (2011) for an excellent and balanced account of the subtle losses (and gains) involved in online learning in relation to face-face.

16. This chapter was completed during the lockdown that followed the coronavirus pandemic (April 2020), a period during which all university teaching staff were faced with the prospect of transferring all teaching online for the foreseeable future. At the time of writing, it is unclear how universities will use the crisis to reshape educational activities in order that they may be available without the risks of human contact.

REFERENCES

Achterhuis, Hans, ed. 2001. *American Philosophy of Technology: The Empirical Turn*. Bloomington: Indiana University Press.

Aldridge, David, and John Tillson. 2018. "Cheating Education: Is Technological Human Enhancement the New Frontier of Learning?" *Educational Theory* 69, no. 8: 589–594.
Anderson, Pamela. 1993. "Having It Both Ways: Ricoeur's Hermeneutics of the Self." *Oxford Literary Review* 15, no. 1–2: 227–252.
Bass, Randall, and J. W. Good. 2004. "Educare and Educere: Is a Balance Possible in the Educational System." *The Educational Forum* 68, no. 2: 161–168.
Biesta, Gert. 2006. *Beyond Learning: Democratic Education for a Human Future*. Boulder: Paradigm Publishers.
Biesta, Gert. 2011. *Good Education in an Age of Measurement: Ethics, Politics, Democracy*. Boulder: Paradigm Publishers.
Biesta, Gert. 2014. *The Beautiful Risk of Education*. London and New York: Routledge.
Biesta, Gert. 2017. *The Rediscovery of Teaching*. London and New York: Routledge.
Borgmann, Albert. 1984. *Technology and the Character of Contemporary Life: A Philosophical Inquiry*. Chicago: The University of Chicago Press.
Bostrom, Nick, and Anders Sandberg. 2009. "Cognitive Enhancement: Methods, Ethics, Regulatory Challenges." *Science and Engineering Ethics* 15, no. 3: 311–341.
Buber, Martin. 1997. *Israel and the World: Essays in a Time of Crisis*. Syracuse: Syracuse University Press.
Burgess, Simon. 2018. "Lessons Learned From Imposing Performance-Related Pay on Teachers." http://theconversation.com/lessons-learned-from-imposing-performance-related-pay-on-teachers-87657. Accessed September 1, 2019.
Dewey, John. 1916. *Democracy and Education*. New York: Macmillan.
Dunne, Joseph. 1997. *Back to the Rough Ground: Practical Judgement and the Lure of Technique*. Notre Dame: University of Notre Dame Press.
Ellul, Jacques. 1973. *The Technological Society*. New York: Random House.
English, Andrea. 2012. "Negativity, Experience and Transformation: Educational Possibilities at the Margins of Experience—Insights from the German Traditions of Philosophy of Education." In *Education and the Kyoto School of Philosophy*, edited by Paul Standish and Naoko Saito, 203–220. Dordrecht: Springer.
Farquhar, Sandy. 2012. "Narrative Identity and Early Childhood Education." *Educational Philosophy and Theory* 44, no. 3: 289–301.
Feenberg, Andrew. 1999. *Questioning Technology*. London and New York: Routledge.
Flint, Kevin, and Nick Peim. 2011. *Rethinking the Education Improvement Agenda: A Critical Philosophical Approach*. London: Bloomsbury.
Freire, Paulo. 2007. *Pedagogy of the Oppressed*. New York: Continuum.
Friesen, Norm. 2011. *The Place of the Classroom and the Space of the Screen: Relational Pedagogy and Internet Technology*. New York: Peter Lang.
Friesen, Norm. 2017a. "Training and *Abrichtung*: Wittgenstein as a Tragic Philosopher of Education." *Educational Philosophy and Theory* 49, no. 1: 68–77.
Friesen, Norm. 2017b. "The Pedagogical Relation Past and Present: Experience, Subjectivity and Failure." *Journal of Curriculum Studies* 49, no. 6: 743–756.
Friesen, Norm. 2017c. *The Textbook and the Lecture: Education in the Age of New Media*. Baltimore: John Hopkins University Press.

Friesen, Norm, and Richard Osguthorpe. 2018. "Tact and the Pedagogical Triangle: The Authenticity of Teachers in Relation." *Teaching and Teacher Education* 70: 255–264.
Gallagher, Shaun. 1992. *Hermeneutics and Education*. Albany: SUNY Press.
Heidegger, Martin. 1977. *The Question Concerning Technology and Other Essays*. New York: Harper and Row.
Hoveid, Martin Honerod, and Halvor Hoveid. 2009. "Educational Practice and Development of Human Capabilities: Mediations of the Student-Teacher Relation at the Interpersonal and Institutional Level." *Studies in Philosophy and Education* 28, no. 5: 461–472.
Kenklies, Karsten. 2020. "Dōgen's Time and the Flow of Otiosity—Exiting the Educational Rat Race." *Journal of Philosophy of Education*. https://doi.org/10.1111/1467-9752.12410.
Klafki, Wolfgang. 1970. "Das pädagogische Verhältnis." In *Erziehungswissenschaft 1. Eine Einführung*, edited by Wolfgang Klafki, et al., 55–91. Frankfurt: Fischer.
Leonardo, Zeus. 2003. "Interpretation and the Problem of Domination: Paul Ricoeur's Hermeneutics." *Studies in Philosophy and Education* 22, no. 5: 329–350.
Lewin, David. 2011. *Technology and the Philosophy of Religion*. Newcastle upon Tyne: Cambridge Scholars Publishing.
Lewin, David. 2014a. "Technological Thinking in Education." In *New Perspectives in Philosophy of Education: Ethics, Politics and Religion*, edited by David Lewin, Alexandre Guilherme, and Morgan White, 29–44. London: Bloomsbury.
Lewin, David. 2014b. "The Leap of Learning." *Ethics and Education* 9, no. 1: 113–126.
Lewin, David. 2019. "Toward a Theory of Pedagogical Reduction: Selection, Simplification, and Generalization in an Age of Critical Education." *Educational Theory* 68, no. 4–5: 495–512.
Lewis, Clive Staples. 2001. *The Abolition on Man*. Michigan: Zondervan Press.
Luhmann, Niklas, and Karls Schorr. 1982. "Das Technologiedefizit der Erziehung und die Pädagogik." In *Zwischen Technologie und Selbstreferenz. Fragen an die Pädagogik*, edited by Niklas Luhmann and Karl Schorr, 11–40. Stuttgart: Suhrkamp.
Marples, Roger. 1999. *The Aims of Education*. London and New York: Routldege.
Masschelein, Jan, and Parten Simons. 2013. *In Defence of School: A Public Issue*. Leuven: Education, Culture & Society Publishers.
Mollenhauer, Klaus. 1972. *Theorien zum Erziehungsprozeß: Zur Einführung in erziehungswissenschaftliche Fragestellungen*. Weinheim: Juventa.
Mollenhauer, Klaus. 2013. *Forgotten Connections: On Culture and Upbringing*. London and New York: Routledge.
Moratalla, Tomas Domingo. 2015. "Application: Between Hermeneutics and Education. Paul Ricoeur's Perspective." *Studia Paedagogica Ignatiana* 18, no. 1: 97–114.
Peters, Richard. 1966. *Ethics and Education*. London: Allen & Unwin.
Peters, Richard. 1983. "Philosophy of Education." In *Educational Theory and Its Foundation Disciplines*, edited by Paul Hirst, 30–61. London and New York: Routledge.

Postman, Neil. 1994. *The Disappearance of Childhood*. London: Vintage.
Ricoeur, Paul. 1984–1988. *Time and Narrative, Volumes 1–3*. Chicago: The University of Chicago Press.
Ricoeur, Paul. 1992. *Oneself as Another*. Chicago: The University of Chicago Press.
Ricoeur Paul. 2005. *The Course of Recognition*. Cambridge, MA: Harvard University Press.
Ricoeur, Paul. 2016. *Philosophical Anthropology*. Oxford: Polity Press.
Ricoeur, Paul, and Jean-Pierre Changeux. 2002. *What Makes Us Think?: A Neuroscientist and a Philosopher Argue About Ethics, Human Nature, and the Brain*. Princeton: Princeton University Press.
Romele, Aalberto. 2014. "Narrative Identity and Social Networking Sites." *Études Ricoeuriennes/Ricoeur Studies* 4, no. 2: 108–122.
Schäfer, Gerd. 2005. *Bildungsprozesse im Kindesalter: Selbstbildung, Erfahrung und Lernen in der frühen Kindheit*. Weinheim und München: Juventa Verlag.
Simard, Denis, and Alain Kerlan. 2011. *Paul Ricoeur et la question éducative*. Laval: Presses de l'Université Laval.
Smeyers, Paul. 2018. *International Handbook of Philosophy of Education*. Dordrecht: Springer.
Stiegler, Bernard. 2011. *Taking Care of Youth and the Generations*. Stanford: Stanford University Press.
Streib, Heinz. 1998. "The Religious Educator as Story-Teller: Suggestions from Paul Ricoeur's Work." *Religious Education* 93, no. 3: 314–331.
Veck, Wayne. 2013. "Martin Buber's Concept of Inclusion as a Critique of Special Education." *International Journal of Inclusive Education* 17, no. 6: 614–628.
Verbeek, Peter-Paul. 2005. *What Things Do: Philosophical Reflections on Technology, Agency, and Design*. University Park: Penn State University Press.
Von Humboldt, Wilhelm. 2015. "Theory of Bildung." In *Teaching as a Reflective Practice: The German Didaktik Tradition*, edited by Ian Westbury, Stefan Hopmann, and Kurt Riquarts, 57–62. London and New York: Routledge.
Westbury, Ian, Stefan Hopmann, and Kurt Riquarts, eds. 2015. *Teaching as a Reflective Practice: The German Didaktik Tradition*. London and New York: Routledge.

Chapter 9

Prostheses as Narrative Technologies*

Bioethical Considerations for Prosthetic Applications in Health Care

Geoffrey Dierckxsens

In recent years, philosophers have been pointing out that one of the future tasks for philosophy of technology, which investigates our experience and use of technology, is to develop an ethical theory of *narrative* technologies. The idea that technologies can have narrative qualities draws on Ricoeur's notion of narrative identity, which can be understood as a synthesis of characters and events of our lived experiences, which can be enacted in the form of stories. Technologies are obviously an integral part of these life stories in today's highly technological world. And our relations with these technologies are often value laden in the sense that they raise ethical questions (e.g., discussions about social media and data protection). Yet, how to understand the normative aspects of narrative technologies is an issue that still remains largely unexamined within the field of philosophy technology.

The aim of this chapter is of course not to develop an entire ethical theory of narrative technologies, which would reach far beyond the length and scope of a book chapter. Yet, my main focus will be to examine certain normative aspects of narrative technologies. I will argue in particular that one way of understanding such normative aspects is by looking at bioethical cases of the use of technological interventions, prosthetic applications in particular, in health care. Bioethics, the ethics of biological and medical research and medical practice, has affinity with philosophy of technology in that both bioethicists and philosophers of technology turn to Ricoeur's phenomenology. In both fields, they argue that looking at a person's embodied experiences and life stories has a significant

---

* The completion of this chapter was made possible by the Lumina Quaeruntur grant of the Czech Academy of Sciences - LQ300092001 and by the international exchange program, SAIA, which I carried out in 2019 and 2020 at the Institute of Philosophy of the Slovak Academy of Sciences in Bratislava

impact on decisions concerning evaluations of technologies, such as prosthetic applications in health. More exactly, whether and to what extent bodies can and should be manipulated by medical technologies involving prostheses and body reconstructions is not only a question of applying ethical, cultural, and legal norms (e.g., respect for body integrity) but also of a patient's actual lived experience of his or her "body biography" (e.g., personal experience of body wholeness embedded in a personal life story) (Slatman 2012, 294).

Although my chapter endorses the idea of body biography, I will argue that it needs to be expanded further by focusing on a concept often overlooked by bioethicists and philosophers of technology: *social imaginaries*. Further elaboration of the notion of body biography in this way will then enable defining normative aspects of prostheses as a narrative technologies. My claim is that by looking at Ricoeur's concept of the social imaginary, understood as the whole of a community's social values and norms, we gain insight in ways patients ethically evaluate prosthetic treatment from within their body biographies. More exactly, whether and to what extent a patient values a prosthesis as good or bad is not only the result of its functionality, comfort, or even how well a patient is capable of engaging with the technology, to use an expression of philosophy of technology (Coeckelbergh and Reijers 2016). Ethical evaluation of technologies (as being valuable or not) always has a social dimension that is not merely individual or rational. Since any kind of evaluation is potentially consciously or unconsciously influenced by different types of social values or norms (whether ideological, religious, political or other), awareness of this influence is crucial in the context of health care; and in particular, as I will demonstrate, in relations between patients and health care professionals that involve prosthetic treatment and evaluation thereof.

This chapter is divided into three parts. In the first part, I will sketch some context by explaining how both philosophers of technology and bioethicists draw on Ricoeur's idea of narrative identity to explain that technologies are an integral part of our life stories, which raises ethical questions. I will argue further that bioethical cases of prosthetic treatment allow developing normative aspects of narrative technologies. The notion of body biography allows understanding how patients, in cases of prosthetic treatment, evaluate prostheses. This evaluation, although embedded in certain ideologies of body wholeness (e.g., beauty ideals), results from a patient's personal values which result from a patient's history of concretely lived experiences within his or her life story (e.g., developing a critical attitude toward oppressing ideologies). In cases of prosthetic treatment, finding ways for patients to identify with a prosthesis and giving it a place in their body biographies is key for the well-being of the patients.

In the second part, I will argue that the notion of body biography nevertheless requires further elaboration on a social level in two ways. First, bioethical theories remain unclear about the different normative meanings that ideologies can express. Oppressing ideologies are of course always problematic. Yet, not all social (nonpersonal) values are always ideological in a negative way, nor is ideology in itself necessarily a negative concept. Values of freedom of expression, for example, support a patient's autonomy. Ricoeur's idea of social imaginaries can help here, in that it explains that a community's social values, norms, principles and symbols can be either ideological (negative or positive) or the expression of a social minority. In particular, the latter can support patients' autonomy in cases of prosthetic treatment. Second, bioethical theories are also unclear about the different ways patients can interact with ideologies (e.g., consciously, unconsciously, and critically). I will argue that the idea of active engaging technologies, which I take from philosophy of technology, allows clarifying *how* patients evaluate prostheses by interacting with social imaginaries: a patient well-being in the use of prostheses requires a continuous active engagement with and awareness of social imaginaries that influence this well-being.

In the final part of this chapter, I will argue that an awareness of social imaginaries increases through stimulating imagination in the relation between health care professionals and patients. I will draw on Ricoeur's idea of *social imagination* in the sense of imagining a different social reality to demonstrate that imagining different values offering an alternative to dominant oppressing ones increases awareness of the influence of oppressing ideologies. This kind of imagination is opposed to phantasy, and there is evidence that it is constructive in open discussion between patients and health care professionals (e.g., Scott 1997). I will then conclude that social imaginaries and social imagination are significant concepts for defining normative aspects of prosthetic applications as narrative technologies.

1. THE CONTEXT

Paul Ricoeur is, strictly speaking, not a philosopher of technology in the sense that we usually understand this term. Certainly, there are references to topics related to technology in his writings (e.g., Ricoeur 1986, 161 ff.; Changeux and Ricoeur 2000, 7, 11, 204, 239, 297, 326, 328). In *What Makes Us Think?* Ricoeur and Changeux discuss technology, for example, in the sense of brain imaging technology in order to discuss the relation between body and mind. Yet, Ricoeur did not develop a theory of technology and how it influences our lifeworld. He also did not develop a bioethical theory (his ethics being mainly contained in the dialectic between self and other in *Oneself as Another* (see Ricoeur 1992).

Yet, philosophers of technology have recently been drawing on Ricoeur's philosophy in order to examine the role of technology in our lifeworld (see Kaplan 2006; Lewin 2012; Coeckelbergh and Reijers 2016; Gransche 2017;

Romele 2019). Coeckelbergh and Reijers (2016) look into Ricoeur's idea that humans use narratives to recount sequences of life events over time (cf. Ricoeur 1984, 1986, 1990), and apply this idea to technology. According to them, understanding narratives in this general sense (i.e., as any kind of recounter of life events, written or in another form) implies that technologies can also be understood as narratives that we use to express life events, and that moreover can have a certain autonomy in expressing such events. For example, they distinguish between "passive" engaging technologies that become part of our social life world, such as bridges over a river, and "active" engaging technologies that are developed to actively engage into a lifeworld (or simulation of it), such as video games.

Furthermore, Coeckelbergh and Reijers (2016) point out that narrative technologies raise ethical questions in the sense of questions concerning moral evaluation of technologies. Thus, they look into "the prospects of deploying the theory of narrative technologies as an ethical theory of technology" (Coeckelbergh and Reijers 2016, 344). In fact, technologies that become part of our lifeworld have social and normative dimensions. Or better: technologies are not value neutral objects, but are part of a lifeworld we share with others and can therefore be ethically evaluated as being right or wrong. For instance, technologies are developed to serve a social function (a bridge), but they can also raise explicit debate about moral values and principles (the debate about violence in video games).

Yet, philosophers of technology remain to a certain extent unclear about what exactly an ethical theory of narrative technologies might look like. As Coeckelbergh and Reijers attest, finding such a theory is one of the future tasks of philosophy of technology. In this chapter, I will explore certain avenues in the direction of developing such a theory by focusing on one particular medical technology: prostheses. Next to philosophy of technology, I will focus therefore on bioethics, in the sense of the ethics of medical research and health care, as well as on Ricoeur's primary work, which contemporary bioethicists often discuss alongside philosophers of technology.

In fact, questions about normative evaluation of prosthetic applications take center stage in recent bioethical and medical discussions (e.g., Slatman 2012; Lindenmeyer 2017). These discussions take different shapes. In line with the more traditional approach to bioethics, bioethical theories formulate guidelines and principles for correct application and use of prosthesis in health care (e.g., Hanspal and Sedki 2018). These theories are normative in the sense that they determine norms that promote the well-being of patients undergoing prosthetic treatment.

Generally speaking, traditional bioethical theories are often so-called "principle-based" theories (I take this notion from Svenaeus 2017, 8). For example, theories investigate formal and legal principles that guide biological or medical research and medical practice, such as principles for the use of germ line editing (e.g., de Miguel Beriain 2018). These principles are clearly important, since biological and medical research as well as medical practice concerns our well-being. Moreover, doing good and avoiding harm is often a matter of life and death.

As far as prosthetic treatment is concerned, much work is done to formulate clear and practical guidelines for medical practitioners and health care professionals. For example, the British Society of Rehabilitation Medicine recently published a revised version on their report of standards and guidelines for amputee and prosthetic rehabilitation (Hanspal and Sedki 2018). Guidelines include giving patients adequate advice, discussing realistic rehabilitation goals, prosthetic options, projected outcomes, as well as providing adequate post-operational treatment. The importance of such standards and guidelines is evident, because of the technological complexity of prosthetic treatments (the surgery is complex, rehabilitation is difficult, prostheses have a precise function and require careful maintenance) and because of the well-being of the patients (amputation is an invasive procedure).

Yet, standards and procedures have their limits, in that the well-being of a patient undergoing prosthetic treatment depends on the patient's personal experience of the treatment itself and on the rehabilitation process (e.g., how he or she is recovering). Medical studies (Gourinat 2019) point to the difficult process of rehabilitation for amputees. In particular, Gourinat points out that while the media often focuses on young dynamic people using prostheses, such as athletes, the majority of amputees are older people who experience difficulties in the use and maintenance of their prostheses. A patient's actual experience of rehabilitation processes after prosthetic treatment should not be overlooked, as well as the exchange of experiences between patients can assist in these processes.

In recent years, bioethical theories therefore increasingly focus on patients' lived experiences. In fact, in response to the limits of the principle-based bioethical theories, phenomenology-based bioethical theories have recently emerged, drawing on the work of Ricoeur and other phenomenologists (e.g., Svenaeus 2017). Although these theories do not deny the importance of moral and legal principles in biomedical research and in health care, they nevertheless criticize the principle-based approach on one point. The main issue being that principle-based theories treat the body mainly as a biological entity or object, and therefore often overlook the patient's actual experience of the body, experiences that constitute his or her life story, and the various possible health problems that relate to these experiences (e.g., distortion of the experience of the body as a "whole" and of prostheses as "alien" in cases of medical amputation) (Bayne and Levy 2005; Slatman 2012).

One important ethical guideline drawn from phenomenology-based bioethical theories is to be wary of the danger of technologies becoming a goal in themselves, blocking or obstructing the well-being of the patient rather than improving it. And it is on this point that technologies raise ethical questions that are relevant for developing a theory of narrative technologies. For example, Svenaeus (2017) argues that modern scientific technologies in general change the ways in which we perceive and understand the world in which we act (Svenaeus 2017, 77). This is of course not necessarily a negative thing. Medical technologies can change, improve or even save our lives. A limb prosthesis, for example, enables a person to walk again. Yet, simultaneously we should be "wary of the technologies that tend to block life-world concerns in order to prolong or even produce life as a goal in itself" (Svenaeus 2017, 86). The point is that the function of medical technologies is to assist life and improve well-being of the narrative identities that we are, while respecting "our personhood and all the human values that go with it" (Svenaeus 2017, 84). The bottom line here is that when prostheses *do not* improve the quality of living of the patient, but instead undermine his or her deeply personal values and or make the body feel alien, the application of the prosthesis is possibly not the best medical treatment for the patient in question.

It is within this context that phenomenology-inspired bioethical theories turn, like philosophers of technology, to Ricoeur's work, in particular to his idea of narrative identity. Slatman (2012), for example, takes cue from Ricoeur's idea that human beings express their embodied experiences in narratives. These narratives are expressions of their identities or how they feel "wholeness" as an embodied being. Slatman focuses on cases of mastectomy after breast cancer and argues that these cases show that the process of re-identification with the body as a whole after breast surgery or mastectomy is far from straightforward and differs based on the patient's personal values and past embodied experiences. In Slatman's words: "The interpretation of patients' stories should be part of counselling prior to interventions such as breast reconstruction, nipple banking, and tattooing" (Slatman 2012, 291).

For certain persons, breast reconstruction, re-identification with the body as a whole can help and for others a reconstruction might feel more alien. According to Slatman, health care professionals should be attentive to interpretation of patients' life stories, because these stories express patients' "body biographies" or how they experience their bodies as an identity over a time of events (Slatman 2012, 293). Slatman demonstrates that Ricoeur's idea, according to which lived existence is a sequence of events that can be narrated, is a timely idea and is significant for real-world questions concerning technological applications in health care (see also Svenaeus 2017, 34).

Ricoeur's idea of narrative identity can thus lead the way to formulate an ethical theory of prosthetic application as a narrative technology. In order for prostheses to contribute to a patient's well-being, to be a "good" technology so to speak, it is important that the patient in question can give the prosthesis a place in his or her body biography, to align it with one's personal values. In a similar way as philosophy of technology, phenomenology-based bioethical theories find their inspiration in the idea that the self has an embodied narrative identity, which is the narrative of one's embodied life experiences and values. To use Ricoeur's own words, narrative identity is a balance between "the perseverance of character and the constancy of the self in promising" (Ricoeur 1992, 124). It amounts to one's personal values, based on one's physical constitution.

What makes a prosthesis normatively valuable is not just its functionality or guidelines for good prosthetic treatment. What makes it normatively valuable also depends on its narrative character, that is, its potential to become part of a patient's life story. One of the potentials of Ricoeur's work is thus its potential to demonstrate that health care and medical practice can concretely improve by looking at, or "reading" to use a hermeneutical concept, embodied *lived* existence in cases of prosthetic applications and not only by examining the physical body as an object. Ricoeur's work therefore contributes to bioethics but also to discussions in philosophy of technology, which raise ethical questions about narrative technologies.

2. BODY BIOGRAPHIES AND SOCIAL IMAGINARIES

In this chapter, it is not my intention to criticize the notion of body biographics. Yet, this notion nevertheless needs further elaboration on a *social* level in order to develop further an ethical theory of prostheses as narrative technologies. While phenomenology-based bioethical theories raise important awareness of the impact of social factors on a patient's experiences, such as the fact that "bodily integrity is always embedded in a certain ideology of wholeness" (Slatman 2012, 283), this social impact is typically not further specified. Bioethical theories often assume that there are only two possible ways to go: personal autonomy and ideology. In fact, phenomenological theories of bioethics do not specify *how* patients interact with social ideologies, and such an elaboration is needed in order to develop further normative aspects of prosthetic treatment as a narrative technology. After all, social ideologies clearly can have a large impact on a patient's well-being in different ways. Patients can acquire and embrace values and react to ideologies in various ways, for example, consciously or unconsciously, rationally or by intuition, with reluctance or spontaneously and enthusiastically.

Moreover, it is not immediately clear how to distinguish between different social ideologies, which exist on different normative levels, both negative and positive ranging from oppressing/harmful ideologies, to ethically informed personal and social frameworks of ideas, to explicit ethically and juridically approved normative frameworks. And these different relations with ideologies influence our normative evaluation of technologies as being "good" or "bad." It is of course not my intention here to downplay the devastating negative impact ideologies often have on the well-being of people, especially minorities, but to better understand the relation between well-being and ideology. Therefore, the question I will focus on in the remainder is how do patients come to embrace, critique, and develop certain personal values and norms by interacting with social ideologies (i.e., how such norms become part of their body biographies or stand in contrast to these biographies).

Let us return first to philosophy of technology, and focus on whether and how philosophers of technology perceive normative evaluation of technologies. Philosophers of technology point out that we are always in a sense involved with technologies. Verbeek (2008), for example, introduces the concept of "cyborg intentionality" and argues that our intentional relationship with the world, the way we experience objects in the world, is always in a sense mediated by technology on several different levels of consciousness, so that we are all "cyborgs" so to speak (390). Yet the normative value of these technologies is obviously not transparent in itself, especially given that technologies are to such a large extent interwoven in our daily lives, so that we are often not even fully conscious of their presence.

Surely, Verbeek distinguishes between several types of cyber intentionality, in particular mediated intentionality, hybrid intentionality and composite intentionality (Ibid.). Mediated intentionality means that technologies mediate our intentional relations with the world, as is, for example, the case when using a pair of glasses while reading. Hybrid intentionality designates, in turn, a relation in which "the human and the technological are merged into a new entity," as is the case for cyborgs: entities that are half human, half machine (Ibid.). In relations of composite intentionality, the artifact itself has a certain intentionality. For example, a sound recorder has different intentional relationship to sound than humans do, in that the recorder registers in more detail background sounds than humans and are incapable of focusing on one particular sound.

However, these categories do not tell us much about normative evaluation of technologies, prostheses in particular. And it is also questionable whether the term "cyborg" is of much use in health care. At first glance, it might seem that Verbeek's second type of intentionality—hybrid intentionality—does provide further insight in patients' relations with prostheses. The use of medical technologies implies merging the intentionality of the artifact with that of the human. Verbeek himself mentions implants, medication, artificial valves and pacemakers. The same can be said for prostheses. Prostheses, much like implants, also become part of a patient's lifeworld to optimize his or her wellbeing (cf. body biography). In the best-case scenario, the patient is mostly unaware of the prosthesis, like in the case of a dental implant, for instance, which enables a person to chew and eat as if he or she would have a natural tooth. The dental implant does not interfere with a person functioning in a physically and mentally fit way. It is designed instead to support well-being.

However, things are far less self-evident for other prostheses. Persons with limb prostheses, for example, are constantly aware of their prosthesis, since the use and maintenance of these prostheses is often a constant struggle (Gourinat 2019). In cases of limb prostheses, the notion of hybrid intentionality does therefore not apply, since in these cases patients do not become one with the technology. Further distinctions need to be made, therefore, to define normative relations between patients and prostheses.

I propose therefore to make a further distinction between a *subjective* and a *social* dimension of hybrid intentionality. While it is important, as I have argued with the theories discussed earlier, that a patient is able, from his or her subjective point of view, to identify with his or her body and the prosthesis, as to make it part of his or her lifeworld, this identification is not only a personal issue. On a social level, there exist several kinds of normative values and norms that influence how different people perceive prostheses, such as values connected to beauty ideals, but also phantasies about bionic powers or improvements prostheses can bring, as well as the actual ethical guidelines that health care professionals use in the treatment of patients with prostheses.

One important question then is how to distinguish between these different kinds of norms and values, since there are clearly significant normative differences between them. Ricoeur's idea of the social imaginary (1986, 2007), in the sense he understands it—as a social group's set of values, norms, institutions, and symbols that have ethical or political meaning—can help finetune these normative differences. The differences between norms and values and how they influence ethical evaluation of prosthetic applications result from differences in social imaginaries. More exactly, different social groups influence in different ways. For example, an oppressing ideology tends to exert its power more on society than a social minority.

The sum of all different social imaginaries consists of a vast network of various values, norms, institutions, and symbols, and to categorize them all would be beyond the purposes of this chapter. However, one important distinction to better understand differences in normative evaluation of prostheses is Ricoeur's distinction between ideology and utopia (Ricoeur 2007, 181). Whereas an ideology reflects the dominant ruling social, ethical and political values, norms, institutions, and symbols of a social group, utopia represents alternative, possibly improved social, ethical and political values, norms, institutions, and symbols. Utopia means telling a different narrative in the social sense of the word of a group resisting or protesting against the ruling ideology, which is potentially oppressing and violent (Ricoeur does not understand utopia in the pejorative or literary sense).

In cases of prosthetic treatment in health care, we should be wary of oppressing ideologies and respect personal autonomy and values, as I pointed out earlier. Yet not all ideologies are negative. For instance, an ideology in the sense of a set of ideas, guidelines and norms for the improvement of prosthetic treatment is not necessarily a bad thing, neither is good democratic political policy for example. The advantage of using Ricoeur's distinction between ideology and utopia is that it underscores the importance of keeping an open dialogue between different voices and dominant voices, as well as the importance of dissent when needed. Moreover, persons whose body biography contains different personal values than the dominant ones can find support with social minorities that have values in common (e.g., prosthetic patients who gather in social protest against ruling beauty ideals). Interaction with social imaginaries is therefore important for both patients and health care professionals, so as to increase awareness of the impact, it can have on the evaluation of technologies in health care, prostheses in particular. Such awareness stimulates critical assessment of harmful oppressing ideologies, and voicing concern on a social level.

To see the place, these values can have in cases of prosthetic treatment, consider again Coeckelbergh and Reijers' (2016) distinction between passive and active engaging technologies. From a purely functional (but also incomplete) point of view, a limb prosthesis, for example, starts as an active engaging technology in that nurses, doctors, physiotherapists and other medical professionals organize sessions with the patient to apply the prosthesis, and learn to use and maintain it. If that goes well, the prosthesis becomes more of a passive engaging technology, like the bridge, as it gradually becomes more of an extension of the body and person that uses it than an external object with which we engage. However, the metaphor of the bridge does not entirely fit here. Unlike the bridge, which we can, for example, use on a daily basis without thinking much of it, a patient's relation with a prosthesis remains an active daily engagement with an invasive object. We know from research that prostheses do not fully "substitute" body parts, as they are designed to perform one function (e.g., a prosthetic limb for walking, swimming or running), but unlike a human body part are not capable of performing all of these tasks (Gourinat 2019).

It seems that prostheses should be understood as technologies somewhere in between active and passive engaging narrative technologies, ideally being more passive than active, integrated into the embodied condition of a patient and its social surroundings. Yet, part of the invasive character of prostheses is also its vulnerability to the impact of social values. A limb prosthesis, for instance, is something that raises social attention, with mixed, both positive and negative consequences, as the media attention on young dynamic prosthetic athletes also attests. Engaging with social values is thus part of the active engagement process patients and healthcare professionals are dealing with when they get involved in prosthetic treatment.

Part of a patient's well-being is thus an active engagement to identify the prosthesis as part of the person one is (at least to a certain extent writing it into one's body biography), and not as an alien or intruding technology. It becomes personal, yet within a social context of different values. Slatman (2012) calls this, drawing on Husserl, being able to identify one's physical body (*Körper*) with one's lived body (*Leib*) (Slatman 2012, 286). According to this view, a disrupted body experience is the result "of an experience of strangeness [that] can dominate and displace the experience of own," potentially caused by a prosthesis (Slatman 2012, 288).

Ultimately, increased awareness of the influence of social imaginaries will then help patients identify with their prostheses as narrative technologies. The question is then how and to what extent awareness of the impact of social values in cases of prosthetic applications can *increase*, that is, how patients and health care professionals can deal with social values to improve the well-being of patients. Answering this question is a crucial step in developing further an ethical theory of prostheses as narrative technologies. In the next section, I will take up this issue.

3. IMAGINATION FOR INCREASED AWARENESS OF THE IMPACT OF SOCIAL IMAGINARIES

"Social imaginary" is, as mentioned earlier, a broad concept referring to many different kinds of values and norms. For the sake of clarity, I will focus in the following in particular socially shared cultural values that particular social groups, majorities as well as minorities, hold and defend, such as the values of a patriarchal society, Western beauty ideals, but also feminist values. These values can thus be both negative and positive in the sense that they can either contribute or obstruct the well-being of patients.

To illustrate how awareness of the influence of social values on patients' body biographies can increase, we consider the following case study, which is itself not a case of prosthetic treatment, but of cancer screening. This study (Gilbar and Banroy 2018) found that there is a positive impact of family presence in cases of genetic counselling for inherited breast cancer. The researchers of this study interviewed clinicians in various health care centers in Israel and found positive aspects of family presence in these cases. First, family presence provides emotional support for making decisions concerning medical treatment (i.e., the genetic testing itself, preventative, oophorectomy and preventative mastectomy). Second, family presence has informational advantages. Not only can the family members help logistically (driving to the hospital, taking over household tasks, etc.), as is the case for other forms of medical counselling (Laidsaar-Powell et al. 2017).

What is more, in the specific case of genetic counselling for inherited breast cancer, the diagnosis of the patient potentially influences the health of family relatives that are genetically related to the patient (both parties may share the inherited risk for breast cancer). This also makes the family more involved in the testing process and diagnosis. The presence of family members therefore positively stimulates the process of spreading information about testing and potential treatments.

Yet, one of the negative aspects of family presence during genetic counselling for inherited breast cancer was the potential of imbalanced power in the relation between the relatives during the counselling. In fact, as Gilbar and Banroy (2018, 386) themselves point out, patients are obviously influenced by their cultural backgrounds when facing these decisions. In the specific context of Israel, for example, patients often come from patriarchal communities, which can pose a problem for the patient's autonomy. This may prevent the patient from making autonomous decisions. Nevertheless, clinicians also have a way of dealing with this issue by openly discussing the power conflict and, in some cases, by talking privately to the patient. In any case, patients should ultimately have the autonomy to reflect on their situation in private and take decisions for themselves.

This particular medical case points out one way for raising awareness of the impact of social values on the well-being of patients. An open discussion between clinicians and patients and their relatives stimulates this awareness and consequently the autonomy and well-being of patients. This particular case is not about prosthetic treatment. Yet, it is easy to see that social values can impact the autonomy of patients in cases of prosthetic treatment in a very similar way. For instance, if ideologies of body wholeness influences cases of breast reconstruction after mastectomy, these ideologies obviously represent patriarchal values, as studies also attest (e.g., La et al. 2019). In decisions about breast reconstruction, the Western beauty ideal of having a whole and healthy body finds to a large extent expression in masculine ideals about the female body, which risks obscuring the well-being of patients.

We can thus safely assume that an open discussion between clinicians and their patients (with and without relatives included) also increases awareness of the pressures of social values in cases of prosthetic treatment. One problem that might occur, however, in open discussions about cultural values is that one or more of the parties involved are incapable of understanding the other's point of view or remain attached to their own cultural values. For instance, a patriarchal family might have too much of an impact on a patient's personal values and obstruct autonomy.

Ricoeur's idea of imagination can demonstrate a possible way out of such a situation, or at least demonstrates a way of increasing awareness of the impact of social values in health care. Of particular interest for our purposes is a passage on what Ricoeur understands as "imaginative variations" (Ricoeur 1992, 148). Inherent to narrative—and also to our narrative identity—is the capacity to make imaginative variations:

> The narrative does not merely tolerate these variations, it engenders them, seeks them out. In this sense, literature proves to consist in a vast laboratory for thought experiments in which the resources of variation encompassed by narrative identity are put to the test of narration. (Ricoeur 1992, 148)

In other words, being able to tell one's own life story (as well as literary stories) implies having the capacity to be creative, to choose phrasings, endorse beliefs or reject certain dominant values. What is more, this "laboratory for thought experiments" is a laboratory for ethical thought experiments or, in Ricoeur's words "explorations in the realm of good and evil" (Ricoeur 1992, 164). What this means is that because of the imagination, we are capable of understanding diverse existing ethical meanings, such as values, beliefs, norms, ideologies, yet we are also capable of criticizing and endorsing them, changing them or plainly rejecting them.

Applying this to health care would then mean that increased imagination in open discussions between patients and their relatives on the one hand, and clinicians on the other hand, would result in increased awareness of the impact of social values and would facilitate sympathy for another's values. In fact, studies in medical ethics have drawn the attention to the relevance of imagination in both nursing and health care (Scott 1997; Smith 2002). Scott (1997), for example, argues that "moral imagination," a concept that she takes from Iris Murdoch, should be stimulated in relations between patients on the one hand, and nurses and medical practitioners on the other hand (Scott 1997, 45). As Scott points out, imagination stimulates "high-quality role enactment and sensitive moral strategy," and is therefore a valuable quality for understanding patient's situations and communicating with them.

For example, being able to empathically imagine a patient's feelings and sociocultural context is important for health care (e.g., the patient is especially vulnerable because she/he is under social pressure at work or at home). This notion of moral imagination is based on Aristotle's idea of phronesis and Hume's concepts of imagination and sympathy. Imagination in this sense is a virtue, which relates to a practical wisdom to understand another person through indirect cues and perceptual activity. Scott adds that reading and listening to narratives further stimulates moral imagination (Scott 1997, 48–50).

To connect this back to the study on genetic counselling discussed earlier, it is this kind of imagination based practical wisdom—or inquiry into the sociocultural backgrounds of the patients—that the clinicians attested to use in their consultations with the patients. In a more general sense, we can conclude that storytelling in the form of exchanging experiences with the patients can add to the emotional support of the patients and creates a "triadic decision-making model" in the relation patient-family-clinician, as Gilbar and Banroy (2018, 379) call it. Yet, part of this exchange is an understanding of the diverse social and cultural meanings (the social imaginaries), in which imagination is involved as an active component of perceiving these different imaginings.

This conception of moral imagination, as an empathic understanding of another's cultural values, is quite close to Ricoeur's ethics in *Oneself as Another*. It is well-known that Ricoeur also models his ethics on the notions of sympathy, inspired by Hume (Ricoeur 1992, 190 ff.) and phronesis, in dialogue with Aristotle (Ricoeur 1992, 158 ff.). Moreover, the kind of sympathy-based practical wisdom Ricoeur has in mind involves narratives as well. That is to say, we learn to actively sympathize with others through narratives and sharing experiences by storytelling. Ricoeur writes following Walter Benjamin, "storytelling is . . . exchanging *experiences;* by experiences, he [Benjamin] means not scientific observations but the popular exercise of practical wisdom" (Ricoeur 1992, 164, original emphasis).

This kind of imagination connects then to social imaginaries, in that it enables sympathizing with a patient's body biography that is embedded in cultural values. Imagination implies being able to take a different point of view, understanding another's story and personal values, and thus raises awareness of the dangers of the influences of cultural ideologies. It can in that sense be of assistance in health care and in prosthetic treatment in particular, in that increased imagination from the part of health care professionals and relatives will give patients a better chance of expressing their autonomous choices and personal values about particular prosthetic applications.

I am suggesting that the stimulation of social imagination can be useful in cases of prosthetic treatment. This is in a sense, of course, a suggestion that is not without risk. Indeed, one might object that there is a thin line between stimulating imagination and negatively influencing or even manipulating a person's imagination. And, one question is how this relates to prosthetic applications that require envisioning realistic rehabilitation goals. One might argue that imagination risks to take the focus away from reality, creating illusionary expectations from prostheses. However, this would be a misinterpretation of Ricoeur's concept of imagination, which expresses, as I am arguing, practical capabilities of innovation, rather than illusions (Ricoeur 2007, 170).

In fact, the problem of unrealistic expectations highlights the importance of legal and moral principles for bioethics, such as the principle to respect a patient's autonomy. Moreover, if we understand imagination not as phantasy, but in the social sense—in Ricoeur's sense—as the capability of imagining different values and thus sympathizing with others, then imagination goes hand in hand with the principle of respect for autonomy.

Imagining another perspective does not necessarily imply a violation of a person's privacy, even though that might sound contra-intuitive. To try and understand the "story" behind the physical condition of a patient is not the same as to invade a patient's privacy in the sense of posing inappropriate questions. To a certain point, posing personal questions can significantly aid in the understanding, diagnosis and cure of diseases, as the study of genetic counselling for breast cancer attests. Respect for the integrity and autonomy of the patient goes hand in hand with asking these questions, as Slatman's analysis discussed earlier (2012) also attests, since having an open discussion implies respecting boundaries and the integrity of the patient (for that reason it is open). This also implies applying a principle of respect for autonomy and respect for personal identity. In fact, patients appeared to be able to still make autonomous decisions, precisely because of the open discussion (which can, e.g., reveal latent cultural pressure of the family or society) (Gilbar and Banroy 2018, 386).

I would therefore propose as a guideline for prosthetic treatment the stimulation of social imagination within the relation between health care professionals and patients. Concretely, this can take the shape of an open discussion about personal values as well as personal history or cultural background. It is insufficient to apply the principle of respect for individual choice and autonomy alone, since the patients themselves face hard decisions, for which emotional support and the exchange of stories and life experiences can help. Moreover, this exchange implies stimulation of imagination in the social, critical sense that enables patients to take a critical attitude toward cultural ideologies or values related to medical technologies (e.g., beauty ideals of physical perfection).

The fact that body integrity is always embedded in a kind of ideology of body wholeness can be explained now in more dynamic terms. Patients do not have personal body experiences that are simply *opposed* to or different from dominating social values of body wholeness. These values influence our body experiences, as they get *dynamically* shaped in these experiences, and patients can actively engage with these values. For example, one patient can be deeply influenced by social values of physical intactness, while another patient who has a strong critical attitude toward social values and ideologies can be more skeptical. Thus, it appears to be important within health care contexts to *read* the biographies of a patient's body, but part of this reading process is to understand the imaginations behind these biographies—to hear a patient's imaginative variations—so that the person in question can properly express personal values. Knowing one's own body values and boundaries is particularly significant in cases when technology has a serious impact on the body.

4. CONCLUSION

To return to the philosophy of technology, we can conclude that one of the normative aspects of prostheses as narrative technologies is a patient's ability to actively engage with the prosthesis and making it a part of his or her embodied lifeworld (cf. hybrid intentionality). However, being able to do that is far from self-evident and there are both personal and social/normative challenges to it. On a personal level, imagination may help patients to understand and express personal values, for example, values about body wholeness, because imagination allows browsing through different social imaginaries. On a social level, there is a societal and normative challenge in that there should be an imagination-inspired open dialogue (in health care and the public domain in general) that allows thinking out of the box and addressing different social imaginaries. This kind of open dialogues helps raising awareness of the potentially oppressing impact of social values that are part of different social imaginaries. Ricoeur's ideas of social imagination and social imaginaries thus contribute to both philosophy of technology and bioethics, as was my intention to demonstrate in this chapter. Part of an ethical theory of prosthetic applications as narrative technologies would then be a description of the relation between patients, relatives, health care professions, and social imaginaries. Moreover, Ricoeur's notion of social imagination helps understanding this relation.

REFERENCES

Bayne, Tim, and Neil Levy. 2005. "Amputees By Choice: Body Integrity Identity Disorder and the Ethics of Amputation." *Journal of Applied Philosophy* 22, no. 1: 75–86.

Changeux, Jean-Pierre, and Paul Ricoeur. 2000. *What Makes Us Think? A Neuroscientist and a Philosopher Argue about Ethics, Human Nature and the Brain*. Princeton: Princeton University Press.

Coeckelbergh, Mark, and Wessel Reijers. 2016. "Narrative Technologies: A Philosophical Investigation of the Narrative Capacities of Technologies by Using Ricoeur's Narrative Theory." *Human Studies* 39: 325–346.

de Miguel Beriain, Iñigo. 2018. "Should Human Germ Line Editing Be Allowed? Some Suggestions on the Basis of the Existing Regulatory Framework." *Bioethics* 33, no. 1: 105–111.

Gilbar, Roy, and Silvia Barnoy. 2018. "Companions or Patients? The Impact of Family Presence in Genetic Consultations for Inherited Breast Cancer: Relational Autonomy in Practice." *Bioethics* 32, no. 6: 378–387.

Gourinat, Valentine. 2019. "La prothèse comme promesse de réparation du corps amputé: du discours à l'expérience." In *L'humain et ses protheses: savoirs et pratiques du corps transformé*, edited by Cristina Lindenmeyer, 155–170. Paris: Éditions CNRS.

Gransche, Bruno. 2017. "The Art of Staging Simulations. Mise-en-scène, Social Impact, and Simulation Literacy." In *Science and Art of Simulation I. Exploring—Understanding—Knowing*, edited by Michael M. Resch, Andreas Kaminski, and Petra Gehring, 33–50. Dordrecht: Springer.

Hanspal, Rajiv, and Imad Sedki. 2018. *Amputee and Prosthetic Rehabilitation: Standards and Guidelines*, 3rd Edition. Report of the Working Party. British Society of Rehabilitation Medicine, London.

Kaplan, David M. 2006. "Paul Ricoeur and the Philosophy of Technology." *Journal of French Philosophy* 16, no. 1–2: 42–56.

La, Jessica, Sue Jackson, and Rhonda Shaw. 2019. "'Flat and Fabulous': Women's Breast Reconstruction Refusals Post-Mastectomy and the Negotiation of Normative Femininity." *The Journal for Gender Studies* 28, no. 5: 603–616.

Laidsaar-Powell, Rebekah, Phyllis Butow, Cathy Charles, and Amiram Gafni. 2017. "The TRIO Framework: Conceptual Insights into Family Caregiver Involvement and Influence Throughout Cancer Treatment Decision-Making." *Patient Education and Counseling* 100, no. 11: 2035–2046.

Lewin, David. 2012. "Ricoeur and the Capability of Modern Technology." In *From Ricoeur to Action: the Socio-Political Significance of Ricoeur's Thinking*, edited by Todd S. Mei and David Lewin, 54–74. London: Continuum.

Lindenmeyer, Cristina, ed. 2017. *L'humain et ses protheses: savoirs et pratiques du corps transformé*. Paris: Éditions CNRS.

Mitchell, David T., and Sharon Snyder. 2001. *Narrative Prosthesis: Disability and the Dependencies of Discourse*. Ann Arbor: University of Michigan Press.

Ricoeur, Paul. 1984. *Time and Narrative*, vol. 1. Chicago: The University of Chicago Press.

Ricoeur, Paul. 1986a. *The Lectures on Ideology and Utopia*. New York: Columbia University Press.

Ricoeur, Paul. 1986b. *Time and Narrative*, vol. 2. Chicago: The University of Chicago Press.

Ricoeur, Paul. 1990. *Time and Narrative*, vol. 3. Chicago: The University of Chicago Press.

Ricoeur, Paul. 1992. *Oneself as Another*. Chicago: The University of Chicago Press.

Ricoeur, Paul. 2007. *From Text to Action: Essays in Hermeneutics II*. Evanston: Northwestern University Press.

Romele, Alberto. 2019. *Digital Hermeneutics: Philosophical Investigations in New Media and Technologies*. London and New York: Routledge.
Scott, Philomena Anne. 1997. "Imagination in Practice." *Journal of Medical Ethics* 23, no. 1: 45–50.
Slatman, Jenny. 2012. "Phenomenology of Bodily Integrity in Disfiguring Breast Cancer." *Hypathia* 27, no. 2: 281–300.
Smith, Benedict. 2002. "Analogy in Moral Deliberation: the Role of Imagination and Theory in Ethics." *Journal of Medical Ethics* 28: 244–248.
Svenaeus, Frederik. 2017. *Phenomenological Bioethics: Medical Technologies, Human Suffering, and the Meaning of Being Alive*. London and New York: Routledge.
Verbeek, Peter-Paul. 2008. "Cyborg Intentionality: Rethinking the Phenomenology of Human–Technology Relations." *Phenomenology and the Cognitive Sciences* 7: 387–395.

Chapter 10

Responsibility, Technology, and Innovation

The Recognition of a Capable Agent

Guido Gorgoni and Robert Gianni

The idea that innovation must be conducted in a responsible manner is not new and has been constantly affirmed at least throughout the past four decades under different labels, such as technology assessment, stakeholder engagement, ethical, legal, and social implications of research(ELSA), "midstream" modulation of science, and, lastly, Responsible Research and Innovation (RRI) (Burget et al. 2017).

Innovation is today the main production system that aims at overcoming the problem of resources and that can better account for the necessary flexibility in a global dimension. Within the paradigm of innovation, we find different modalities such as disruptive or incremental innovation, according to the degree of novelty expressed by a certain product or process. We can also distinguish innovation according to the moment at which is implemented in the pipeline (Bessant 2013). Benoit Godin has made an extraordinary account of the history of innovation, as well as the challenges that innovation brings to our societies (Godin 2006). However, due to a number of negative events and given its actual holistic nature (Pavie 2018; Prahalad and Ramaswamy 2004), in the past decade, the often-implicit normative objectives of innovation have been brought to light more evidently. Such objectives vary depending on the geographical context and disciplinary domain. Among the various examples, it is worth mentioning the model of frugal innovation, which is becoming increasingly important in middle- and low-income countries (Pansera and Owen 2018; Schroeder and Kaplan 2019). Frugal innovation aims at providing disadvantaged parts of the population with goods that might not be considered essential but that are nevertheless important to guarantee a fair access to technological products. Accordingly, materials like, for instance, phones and fridges are deprived of their unnecessary trimmings and sold at a lower price.

Technological artifacts have for a long time raised concerns with regard to their ethical and societal impact. More recently, the paradigm of innovation, which has become the predominant productive model, has increased these concerns because of its highly flexible and contextual and often disruptive nature. Since a decade, therefore, governance approaches in Europe like the one adopted by the European Commission have been supporting soft regulatory approaches promoting responsible postures (EC 2012). Although the debate about the understanding of the practical meaning of a framework such as RRI has not yet found a sufficiently broad consensus (Burget et al. 2017), it is possible to denote its peculiarity in opening the decision-making process to forms of inclusion based on stakeholder engagement (Pansera and Owen 2019). In this sense, RRI extends the capillarity of a participatory approach to spheres that could hardly be efficiently integrated in the decision-making process (Fisher and Rip 2013). In other words, it is possible to denote how RRI aims to represent a sort of contextual translation of the major guiding principle of European public policies, the precautionary principle, which recommends an attitude of care and prudence when designing innovative products and processes.

According to Bernard Reber (2016), the precautionary Principle is often understood as a form of inaction or renunciation to decisions in the name of precaution (Reber 2016). However, he argues that the precautionary Principle always implies an active decision, which ultimately relies on political assumptions. What is different in the governance inspired by the precautionary Principle is how the decision is reached. In this sense, along with Reber, we can distinguish two basic forms. The first, which is mainly used in the United States, is based on a utilitarian calculation of risks and benefits. The second, on the other hand, which is more common in the European context, is based on deliberative forms of consultation on the basis of rights (Reber 2016, 132–133).

Drawing a parallel with RRI, and following Paul Ricoeur, it is to this second communicative model that we must look if we want to design an innovation that is responsible. But while the claim to responsibility is widespread, causing a "responsibility overload," the almost intuitive meaning of the word "responsibility" is not always fully substantiated as if it was self-evident. On the contrary, responsibility is an eminently contested subject as regards its precise meaning and its normative content (and subsequent implications). In order to better understand the implications of the constant appeal to responsibility, we could move from the reflections made on *The Just*, where Ricoeur reconsiders the semantics of the idea of Responsibility (Ricoeur 2000).

1. REVISITING THE SEMANTICS OF RESPONSIBILITY

Ricoeur's contribution is important for the framing of responsibility in technological innovation because it enables moving away from focusing only on the negative consequences of an event. In particular, in his study on responsibility, Ricoeur proposes to reconnect the idea of responsibility to the semantics of moral imputation, thereby regaining the sense of the connection between the agent and the action more than that of the attribution of its (typically negative) consequences.

This way Ricoeur wants to depart from the traditional legal approach, where the designation of a "responsible" person is framed formally as a function of the imputation of the consequences to a legal subject (not necessarily the actual agent nor a human one). By contrast, in Ricoeur's view, linking back the idea of responsibility to the relationship with the agent helps in distinguishing more clearly two aspects: that of the obligation to answer to the victims of damages and that of the designation of the responsible subject. What Ricoeur has in mind is to bring back the active meaning of the term "responsible" and to unveil the often-paradoxical nature of a notion that links individual and collective agency. This conceptual distinction opens the theoretical possibility of re-articulating the semantics of legal responsibility out of the retributive logic.

In contrast with the legal-positivistic stance inherited by Hans Kelsen's *Pure Theory of Law*, where responsibility is constructed as a purely formal concept (Kelsen 2009), Ricoeur indicates the primary root of the idea of responsibility in the idea of moral imputation, stating that the founding concept has to be sought outside the semantic field of the verb "to respond," by looking closely in the semantic field of the verb "to impute" (Ricoeur 2000, 13).

The root meaning of the verb *imputare*, recalls Ricoeur, is to attribute an action to somebody, who is considered as its authentic author: it is a sort of moral calculation (*putare*) in which the action is placed on one's moral account. In order to attribute an action to somebody as its author, it is necessary to explore his or her role in producing the event and therefore recognizing him or her as the author of the action.

The reasons behind this ascription, therefore, go back to the entanglement of the subject with the action, which is not only externally attributed but also self-reflexive, as part of the self-comprehension driven by the attestation of the capacities at the various levels of the constitution of the Self (Ricoeur 1992). This is an important philosophical move as the self-understanding of the subject as a responsible agent is an essential element of the responsibility idea (Dierckxsens 2017, 584).

This semantic renewal is particularly relevant when talking about responsibility in innovation, as it implies that the responsibility of the innovators shall be engaged in the first person. This means that responsibility could not be identified simply with a procedure or a technical device but instead holds also an unavoidable element of personal commitment, which is the element ensuring the orientation of responsibility toward the future.

Indeed, many uncertainties surround the attribution of responsibility in the context of scientific and technological innovation, mainly due to the long-term impacts in the future of the choices made. In particular, it is hard identifying the author of the action, or determining the extents of the effects for which one is made responsible; moreover, the relation of responsibility is troubled by the loss of reciprocity between the agent and the victim of the action (Ricoeur 2000, 28–31).

For these reasons, the orientation of responsibility toward the future cannot rely only on the anticipated calculation and allocation of the negative consequences of innovation along a risk-based model but has to embrace more frankly a virtuous commitment toward commonly shared societal values, in particular by defining standards of the acceptable risks and societal benefits of innovation.

2. TURNING RESPONSIBILITY TOWARD THE FUTURE

In order to adequately take into account the societal issues posed by technological innovation, we need, says Ricoeur, to orient the very idea of responsibility toward the future so that the idea according to which:

> we are eminently responsible for what we *have done* (a retrospective orientation that the moral idea of responsibility has in common with the juridical idea), must instead be substituted with a more deliberately prospective orientation, where the idea of prevention of future harm has to be added to that of reparation for harm already done. (Ricoeur 2000, 31)

This idea would not be fully clear, nor very original, if it was considered from the perspective of the "outputs" of the mechanisms of legal responsibility, where it has been implemented under the form of indemnization through the mechanisms of insurance. Instead, however, what Ricoeur has in mind is that of reconnecting the idea of responsibility to that of an agent (through the semantics of moral imputation) in order to recover a central role for the acting subject. In the following, we will examine more closely the contribution of Ricoeur's philosophy to this relevant shifting of the meaning of responsibility in a prospective sense.

3. RETROSPECTIVE AND PROSPECTIVE RESPONSIBILITY

Responsibility can be directed both to the past, as it is usually understood in legal terms, and toward the future (which is more frequent in ethical terms). In order to understand the difference between these two orientations, we could follow Peter Cane's suggestion (2002) that when we speak of responsibility in terms of accountability, answerability, and liability, we look backward to conduct and events in the past. These concepts are at the core of what we could call "historic responsibility." By contrast, when we refer to responsibility through ideas such as roles or tasks, we look to the future and in this sense we sketch "prospective responsibilities" (Cane 2002, 31).

The difference between the two lies in that retrospective responsibility is substantiated by an *ex post facto* judgment over a given situation; it is mainly linked to ideas such as liability or damage and is therefore characterized essentially in reactive terms. By contrast, the idea of a prospective responsibility refers to the notion of going beyond the perspective of complying with some pre-established duties, and to proactively assuming responsibilities for a certain state of activities even when specific duties are not (or cannot) be established in advance (Cane 2002, 48).

This way of thinking about prospective responsibility is relevant in the context of a reflection on technological innovation since it allows to situate the discussion beyond the reference to the established paradigms of *fault*, which constitutes the "standard" model of responsibility, based on the idea of liability, and that of *risk*, based on indemnization. It points more clearly in the direction of the idea of *precaution*, exemplified by the precautionary Principle, which aims at articulating responsibility in situations that are not adequately covered by the usual means of risk management in reason of the radical uncertainties they generate, involving fundamental value judgments, as is the case of contemporary technological innovations (Stirling 2017).

In more recent times, the well-established liability model and the risk model have been integrated by the precautionary principle, since the unpredictable long-term effects of technological innovation defy the possibility to find an identifiable author of the action (by attributing the fault) or the possibility to rely on knowledge in order to anticipate its possible outcomes (through risk management measures).

The precautionary Principle is a normative reference invoked precisely in the situations of radical uncertainty generated by the processes of techno-scientific innovation. In particular, precaution is invoked "where there are threats of serious or irreversible damage, lack of full scientific certainty shall not be used as a reason for postponing cost-effective measures to prevent environmental degradation" (United Nations Conference on Environment and Development 1993).[1]

It emerges clearly that the precautionary approach is fundamentally different from the standard legal approaches to responsibility as it leads to incorporate considerations going beyond the textual dictates of positive law (Boisson de Chazournes 2009, 163), disrupting the logic of the two former models of legal responsibility. Instead, the precautionary logic configures responsibility along the ethical idea of a decision in a given situation (Ewald 2001), which reconnects responsibility to the agent and more fundamentally focuses its semantics on the engagement of the agent. This element makes the case for the relevance of the reflections of Ricoeur on a responsibility strongly connected with the agent.

4. RESPONSIBILITY AS RESPONSIVENESS

The qualifying feature of this sense of responsibility is the distinctive element of personal commitment, which goes beyond the "morality of duty" typical of the law and rather aligned with a "morality of aspiration" distinguishing ethics (Fuller 1969). Along with Graham Haydon, we could designate this as "virtue-responsibility," which refers to some personal quality of the agent, be it a capacity or a disposition (Haydon 1978, 46). This implies the idea of "taking role-responsibility seriously" (Cane 2002, 32) as it makes reference to a proactive engagement that extends further than simple compliance with an obligation.

This declination of responsibility closely echoes the prospective idea of responsibility evoked by Ricoeur, in particular as it is eminently proactive rather than reactive, and is essentially characterized as a disposition of the agent, since it implies "a willingness to understand and confront the other's commitments and concerns with ours, to look for a possible terrain of sharing. It entails readiness to rethink our own problem definition, goals, strategies, and identity" (Pellizzoni 2004, 557).

In its prospective declination, responsibility counts more as an attitude than as an obligation, and therefore it is strictly linked with the identity of the legal subject, in the sense of his or her self-understanding as a responsible agent in the multiple relations with the others. As a capacity or disposition, responsibility is more relevant in that it is assumed in an active or even proactive manner, more than as something that is ascribed to the subject afterward and from the outside.

In contrast with liability or accountability, responsiveness implies behaviors and practices that extend over and above legal requirements and which therefore have to be fulfilled with voluntary, extra-legal engagements: this takes responsibility far from the logic of responding to a charge (reaction) typically associated to legal responsibility, and gets it closer to the logic of responding to a call (response) not linked to legal duties and obligations.

Ricoeur's reflections can help in disentangling yet another declination of the proliferating responsibility rhetoric, and in distinguishing what are the conditions for unfolding the potentialities behind the idea of "responsible" innovation, in particular figuring innovation in other terms than a purely technical advance and introducing the reference to wider historical and societal dimensions (what once was included under the idea of "progress," which includes a wider societal vision, richer than the idea of "innovation," which is more technically framed).

5. RECOGNIZING RESPONSIBILITY TOWARD THE OTHER

This prospective declination of the idea of responsibility requires a subject recognizing its responsibility toward the Other. This way of framing responsibility attracts into the discussion the ideas of identity and recognition, on both of which Ricoeur's reflection offers significant contributions, since it calls into consideration the self-comprehension of the agent and its relationship to the others.

In particular, considering responsibility and its subject through the idea of recognition leads to frame responsibility within intersubjective dynamics; in doing so, Ricoeur follows closely Honneth's "struggles for recognition" but he draws different conclusions, as for him recognition does not necessarily lead to conflict, but contemplates also, besides the conflict, the possibility of a pacific confrontation.

In *The Course of Recognition*, Ricoeur states that recognition involves simultaneously the other person and the norm: "as regards the norm, it signifies, in the lexical sense of the word, to take as valid, to assert validity; as regards the person, recognition means identifying each person as free and equal to every other person" (Ricoeur 2005, 197).

This implies that the responsible subject cannot be entirely defined as the designatory of the formal legal imputation of responsibilities, since this way the legal subject, in Ricoeur's words, acquires a "dialogical and institutional structure" marked by the triadic relation "I-you-third person," which is mediated by different "orders of recognition" (Ricoeur 2000, 5–6).

Following Ricoeur's explicit suggestion, the capacities of the self need to be actualized through the "continual mediation of interpersonal forms of otherness and of institutional forms of association in order to become real powers to which correspond real rights" (Ricoeur 2000, 6). The connection between universal rights and the capacities of the subject of rights is the product of a concrete "struggle for recognition" (Ricoeur 2005, 152–153) which is mobilized by the indignation subsequent to mis-recognition. Here it emerges clearly how responsibility is to be understood primarily in terms of a capacity of the self, which is both reflexive and intersubjective, rather than in the purely legal terms of imputation: The term *responsibility* therefore covers self-assertion and the recognition of the equal right of others to contribute to advances in the rule of law and of rights (Ricoeur 2005, 200).

In this perspective, responsibility appears as a specific form of capacity (responsiveness) characterizing a fully-fledged subject of rights, namely the idea of responsiveness, which is strictly interconnected with the narrative identity of the Self. In this way, the idea of responsibility is taken beyond the limits of the strict morality of the duty, in the direction of an ethics of care.

6. THE SEMANTIC AND PRAGMATIC OF A CAPABLE AGENT

Ricoeur has repeatedly stressed the need to understand responsibility as the last stage of a triadic ethical concept that is triggered by the linguistic relationship between two actors. It is precisely language that represents the first stage of the process of making an individual responsible, since it is through the recognition of one's ability to designate oneself that one becomes a person. On the basis of Strawson's analyses, Ricoeur adds to this basic process of individuation the abstract form in the semantics of psychic predicates, that is, the fact that certain characteristics can be recognized beyond their concretization in a specific person. Through the institution of language that entails universal characteristics, Ricoeur aims to transpose to the "other" the recognition of the characteristics of an individual. Accordingly, a subject can assume that a certain type of property that he attributes to himself can be equally found in other subjects. This construct is formed on a semantic basis of recognition that emancipates itself from the I and the You in favor of the Self. The figure of the Self makes it possible to formulate a communicative reflection between the I and the other as a unifying figure of differences. This is how the Self plays a crucial role in the construction of the "ethical person." Only through the recognition of this initial but essential link can, according to Ricoeur, the ethical personality be developed in the direction of a responsible approach.

The next step, for Ricoeur, is the one that brings the subject into the practical sphere where relations with the other become the domain of the legal and moral sphere. It is here that responsibility is formed as the capacity to discern the right and the good. Responsibility is in fact triggered by the passage from the locutory act (I am) to an action aimed at exerting an influence externally. Inasmuch as one individual acts, he wields power over someone else. An action, according to Ricoeur, is always done by someone and suffered by someone else. It is to this dissymmetry that the action triggers, and to the violence potentially inherent in it, that responsibility is called to answer as a model of safeguard.

However, in the Ricoeurian architecture, with the advent of moral duties, the cruciality of the intersubjective linguistic dimension does not disappear, but rather passes from an abstract and semantic dimension to a pragmatic one. The bi-univocal relationship between language and action that Ricoeur never stops highlighting is of particular relevance for our analysis. On the one hand, Ricoeur urges a more substantial use of language to understand agency because, if not everything is language, everything in experience does not access the sense but under the condition of being brought to language (Ricoeur 1994, 82). On the other hand, language must be solicited in its pragmatic dimension, where the illocutionary side represents its true nodal force and its ethical value. Affirmations such as "I promise" or "I will be responsible" imply a series of behaviors and actions that, while generated in a linguistic dimension, transcend it in the direction of a pragmatic ethics. The grammar of the Self applied to the sphere of practice must be concretized in a dialogue that builds on the concrete and specific challenges of an interlocutory context (Ricoeur 1994, 84).

7. THE ROLE OF INSTITUTIONS IN THE PRAGMATIC OF RECOGNITION

The difficult relationship between responsibility in a retrospective and fundamentally juridical sense, with the ability to go further, in the direction of care, of a virtuous approach, is what Ricoeur underlines with lucidity and to which he tries to offer a solution through a modular structure of the ethical personality. Similar to the linguistic relationship, the ethical process cannot be limited to the I/You relationship of the friendship model. First of all, this is the case because this model risks dispersing the ethical unity into fragmented forms of relationship that would not be immune from criticisms of relativism. Second, a model of ethical relationship based on the I/You couple requires a physical presence that today's global and plural societies make unrealistic. How can an I relate to the myriad of You present in the external reality in the same way in which I interact with a friend?

Ricoeur understands very well that the relationship between You and I needs a mediation in the practical dimension as well as it did in the linguistic one. The solution for him is to transpose his linguistic construction, where the semantic relationship between the "You" and the "I" is sublimated in the Self, to the practical interactions regulated by the moral and juridical sphere. To the relationship between You and I on the basis of a friendship model, Ricoeur then adds a third vector. In this way, he manages to respond to the challenges of pluralism and allows us to think of a concept of responsibility capable of emancipating itself from its legal constraints without getting lost in the ineffectiveness of relativism. It is once again through the role of the institution that Ricoeur attempts to resolve the relationship between the I and the infinity of the You. According to Ricoeur, the institution plays different roles and somehow exemplifies the role and complexity of responsibility as an agency. The institution is a sort of field where the "I" and the "You" meet and can interact following a series of rules that outline their space of action. A clear example that Ricoeur offers is language that, although used by two or more actors, also exists beyond its users. The rules of the language are therefore helpful to regulate the interaction and to delimit the space of *maneuver* within which issues must be evaluated. As in the game of chess, the use of rules does not predetermine the outcome of the game. As well in language, rule-based communicative exchanges cannot predict whether they will take the form of an altercation or an agreement (Ricoeur 1994, 89).

8. BETWEEN TRADITION AND INNOVATION

However, institutions, by providing the actors with a set of rules where they can interact, express also another implicit function. In fact, the institution takes on the role of a link between tradition and innovation. On the one hand, it exemplifies all the innovations that have been made before and that have found a consensus in a given historical-geographical context. As Ricoeur sums it up beautifully, "To be born is to appear in an environment where words have spoken before us" (Ricoeur 1994, 86). On the other hand, original situations may require an innovation of the institution itself in order to represent an up-to-date instrument of dispute resolution. A process that respects tradition must be closely linked to innovation. An innovation that for Ricoeur must also turn to the past, to those promises that have not been kept, cannot disregard the use of this critical posture for the future and the challenges it brings with it (Ricoeur 1994, 101).

It is therefore not difficult to draw a parallel between the role of the institution *tout court* and that of responsibility as a specific institutional arrangement. In fact, Ricoeur reminds us, institutions should not be understood in their purely legal or political sense but as pre-ethical entities that give meaning to human praxis (Ricoeur 1990, 125). Just as language, while maintaining more or less stable structures, is enriched by the contextual use of it, so too the concept of responsibility, clear in its restrictive dimension, develops new senses according to the context of its use. Even if two subjects or a subject faced with uncertainty know the rules within which to move, they will be responsible for translating them into ethically appropriate actions (Gunther 1993).

In this sense, it is important to remember that for Ricoeur, the institution, or in its most recent variant, the symbolic order, is not immediately accessible to everyone. The vulnerability of those who are unable to access the symbolic order, to achieve their own autonomy is therefore for Ricoeur the objective of the principle of responsibility (Ricoeur 2007). Ricoeur tells us: "To be able of entering into a symbolic order is to be capable of entering into an order of recognition, of inscribing oneself in a 'we' that distributes the and apportions the authority of the symbolic order" (Ricoeur 2007, 88). Responsibility is always intersubjective, but this relationship must be integrated in an impersonal structure involving anyone (*chacun*) living under the same laws. And it is the dimension of politics that can best guarantee the necessary stability to the system as well as the access to this symbolic order. And it is here that the circle, which started from the semantic use of recognition, developed within the legal-moral sphere closes with the formation of the ethical person through a cooperative basis of recognition.

The loss of credibility of traditional forms of authority requires "a patient reconstitution of a consensus of a different, less dogmatic, less univocal and hence more deliberately pluralistic type, meant to knit together tradition and innovation" (Ricoeur 2007, 87). The public space is what for Ricoeur, on the basis of the insights of Hannah Arendt, expresses the condition of plurality that results from the extension of interhuman relations to all those whom the relationship between the I and the you leave out (Ricoeur 2007, 73).

9. THE PARADIGM OF GIFT AS INNOVATIVE RESPONSIBILITY

We underline one last aspect that once again highlights the virtuous, subjective and active characterization of Ricoeurian responsibility, which translates the power of creative subjectivity of phenomenological origin into a stable structure. We have seen how responsibility has a hard basis of objective regulation of relationships. In this sense, responsibility in the sense of imputability or accountability is a relationship of reciprocity based on the principle of equality. However, according to Ricoeur, this understanding is reductive because it risks ousting the creative contribution of the subject and his ability to go beyond the set of established rules. Accordingly, it risks limiting not only the practical effectiveness of responsibility in responding to new challenges but also the confidence in the positivity of its role. According to Ricoeur, therefore, in order for responsibility not to be reduced to an objective or repetitive relationship of reciprocity, and instead to be able to motivate creative and positive approaches, it needs to preserve its spontaneous and innovative nature. Within the analysis of recognition as an engine of interaction, in order to understand what can guarantee to it a sense and protect it from the criticism of being nothing more than a power struggle, Ricoeur proposes to take an alternative example to the logic of reciprocity.

The role of the gift with its paradigm focused on mutuality represents in his eyes this example. The gift is what for Ricoeur exemplifies a state of peace in the struggle for recognition because in some way, it represents a model of recognition that has taken place. In this sense, the necessity of the gift in the architecture of recognition lies in its character of moral motivation. In other words, the tangibility of the gift and its alternative to the competitive model of commercial exchange makes it a source of inspiration and a moral impetus in the dynamics of recognition.

But what represents an interesting point for our analysis of a responsible agency architecture is the need, not only for there to be positive examples, but for responsibility to survive and respond to its nature via means of spontaneous actions, not linked to the logic of reciprocity, but inserted in the dimension of mutuality. These actions, denotes Ricoeur, are crucial if we want to be bring back the unpredictable subjectivity, the *agape* within the impersonal relationship of global responsibility. This does not imply that we should disregard the objectivity aimed at maintaining an adequate level of justice. The gift, as well as charity and forgiveness, does not eliminate the reciprocity inherent in exchange, justice, and the regulation of debts. It represents a surplus that somehow makes it possible for reciprocity to maintain its value necessary for reciprocity to remain alive. The logic underlying gift in Ricoeur's view enables justice to be given "its boldness and momentum." The gift may or may not be reciprocated and no matter whether the attention is placed on the gift as an object or as a relationship, what takes on a nodal value is the guarantee, as Ricoeur suggests, that taking a responsible attitude of care always has value even if it does not necessarily have a price (Ricoeur 2005, chapter 5).

10. CONCLUSION

Ricoeur's reflections on responsibility are highly fruitful in understanding the challenges that innovation poses to the agency already proven by the tensions of pluralism. How can we maintain, Ricoeur asks, a right balance between the necessary assumption of responsibility and its radicalization, which risks dispersing its motivational force? (Ricoeur 2000, 33). How can we use the instrument of responsibility in a context of uncertainty such as that of technological innovation? The answer that the French philosopher offers us at the end of his reflections is one that suggests understanding the ethical person as the result of a relationship of recognition and interaction within the right institutions in an illocutionary manner. The resulting conceptual framework is one that promotes the linguistic interaction within structures that regulate relationships of recognition and stimulate the efforts of care toward himself and the others that every subject can put in place (Ricoeur 2000, 90).

It is clear that technical processes and thus innovation cannot be considered intrinsically responsible only through the concrete mediation of interpersonal relationships, in which the crucial role of the "responsible" subject cannot be substituted by the mechanisms of a formal framework, that technology can be evaluated in its responsibility. It is therefore clear that the concept of responsibility developed over the years by Ricoeur is embedded in an architecture of the capable agent, in a relationship with other individuals within the appropriate institutions. For Ricoeur, responsibility must be able to hold its various dimensions together and place them in a narrative relationship with other individuals (Gianni 2019). In this sense, Ricoeur's concept of responsibility can be well reconciled within the framework of RRI, which is able to combine respect for existing rules with the challenges arising from the uncertainty of innovation in a pluralist context. It is through the responsibility to mitigate the vulnerability of others and the relentless attempt to include the weakest or excluded ones that "responsible innovation" frameworks, such as RRI, will be able to respond to the challenges to which they are called.

The renewal of the concept of responsibility proposed by Ricoeur acquires a deeper meaning with reference to the subject of this responsibility, a subject assumed in its ethical and phenomenological constitution and not confined to a pure legal abstraction. In this respect, the renewal of the semantics of responsibility under the sign of the idea of capacity and imputation, with the contemporary revision of the fundamental legal anthropology under the sign of attestation proposed by Ricoeur, has a crucial theoretical relevance in articulating the legal idea of responsibility beyond the idea of obligation and reaction, along the idea of a projection of responsibility over the future, which is strictly intertwined with the idea of a capable subject recognizing and assuming its responsibility toward the Other.

NOTE

1. Principle 15 of the Rio Declaration on Environment and Development (Rio de Janeiro, 3-14 June 1992). UN Doc. A/CONF.151/26, vol I, annex I, 1992. After having being consecrated as a general principle of European law by the EU Court of

Justice, despite being originated in the context of environmental regulation, the precautionary principle is now enshrined in article 191 of the Treaty on the Functioning of the European Union (under the title dedicated to the Environment).

REFERENCES

Bessant, John. 2013. "Innovation in the Twenty-First Century." In *Responsible Innovation: Managing the Responsible Emergence of Science and Innovation in Society*, edited by Richard Owen, Bessant John, and Heintz Maggie, 1–25. Hoboken: John Wiley and Sons.

Boisson de Chazournes, Laurence. 2009. "New Technologies, the Precautionary Principle, and Public Participation." In *New Technologies and Human Rights*, edited by Thérèse Murphy, 161–194. Oxford: Oxford University Press.

Bovens, Mark. 1998. *The Quest for Responsibility: Accountability and Citizenship in Complex Organizations*. Cambridge: Cambridge University Press.

Burget, Mirjam, Emanuele Bardone, and Margus Pedaste. 2017. "Definitions and Conceptual Dimensions of Responsible Research and Innovation: A Literature Review." *Science and Engineering Ethics* 23, no. 1: 1–19.

Cane, Peter. 2002. *Responsibility in Law and Morality*. Portland: Hart Publishing.

Dierckxsens, Geoffrey. 2017. "Responsibility and the Physical Body: Paul Ricoeur on Analytical Philosophy of Language, Cognitive Science, and the Task of Phenomenological Hermeneutics." *Philosophy Today* 61, no. 3: 573–593.

European Commission. 2012. *Responsible Research and Innovation: Europe's Ability to Respond to Societal Challenges*. Brussels: European Commission.

Ewald, François. 2008. "Le principe de précaution." In *Le principe de précaution*, edited by François Ewald, Nicolas de Sadeleer, and Christian Gollier, 27–42. Paris: PUF.

Fisher, Erik, and Arie Rip. 2013. "Responsible Innovation: Multi-Level Dynamics and Soft Interventions." In *Responsible Innovation: Managing the Responsible Emergence of Science and Innovation in Society*, edited by Richard Owen, John Bessant, and Maggie Heintz, 165–183. Hoboken: John Wiley and Sons.

Gianni, Robert. 2019. "The Discourse of Responsibility: A Social Perspective." In *Responsible Research and Innovation: From Concepts to Practices*, edited by Robert Gianni, John Pearson, and Bernard Reber, 11–34. London and New York: Routledge.

Godin, Benoit. 2006. "The Linear Model of Innovation: The Historical Construction of an Analytical Framework." *Science, Technology, & Human Values* 31, no. 6: 639–667.

Gunther, Klaus. 1993. *The Sense of Appropriateness*. Albany: SUNY Press.

Hart, Herbert, and Lionel Adolphus. 1968. *Punishment and Responsibility: Essays in the Philosophy of Law*. Oxford: Oxford University Press.

Haydon, Graham. 1978. "On Being Responsible." *The Philosophical Quarterly* 28, no. 110: 46–57.

Kelsen, Hans. 2009. *Pure Theory of Law*, 5th ed. New Jersey: The Lawbook Exchange.
Pansera, Mario, and Richard Owen. 2018. *Innovation and Development: The Politics at the Bottom of the Pyramid*. London and New York: ISTE/Wiley.
Pavie, Xavier. 2019. "From Responsible-Innovation to Innovation-Care. Beyond Constraints, a Holistic Approach of Innovation." In *Responsible Research and Innovation: From Concepts to Practices*, edited by Robert Gianni, John Pearson, and Bernard Reber, 245–267. New York and London: Routledge.
Pellizzoni, Luigi. 2004. "Responsibility and Environmental Governance." *Environmental Politics* 13, no. 3: 541–565.
Prahalad, Coimbatore Krishnarao, and Venkat Ramaswamy. 2004. *The Future of Competition: Co-Creating Unique Value with Customers*. Cambridge, MA: Harvard Business School Press.
Reber, Bernard. 2016. *Precautionary Principle, Pluralism and Deliberation*. London and New York: ISTE/Wiley.
Ricoeur, Paul 1990. "Approches de la personne." *Esprit* 160, no. 3–4: 115–130.
Ricoeur, Paul. 1992. *Oneself as Another*. Chicago: The University of Chicago Press.
Ricoeur, Paul. 2005. *The Course of Recognition*. Cambridge, MA: Harvard University Press.
Schroeder, Doris, and David Kaplan. 2019. "Responsible Inclusive Innovation: Tackling Grand Challenges Globally." In *International Handbook on Responsible Innovation*, edited by René von Schomberg, and Jonathan Hankins, 308–324. Cheltenham: Edward Elgar.
Stirling, Andrew. 2017. "Precaution in the Governance of Technology." In *The Oxford Handbook of Law, Regulation and Technology*, edited by Roger Brownsword, Eloise Scotford, and Karen Yeung, 645–669. Oxford: Oxford University Press.
United Nations Conference on Environment and Development. 1993. *Rio Declaration on Environment and Development*. Rio de Janeiro: United Nations Conference on Environment and Development.
Villey, Michel 1977. "Esquisse historique sur le mot responsable." *Archives de Philosophie Du Droit* 22: 45–58.

Part III

RICOEUR AND TWENTY-FIRST-CENTURY TECHNOLOGY

Chapter 11

Ricoeur and E-health

Alain Loute

As the title of this chapter indicates, I would like to confront Paul Ricoeur's hermeneutics with e-health. What is e-health? This term refers to the use of information and telecommunication technologies in the field of health and well-being: "e-health is an emerging field in the intersection of medical informatics, public health and business, referring to health services and information delivered or enhanced through the Internet and related technologies" (Eysenbach 2001). Following Eysenbach, the transformations in health brought about by digital technology are multiple and profound, which is why he advocates an extensive and inclusive definition of e-health.

> In a broader sense, the term characterizes not only a technical development, but also a state-of-mind, a way of thinking, an attitude, and a commitment for networked, global thinking, to improve health care locally, regionally, and worldwide by using information and communication technology. (Eysenbach 2001, 118)

The purpose of this chapter may be surprising. Why, first of all, confront Ricoeur's hermeneutics with the field of health? Narrative medicine and clinical ethics have been advocating for many years to value narrativity in clinical practice and ethics. In this movement of narrative medicine, Paul Ricoeur's hermeneutics occupies a key place.

I have decided to focus more precisely here on e-health, because it confronts narrative medicine and narrative clinical ethics, and more fundamentally Ricoeur's hermeneutics, with a paradox. One might think that the latter, and hermeneutics in general, are particularly well equipped to deal with the epistemological and ethical challenges of new information and communication technologies. Does not its attention to interpretation, narrativity, and dialogue provide hermeneutics with a highly relevant perspective for questioning the use of information and communication technologies in health? The paradox is that, although they value the voice of healthcare stakeholders, including patients, and they paid attention to communication, they do not really allow us to position ourselves in relation to the emergence of what is called e-health.

The thesis I would like to defend in this chapter is that e-health destabilizes hermeneutics on a threefold level, and that, as a result, it is difficult for hermeneutics to identify the ethical and epistemological issues of e-health. I will show that the problem with hermeneutics is, first of all, its *idealism*, whereas it would be necessary to consider the materiality of meaning, that is, the digital as a new support for inscription. Secondly, e-health questions the *semantic optimism* of hermeneutics, by this expression, we mean the presupposition of a continuity of meaning, or to put it in Ricoeurian terms, of a continuity of "traditionality," the interplay between sedimentation and innovation, whereas digital technology may constitute a significant discontinuity in the way it processes data in relation to the dynamics of interpretation. Finally, hermeneutics favors temporality to the detriment of spatiality, while e-health is perhaps a new form of spatialization of care, to borrow the Foucaultian expression used in *The Birth of the Clinic*.

Before I begin, I would like to make one more methodological clarification. First of all, it must be noted that Ricoeur has already been brought to the attention of philosophers of technology. Among other motivations, it was a question for these authors of responding to a limit of the "empirical turn" (Achterhuis 2001) in the philosophy of technology, namely the "neglect of language in such inquiries about technology" (Coeckelberg and Reijers 2016, 326). The "empirical turn" would have led to a sacrifice of language for an analysis of the material dimension of technologies. Numerous authors have tried to find in Ricoeur's work a way to renew a philosophy of technology that allows us to investigate the ways technology and language interact.

This chapter does not pretend to give an overview of all this work, which, according to some, constitutes "a sort of 'hermeneutic turn' in philosophy of technology" (Romele 2020, 71). Moreover, my aim in this chapter is different from that of many authors who claim to be part of this "hermeneutic turn." For many, the objective is to use Ricoeur's work to develop a new philosophy of technology.

In a sense, this chapter would like to take the opposite path. It seems to me that these works have not sufficiently taken into account the fact that the philosophy of technology can question fundamental presuppositions of Ricoeurian hermeneutics. My objective in this chapter is therefore to critically questioning the latter. Such a research movement is not radically new. David Kaplan began his 2006 article by stating that "Ricoeur's works would be enhanced if he read in the philosophy of technology just as much as the philosophy of technology would benefit if Ricoeur were to join in the conversation" (Kaplan 2006, 42). I would like to radicalize this perspective. Kaplan limits himself to ask two questions: "What Ricoeur Adds to the Philosophy of Technology" and "What the Philosophy of Technology Adds to Ricoeur" (Kaplan 2006). The hypothesis I would like to investigate is that the philosophy of technology has a heuristic power: it allows us to uncover and question some assumptions of Ricoeurian hermeneutics.

First, I will begin by presenting what constitutes a rise in narrativity and the hermeneutical paradigm in contemporary clinical ethics. Thereafter, I will focus here in particular on Paul Ricoeur's philosophical hermeneutics. The second point of this chapter will confront this hermeneutical paradigm with the specific field of digital health by showing the triple destabilization it induces.

1. HERMENEUTICS AND NARRATIVE ETHICS IN THE FIELD OF HEALTH

Since the 1980s, more and more ethical approaches have been developed in opposition to or in addition to the classical bioethical approach—"Principlism"—as established by Beauchamp and Childress (*Principles of Biomedical Ethics* 1979). Among other developments, a so-called clinical ethics movement has developed, seeking to reconnect with clinical practice and recontextualize ethical reflection.

What interests me particularly here is that many clinical ethics committees organize the ethical discussion based on one or more narratives of the situation. According to the proponents of such a clinical ethic, the benefits of such an approach are plural. According to Hubert Doucet, in *Au pays de la bioéthique*:

> The value of the narrative approach to health care lies in the fact that the narrative is rich in an amplitude that is lacking in information, to use Walter Benjamin's idea. It is, in fact, the recognition that a patient's life is a story, and that it has a narrative coherence that reveals the particular meaning of the human events in that person's life. (Doucet 1996, 151, my translation)

There would be, Doucet tells us, a "darkness" specific to life that only logical reason could not dispel. The imagination at work in narrative is essential to "pierce the depths of a life's history" (Doucet 1996, 151, my translation).

Another contribution of narrativity to clinical ethics is that it would allow us to ask and problematize the question, "Who formulates the ethical problem?" For Doucet, "proponents of the narrative approach insist that case studies are formulated in an objective manner, resembling impersonal reports written by seconded observers" (Doucet 1996, 154, my translation). A narrative approach can enrich the formulation of a problem by way of multiple voices, each of the stakeholders being able to narrate, from their point of view, the situation.

This is why Didier Caenepeel and Guy Jobin write that "it is also stressed that narrative epistemology has a 'political' dimension, which is the relativization of the monopoly of medical discourse in the definition of ethical issues. . . . According to Brody (1999), the ethics committee adopting an informed approach through narrativity is a manifestation of the democratization of healthcare institutions" (Caepeneel and Jobin 2005, 111, my translation).

In order to conclude this overview of narrative clinical ethics, I would like to stress that for some authors, narrativity, in addition to its mobilization in clinical ethics committees, can also transform the medical relationship, the daily clinical practice. Marie Gaille, for example, considers that narrative ethics can "restore to medical practice a humanistic dimension that is often considered to be misguided or even lost in a practice in which the use of biotechnology and evidence prevails" (Gaille 2012, 213–214, my translation). In this perspective,

> the use of narrative ethics is conceived of as a tool for the diagnosis itself, the patient's narrative can contribute, by way of the elements it delivers to the doctor, to an understanding of the pathology from which he is suffering. The subjective dimension of health is highlighted by this ethical trend. (Gaille 2012, 213–214, my translation)

If the question of the subjective dimension of health is pertinent here, it is because in allowing the patient to recount his life, in a sense, he is also allowed to be a subject again. Such an affirmation is based on Paul Ricoeur's reflections on narrative identity, for whom subjectivity is purely narrative. For the French philosopher, the subject, in living experience, has only a discordant and chaotic intuition of himself. The relationship he has to himself is opaque. What allows the subject to understand himself, to grasp his ipseity, is the mediation of narrative imagination. It is by "emplotting" his life, by looking back at the different episodes of his life and putting them together in the form of a story, that the subject comes to an understanding of himself (Loute 2012).

In connection with this valorization of the narrative of the patient's life, it is a matter, for an author like Rita Charon,[1] of training professionals to become "attentive readers" of their patients' stories, to develop through "narrative training" a better ability to read patients' stories, as for Rita Charon, a doctor is not naturally and spontaneously an attentive reader of the stories of these patients. "Although everyone grows up listening to and telling stories, sophisticated knowledge of how stories work is not attained without considerable effort and commitment" (Charon 2006, IX). Rita Charon thus insists on the importance of integrating literature theory courses or writing workshops into the initial training of health professionals, in order to develop what she calls their "narrative competence," a "competence to recognize, absorb, interpret, and be moved by the stories of illness" (Charon 2006, VII).

2. HERMENEUTICS AND E-HEALTH

I would like to confront this movement of valuing narrativity in the field of health with the emergence of what is called "e-health." Defining precisely what is meant by e-health is not an easy task. Reading the literature on the subject, a terminological blur surrounding this term permeates the field. It is not uncommon to see the terms "digital health," "telemedicine," "connected health," "e-health," and so on, sometimes appearing in a way that is not always precise or clearly distinguished. Authors who have carried out a systematic review of the literature (Oh et al. 2005) have identified up to fifty-one different definitions of e-health, depending on whether the focus is on the process of executing the medical act at a distance, the technology used or the service relationship. Such vagueness reflects the uncertainty of the evolution of the health field. It could also be the trace of an attempt to frame it and regulate it normatively. Thus, in France, "telemedicine" strictly defines acts of medicine performed at a distance, unlike "e-health," which covers the use of information and telecommunication technologies in the field of health and well-being. In this chapter, I will not dwell on the stakes of a precise delimitation of this field of e-health, but will question the ability of hermeneutics and narrative ethics to identify e-health issues. Do they allow actors to position themselves ethically in relation to e-health devices?

Such a question is necessary because the development of health technologies has often provoked very strong and opposing reactions. Some see it as the worst of recent developments, others as the best. For (Rialle et al. 2014), the field of e-health is not immune to this phenomenon. For them

> on the one hand, the fear of dehumanization of support and care for people experiencing loss of autonomy and, on the other hand, the hope and desire to have effective technological tools at their disposal as soon as possible that live up to society's pressing issues are simultaneously and increasingly expressed. (Rialle et al. 2014, 135, my translation)

Does hermeneutics allow us to orient ourselves in these debates? Does its attention to interpretation, narrativity, and dialogue provide hermeneutics with a highly relevant perspective for questioning digital health? The thesis I would like to defend in this chapter is that e-health destabilizes hermeneutics on a triple level, and that, as a result, it is difficult for the latter to be able to identify the epistemological issues of e-health and its ethical implications. Let us begin by showing how e-health questions a certain form of hermeneutic "idealism," by which we mean here a neglect of the materiality of meaning.

3. HERMENEUTIC IDEALISM

Why label hermeneutics as idealism? I use this expression following Alberto Romele who speaks on this subject "of idealism of matter" (Romele 2020). For him, authors such as Heidegger, Gadamer, and Ricoeur are interested in language but do not care about its materiality, the means of transmission, or the generation of meaning. In this observation, he underlines the figure of Paul Ricoeur, which he considers to be a kind of exception. "Among the hermeneutists, Ricoeur is indeed the one who has most insisted on the externalizations and materializations of language. Ricoeur has focused his attention on written and fixed forms of language: symbols, signs, narratives, et cetera" (Romele 2020, 6). The reason, according to Romele, is that he was primarily interested in written language. In this respect, it is interesting to recall here the important role that the concept of "text" actually occupies in Ricoeur's philosophy of narrativity. To follow Ricoeur, the text would be the paradigm *par excellence* of the narrative. The philosopher writes as follows:

> In my view, the text is much more than a particular case of intersubjective communication: it is the paradigm of distanciation in communication. As such, it displays a fundamental characteristic of the very historicity of human experience, namely, that it is communication in and through distance. (Ricoeur 1991b, 76)

Because of this power of distancing, the written narrative would have more power to reveal and transform our living experience than an oral narrative. The power of the narrative imagination is therefore at its strongest in written discourse. "Mediation by texts seems to restrict the sphere of interpretation to writing and literature to the detriment of oral cultures. This is true. But what the definition loses in extension, it gains in intensity. Indeed, writing opens up new and original resources for discourse" (Ricoeur 1991a, 17).

What explains this power of the text is the fact that, in its written form, discourse gains in autonomy. Thanks to writing, the narrative gains in power; it becomes autonomous with regard to the limits of dialogue: it can be received by a multiplicity of readers; it can be read even if the context of its production has changed (we continue to read Aristotle despite the disappearance of the ancient Greek world); it also becomes autonomous with regard to the author's intentions. Finally, while in spoken speech the reference is ostensive, the text frees its reference from the limits of the ostensive reference; it opens references.

For Romele, the Ricoeurian attitude is paradoxical. "On the one hand, Ricoeur externalizes and materializes language and interpretational processes, in general, more than all other ontological hermeneutists. On the other hand, however, his understanding of the materialized language remains paradoxically idealistic." Ricoeur is particularly interested in rules that underlie the functioning of symbols, metaphors, and narratives. But, "he is not interested into the materialities of these means for transmitting meaning" (Romele 2020, 6).

I share Romele's view, but I would radicalize it. I think that even writing remains approached by Ricoeur in an idealistic way and without taking into account its materiality. Indeed, symptomatically, it should be noted that, despite the fact that the text occupies a privileged place in his philosophy of narrativity, Ricoeur does not have any specific reflection on the "book" object. Everything happens as if the book were only a support for meaning, a neutral vehicle whose form would in no way influence the meaning of the text. Generally speaking, it seems that Ricoeur lacks a thorough reflection on the materiality of meaning, on the technical and material media of the work of imagination. Perhaps Ricoeur has too much faith in the power of language to emancipate itself from its materiality?

Similarly, does the phenomenon of semantic innovation become totally autonomous from sensibility, as Ricoeur seems to indicate? It should be noted that he does not deny that semantic innovation can have effects on the sensory level. "In schematizing metaphorical attribution, imagination is diffused in all directions, reviving former experiences, awakening dormant memories, irrigating adjacent sensorial fields. . . . The poet is this artisan of language who engenders and shapes images through language alone" (Ricoeur 1991c, 173). In reading, semantic innovation can have repercussions on the sensory level. Nevertheless, for Ricoeur, the "work of imagination is to schematize metaphorical attribution. Like the Kantian schema, it gives an image to an emerging meaning. Before being a fading perception, the image is an emerging meaning" (Ricoeur 1991c, 173). While he does not deny that imagination can echo the senses, Ricoeur insists that it is in language that imagination is creative: "we see images only insofar as we first hear them" (Ricoeur 1991c, 174).

The semantic innovation produced by the narrative imagination therefore seems to owe nothing to the materiality of the text itself, nor to the sensory experience or to any relationship with a technical object. However, as a historian like Roger Chartier reminds us, "when the 'same' text is apprehended through very different mechanisms of representation, it is no longer the same" (Chartier 1995, 2).

The material forms and mediations of texts as well as the way in which they communicate have an impact on narrativity. This point was raised in the field of new health technologies by a report of the "European Group on Ethics in Science and New Technologies" (European Group on Ethics in Science and New Technologies 2015). The report highlights the ambivalence of new technologies in the health sector in terms of sharing experiences and communication. On the one hand, social networks, mobile health, digital health, and so on provide many opportunities to exchange experiences about disease. On the other hand, these information technologies run the risk of transforming a rich and diverse experience into discrete, isolated information, detached from the social context and from the biography of the subject. In other words, certain devices, techniques, and mediations can constitute conditions that allow the subject to return to his experience through self-reflection. But other devices may fragment the work of the imagination, so that narrativity is more likely to break down subjectivities than to constitute them.

The effect of many of these devices should be understood less as a form of imposing a specific semantic content, than as subjecting the imagination to an economy of attention.[2] For many thinkers, cognitive capitalism[3] and the development of cyberspace have led to the massive development of techniques for capturing our attention: audience measurement, algorithmic profiling, targeting, and so on. For Jonathan Crary, "Most important now is not the capture of attentiveness by a delimited object—a movie, television program, or piece of music . . . but rather the remaking of attention into repetitive operations and responses that always overlap with acts of looking or listening" (Crary 2014, 68). The repeated and mechanical movements operated by digital technologies that capture attention have the effect of fragmenting the work of the imagination.

In his latest book, Yves Citton proposes the concept of "mediarchy" to designate the power exercised by the media as environments.

> We live in mediarchy as soon as our communication devices structure from within our attentional dispositions . . . and therefore our orientation capacities, by organizing the environments in which we act . . . in a way that always slightly exceeds our intentional control. (Citton 2017, 49, my translation)

For Citton, paying attention to this mediarchy requires an effort, because "we have trouble seeing the media for the good reason that they are made not to be seen" (Citton 2017, 25, my translation). Hermeneutics may then reinforce this occultation of the media.

Narrative clinical ethics would benefit from exploring this issue of the material mediations of narrative work. Rita Charon herself is inaugurating such a work site by taking an interest in the patient file. She critically analyzes the codes that structure these medical records. For example, in these cases, doctors are encouraged to delete their "I," their own voice. According to her, it would be necessary to break these codes and develop a "more direct and transparent ways of representing the full range of what we doctors, nurses, and social workers come to know about our patients and ourselves" (Charon 2006, 146). Nevertheless, Charon seems to be particularly interested in the medical record as it constitutes a text of a particular kind, a text with an impersonal narrator. But isn't a file also a technical object? Who has access to it? Does the materiality and form of this object favor a concentration of power or does it make possible strengthened interprofessional collaboration? Hospital management necessarily involves the task of managing the attention of the professionals who work in the hospital. Doesn't the way the patient file disciplines and focuses the attention of professionals have an effect on how carefully they can then read their patients' stories?

Finally, if there is a need to focus on the technical objects and devices of digital health, it is to address an issue that falls within the scope of hermeneutics, but that remains unsatisfactorily addressed. These are the materials that allow for this "traditionality."

In order to clarify my point, I would like to quote Maurizio Ferraris for whom before being an instrument of communication, the internet is a recording tool. This idea of recording goes beyond the simple idea of a material for meaning. It leads us to think about what can sustain and even sometimes empower the strength of our discourse. "The social object is dependent on minds for its beginning to exist, but once it has been recorded it acquires an independent existence" (Ferraris 2014, 210). By social object, he means "the result of (1) a social act (which means, an act that involves at least two people) that is (2) characterized by being recorded" (Ferraris 2014, 209). He also uses the term "documentality" to refer to the sphere in which social objects are generated.

These concepts are interesting to reflect on the historicity of the narrative schematism at the heart of Paul Ricoeur's hermeneutics. Ricoeur brings together with Kantian schematism the work of imagination at work in the metaphorical use and the narrative use of language. Schematism places "an intuitive manifold under the rule of a concept" (Ricoeur 1984, 66). The narrative imagination, in a similar way, "takes together" incidents that it places under a story taken as a whole. In both cases, the imagination synthesizes the heterogeneous. The result is a unity, a link that would not exist if it remained focused solely on the events that make up the story. In the same way, the scheme produces, by synthesis, a bond that does not exist in intuition alone.

This reference Ricoeur makes to Kantian schematism calls for a first clarification. If the Kantian schemes are a priori, as an a priori determination of time, the nature of the schemes in Ricoeur is different. The schemes are of a linguistic and historical nature. "This schematism . . . is constituted within a history that has all the characteristics of a tradition" (Ricoeur 1984, 68). Ricoeur calls this process of constitution "traditionality." He reports of a double phenomenon of sedimentation and innovation.

It remains to be seen how this interplay of sedimentation and innovation is constituted. Should we postulate a collective memory or a social imagination in which past narrative schemes are stratified and sedimented? Is the notion of an archive approached to account for historiographical work sufficient to account for traditionality? Is the latter apprehensible without taking into account the technical objects and devices that ensure the recording and transmission of speech? Doesn't the field of research—opened up by documentality—make it possible to investigate all these questions?

Documentality is also essential to capture some of the dynamics generated by e-health, more specifically in the context of remote medical surveillance. *Télésurveillance* is one of the telemedicine acts recognized by French law.[4] Remote monitoring is when a patient with a chronic disease is monitored at home by clinical or biological indicators chosen by a medical health professional, collected spontaneously by a medical device or entered by the patient or a medical assistant, and then transmitted to the medical professional via telemedicine services. Might one of the risks of these connected devices and objects be that they do not "speak" for the patient? For example, it could play a probative role in a trial, producing evidence of good patient compliance, discriminating between "good" and "bad" patients. In his book on telemedicine in France, Pierre Simon reported on this possibility of controlling patients from a sleep apnea remote monitoring program that aimed to deploy continuous positive airway pressure machines to more than 800,000 patients. These machines had to be connected so that the healthcare provider could inform the French National Health Insurance Agency (l'Assurance Maladie) of cases of non-compliance (use of the devices less than three hours a day). This telemonitoring program was abandoned following an appeal to the French Council of State (Conseil d'État) by Patient Representatives (Simon 2015, 24).

4. THE SEMANTIC OPTIMISM OF HERMENEUTICS

The second characteristic of Ricoeurian hermeneutics that makes it difficult to identify the challenges of e-health is its "semantic optimism" (Loute 2017).

What does such an expression mean? By semantic optimism, I mean the fact that Ricoeur's philosophy is based on too much faith in the powers of imagination. A semantic optimism, not a political one, Ricoeur having reminded us on many occasions that it is always possible to "do wrong." Ricoeur has "confidence in the powerful institution of language" (Ricoeur 1985, 22). The question is rather to ask whether Ricoeur does not grant too much power to "meaning." Everything happens as if, for him, the human experience was always tellable.

According to Ricoeur, shared stories have the power to strengthen social ties, to give us back reasons to live together. But does Ricoeur consider in its full measure the fact that these stories can allow for the exercise of a form of domination, that texts can be imposed by a ruling class as the only legitimate texts, and that the mediation of these stories can lead to a real experience of alienation? As Enrique Dussel points out, in the same way that capital can constitute the producer as a mediation of capital valorization, the text can constitute "the reader as a mediation of the 'thing of the text' " (Dussel 1996, 85).

Ricoeur is in fact well aware that discourse can be the occasion for the exercise of symbolic and epistemic violence. To put it in terms of an expression taken from *History and Truth*, the narrative is always confronted with the risk of the "faux pas du total au totalitaire" (Ricoeur 1964, 191). The narrative, in its aim of a totality of meaning, contains the risk of closing off the possible and imposing with force the concordance and coherence of meaning. To this risk, Ricoeur would answer that it is always possible to tell otherwise. The solution is therefore to be found in the language itself. Any story, even the most succinct, has a free semantic potential that can always be revealed.

In *Time and Narrative*, Ricoeur examines another possible limit of the imagination, a risk that would no longer be sought this time in the imposition of a concordance, but in the discordance and fragmentation of meaning. Are there not certain discourses that not only resist but also seem to oppose any rule of narrative composition? It should be remembered that narrative schematism is historicized. Then, can't we imagine that one day this schematism might disappear? Don't some contemporary works push the distortion of the rules to such an extent that they indicate the possibility of the death of the story, revisiting the theme, already addressed by Walter Benjamin in *The Storyteller*, of an end to the art of narration?

For Ricoeur, however, "innovation remains a form of behavior governed by rules. The labor of imagination is not born from nothing" (Ricoeur 1984, 69). Imagination is always a game with rules. "I must admit, the assumptions on which I will expand at leisure further do not allow me to think of a radical anomie, but only a play with rules. Only a regulated imagination is conceivable" (Ricoeur 1986, 16, my translation)[5]. Even the most schismatic work cannot free itself from play with the sedimented patterns of a narrative tradition. Ricoeur quotes Frank Kermode on this subject: "Schism is meaningless without reference to some prior condition; the absolutely New is simply unintelligible, even as novelty," for "novelty of itself implies the existence of what is not a novel, a past" (Kermode 1967, 117)[6].

Ricoeur even writes that "productive imagination is not only rule-governed, it constitutes the generative matrix of rules" (Ricoeur 1984, 68). Indeed, a semantic optimism runs through Ricoeur's work, as the following quotation indicates: "Beyond every possible suspicion, we must have confidence in the powerful institution of language. This a wager that brings its own justification" (Ricoeur 1985, 22). To put it another way, "The search for concordance is part of the unavoidable assumptions of discourse and communication. . . . The Universal pragmatics of discourse says what amounts to the same thing. Intelligibility always precedes itself and justifies itself" (Ricoeur 1985, 28).

Nevertheless, Ricoeur himself moderates this optimism by indicating that even if the end of the story is improbable, it must nevertheless be considered conceivable: "one may always refuse the possibility of coherent discourse" (Ricoeur 1985, 28). "Either discourse or violence, Eric Weil has said in his *Logique de la Philosophie*" (Ricoeur 1985, 28). The productive imagination therefore depends on something that is not it and that makes it possible, namely a motivation to give meaning to its actions. This is what leads Ricoeur to write that "the search for concordance stems from a concern that maintains, elsewhere than in literature, its motivations . . .: it is . . . in the judicial domain—great archipelago of coherence—but also in the ethical relationships with others and in the reflection on the foundations of democracy, that the demand for intelligibility is inexpungable" (Ricoeur 1990, 199, my translation).

In terms of the narrative schematism that constitutes traditionality, this "semantic optimism" means that it presupposes a form of continuity in the interplay of sedimentation and innovation, a continuity in the work of interpretation. On this point, Ricoeur's analysis of Foucault's *Archaeology of Knowledge* in *Time and Narrative vol. III* is very instructive. According to him, Foucault favored discontinuity over continuity in history, the latter being associated with the ambition of a constituent consciousness and mastery of meaning. Ricoeur, on the contrary, legitimizes the presumption of continuity in history, discontinuity being related to the innovation pole of the dialectic of traditionality between sedimentation and innovation.

However, many current reflections on the digital tend to nuance or at least question this optimistic presupposition of the continuity of traditionality. This questioning of the continuity of meaning is linked to the "measurability" and "calculability" that digital technology introduces. As Bruno Bachimont writes, "The digital is therefore this art of sign asceticism, which consists in de-semantizing what is spoken of in order to reduce it to formal and blind symbols, thus gaining the ability to manipulate them via a machine" (Bachimont 2012, my translation). There have been many critiques of new attempts to measure the dynamics of care and to subject it to the reign of calculation. In a recent article, we examined different forms of attempts to defend—epistemologically, and inseparably, ethically—the disproportion of care (Loute 2016). It would fall entirely within the scope of what Ricoeur calls the logic of excess or the logic of superabundance, an antinomy of the logic of equivalence to which justice belongs. But it seems to me that it is not correct to understand the digital as a form of "one-dimensional rationality" that might dominate the reign of numbers at the expense of meaning. Indeed, calculations carried out through digital technology make it possible to feed the phenomenon of semantic innovation. They could be understood as feeding this semantic productivity. To understand this point, we must follow Bachimont in the distinction he makes between two dimensions of the digital. Digital technology is not just the ideal level of computation that Bachimont still calls "computational." For Bachimont, "for these calculations to be of any interest, the signs must then be given meaning according to an external and arbitrary convention, totally unmotivated by the signs themselves" (Bachimont 2012, my translation). The digital must also be understood in a second metaphorical or schematic dimension that aims to translate the calculated results into perceptual dimensions; "despite the novelty of the calculation and algorithms used, we still approach spaces to read, sounds to hear, images or representations to see, gestures to perform, returns of effort to feel, etc." (Bachimont 2012, my translation). It could be said that it is a matter of allowing for the semantic reappropriation of the results of the calculation.

Semantic reappropriation of the results of the calculation can indeed be an opportunity for real breakthrough in interpretative work. Bachimont points out that there is a tension between the universality of the computational, dealing with abstract information and its calculation, and the diversity and contingency of its achievements. "In general, we get by adopting a formalized theoretical framework, in which processing operations belong to the same theoretical perspective as those that made it possible to digitize the content in question: we start with sound signals that we record, digitize and transform as sound signals" (Bachimont 2012, my translation). But, he adds, "it is more complicated when addressing areas in which such a theoretical framework is lacking" (Bachimont 2012, my translation).

Isn't this the case for many uses of big data in health, for example? Are we not faced with the difficulty of interpreting a great many results? According to some authors, big data is an epistemological revolution. As Antoinette Rouvroy writes, "Unlike conventional statistical processing, in which statistical hypotheses or categories precede and govern the collection of data, in Big Data-type processing, the exact opposite occurs: data collection and processing come first and give rise to hypotheses or categories from among the mass of data" (Rouvroy 2016). Rather than subsuming data into pre-constituted categories, big data inductively "brings out" the categories of the mass of data itself.

Historian Alain Desrosières taught us that "quantifying" reality is only possible through the work of qualification of beings and phenomena, the application of an equivalence convention allowing for the work of measurement. It is as if big data could do without this conventional process and make reality speak immanently, without any mediation. Some have seen this as the announcement of the "end of the theory," to use the well-known title of an article by Chris Anderson (2008). It would be a matter of inferring correlations, without further attempting to construct a model that could give them a causal explanation.

In the health sector, big data are used in what some call personalized medicine, among other things. Xavier Guchet points out in this regard that we often speak of personalized medicine as a medical technology: an approach to medicine more linked to technological advances, rather than to conceptual advancements. "Personalized medicine is generally not described as the clinical application of prior biological knowledge, as a medicine preceded by theory; it is rather defined as a theoretical medicine, without prior assumptions, supposedly guided only by the positivity of the data delivered by techniques, and processed by algorithms" (Guchet 2016, 139, my translation). According to Guchet, the great hope placed in the development of this medicine must not hide the lack of biological knowledge concerning the etiology of multifactorial diseases such as cancer. Because of its semantic optimism, the presupposition of the continuity of the work of interpretation, is hermeneutics really able to identify the epistemological issues related to these latest technological developments?

5. THE OCCULTATION OF SPATIALITY

Finally, the last characteristic of hermeneutics, at least in its Ricoeurian version, which undermines its ability to identify the epistemological and ethical issues of digital health, is the privilege granted to temporality and the occultation of spatiality. One of Ricoeur's main hypotheses at the beginning of *Time and Narrative* is that narrative can provide a poetic solution to the aporias with which any speculative approach to time is confronted; "time becomes human time to the extent that it is organized after the manner of a narrative" (Ricoeur 1984, 3). To take up Ricoeur's case, unless I am mistaken, the triple mimesis model is applied only to space on the occasion of a reflection on architecture, notably *in Memory, History and Forgetting*, and in an article entitled "Architecture and Narrativity."

Such an omission of spatiality is harmful because several "studies (notably Cartwright 2000; Nicolini 2007; Dyb and Halford 2009; Petersson 2011; Oudshoorn 2012) have tried to show that telemedicine does not abolish borders or spaces. This runs against a common-view often implicit in public policy" (Mathieu-Fritz and Gaglio 2018, 4).

On the contrary, telemedicine is transforming the places and spaces of care.[7] It requires a certain reorganization of the home, induces a reorganization of organizations, blurs the boundaries between institutions, makes it possible to visualize a territory for the implementation of public health policies, and so forth. Digital technology, for example, is changing the very notion of territory and habitat. "The material and visual control of connected objects is important. With these technologies, the question arises of transforming the home into a hybrid place, combining the private and public spheres, with the intrusion of indiscreet technologies" (Mayère 2018, 210, my translation). Some people talk about a "medicalization of the home" (Arras and Neveloff Dubler). In her "technogeographic" approach to telecare, Nelly Oudshoorn (2011) reports on the transformation of the geography of care by way of these technical devices.

We could even hypothesize that e-health makes it possible to constitute a new form of spatialization of care, in the sense of Foucault's ternary spatialization emphasized in *The Birth of the Clinic*. "Let us call tertiary spatialization all the gestures by which, in a given society, a disease is circumscribed, medically invested, isolated, divided up into closed, privileged regions, or distributed throughout cure centres, arranged in the most favorable way" (Foucault 1973, 16). The point is quite obvious to me in the case of telemedicine, as it is becoming such a way for public authorities to act on their respective territories. Telemedicine

> refers to any medical procedure remotely using ICT, and it is marginally practiced until 2009. Informal networks intend to reduce the impact of spatial inequalities in medical demography. But legal recognition of telemedicine and its introduction into the HPST law enable it to become a tool for public health policy (Rauly 2015).

Telemedicine supports real public policy objectives.

Amandine Rauly defends the idea that, in the context of telemedicine, technology does not only play the role of supporting medical practice; for her, telemedicine is mobilized to apply public policies based on a "market reference framework."

Starting in 2010, telemedicine will become a tool of French public policy (HAS 2011) which should allow a better control of health expenditure and a rationalization of it. The telemedicine tool can then become "new eyes" for the social state, supposed to be "one-eyed," within an asymmetric informational relationship. The technical object makes it possible to standardize medical protocols, to circulate private information between the actors of the health system and thus to limit moral hazard. (Rauly 2015, my translation)

By favoring a consideration of temporality, can hermeneutics take into account this new form of political management of space?

6. CONCLUSION

In this chapter, I have sought to confront Ricoeur's hermeneutics with the field of e-health. In my opinion, the interest in doing so goes beyond the simple application of this philosophy to a new object. The confrontation with e-health has the more fundamental effect of questioning hermeneutics on a threefold level. I have tried to show that, at a fundamental level, e-health destabilizes hermeneutics and that, as a result, it is difficult for the latter to be able to identify the ethical and epistemological issues of digital health. It questions its neglect of materiality, its presupposition of a continuity of meaning and its occultation of spatiality.

What avenues for research can we identify to continue the work begun in this chapter? A first line of research could consist in exploring the link between imagination and materiality from the Simondonian philosophy of imagination. As Vincent Beaubois points out, in Simondon, the scheme is never the pure product of a human consciousness, but the result of a certain frequentation, of a certain participation in an external reality (Beaubois 2016). Simondonian philosophy of imagination invites us to conceive of schematization as a collaboration between human and technical, and not, as in Ricoeur's work, as a work internal to the language field.

A second line of research could be to reread the pages that Ricoeur devoted to the concept of "trace." Ricoeur devotes several pages to the concept of trace in the third volume of *Time and Narrative*. The problem of the trace is discussed in a section devoted to historical time. "Traces" are presented as "reflective instruments." They play "the role of connectors between lived time and universal time" (Ricoeur 1988, 104). Could not one of the interests of this concept be to allow us to understand what can ensure continuity or create discontinuities of traditionality? Doesn't the continuity of meaning depend on the materiality of the reflective instruments?

Finally, a final line of research would consist in exploring the avenues for research opened up by the few pages devoted by Ricoeur to the link between space and narrativity. Recent publications in the field of Ricoeurian studies have returned to the few pages devoted by Ricoeur to the intersection between space and narrativity (D'Alessandris 2019), (Lelièvre and Inizan 2016). The translation in English of the article "Architecture and Narrativity" has been published in 2016 (Ricoeur 2016). In this article, Ricoeur

> would like to put an analogy in place, or rather something that appears at first sight to be only an analogy: a narrow parallelism between architecture and narrativity, in that architecture would be to space what narrative is to time, namely a "configurative" process; a parallelism between on the one hand constructing, that is, building in space, and on the other hand recounting, emplotment in time. (Ricoeur 2016, 31)

Architectural *making* would also produce "a spatial synthesis of the heterogeneous" composing between several variables: units of space, massive forms, and the boundary surfaces. Despite the fact that Ricoeur makes it clear that this is only an "analogy" between architecture and narrativity, could these reflections nevertheless be useful in explaining the way in which technologies in the field of e-health transform spaces?

NOTES

1. For a more in-depth analysis of narrative medicine, the reader may refer to (Loute 2019) and (Le Berre and Loute 2018).
2. On this point, see also Loute (2017).
3. On this point, see also Cavazzini and Loute (2017).
4. For a more detailed presentation of this legal framework, see Williatte and Loute (2017).
5. This part of the article is not included in the following translation: "On Interpretation," translated by Kathleen Blamey, in *From Text to Action, Essays in Hermeneutics II*, Evanston, Northwestern University Press, 1991, pp. 1–20.
6. Quoted by Ricœur (1985, 26).
7. For further developments on this issue of spatialization of care induced by telemedicine, the reader may refer to Loute (2019).

REFERENCES

Achterhuis, Hans, ed. 2001. *American Philosophy of Technology: The Empirical Turn*. Indianapolis: Indiana University Press.
Anderson, Chris. 2008. "The End of Theory: The Data Deluge Makes the Scientific Method Obsolete." *Wired Magazine*, June 23, 2008. https://www.wired.com/2008/06/pb-theory/.

Bachimont, Bruno. 2012. "Pour une critique phénoménologique de la raison computationnelle." https://www.ina-expert.com/e-dossier-de-l-audiovisuel-l-education-aux-cultures-de-l-information/pour-une-critique-phenomenologique-de-la-raison-computationnelle.html.

Beaubois, Vincent. 2016. "Un schématisme pratique de l'imagination." *Appareil* 16. http://appareil.revues.org/2247.

Caepeneel, Didier, and Guy Jobin. 2005. "Discursivité et co-autorité en éthique clinique: regard critique sur le rôle et les fonctions de la délibération éthique en comité." *Journal International de Bioéthique* 16 (3): 105–133.

Cavazzini, Andrea, and Alain Loute. 2017. Éditorial. *Cahiers du GRM* 11. http://journals.openedition.org/grm/997.

Charon, Rita. 2006. *Narrative Medicine: Honoring the Stories of Illness*. Oxford: Oxford University Press.

Chartier, Roger. 1995. *Forms and Meanings, Text, Performances, and Audiences from Codex to Computer*. Philadelphia: University of Pennsylvania Press.

Citton, Yves. 2017. *Médiarchie*. Paris: Seuil.

Coeckelbergh, Mark, and Wessel Reijers. 2016. "Narrative Technologies: A Philosophical Investigation of the Narrative Capacities of Technologies by Using Ricoeur's Narrative Theory." *Human Studies* 39 (3): 325–246.

Crary, Jonathan. 2014. *24/7: Late Capitalism and the Ends of Sleep*. London: Verso.

D'Alessandris, Francesca. 2019. "La durée dans la dureté. Espaces de la mémoire et mémoires de l'espace chez Paul Ricoeur." *Etudes Ricoeuriennes/Ricoeur Studies* 10 (1): 58–72.

Doucet, Hubert. 1996. *Au pays de la bioéthique, L'éthique biomédicale aux Etats-Unis*. Geneva: Labor et Fides.

Dussel, Enrique. 1996. *The Underside of Modernity. Apel, Ricoeur, Rorty, Taylor and the Philosophy of Liberation*. Translated by Eduardo Mendieta. Atlantic Highlands, NJ: Humanities Press.

European Group on Ethics in Science and New Technologies. 2015. *Ethics of New Health Technologies and Citizen Participation*. Opinion 29. https://ec.europa.eu/research/ege/pdf/opinion-29_ege.pdf.

Eysenbach, Gunther. 2001. "What Is E-Health?" *Journal of Medical Internet Research* 3 (2).

Ferraris, Maurizio. 2014. "Total Mobilization." *The Monist* 97 (2): 200–221.

Foucault, Michel. 1973. *The Birth of the Clinic*. Translated by Alan M. Sheridan. London: Routledge.

Gaille, Marie. 2012. "Des mots et des maux: Que peut-on espérer des récits de vie dans la relation de soin à travers les âges." *Médecine/Sciences* 28: 213–214.

Guchet, Xavier. 2016. *La médecine personnalisée: Un essai philosophique*. Paris: Les Belles Lettres.

Kaplan, David M. 2006. "Paul Ricoeur and the Philosophy of Technology." *Journal of French and Francophone Philosophy* 16 (1–2): 42–56.

Kermode, Frank. 1967. *The Sense of an Ending: Studies in the Theory of Fiction*. New York: Oxford University Press.

Le Berre, Rozenn, and Alain Loute. 2018. "Raconter la souffrance en soins palliatifs: les usages multiples du récit." *Médecine Palliative* 17: 208–217. https://doi.org/10.1016/j.medpal.2018.06.004.

Lelièvre, Samuel, and Yvon Inizan. 2016. "Introduction." *Etudes Ricoeuriennes/ Ricoeur Studies* 7 (2): 8–13.

Loute, Alain. 2012. "Normative Creativity in Paul Ricoeur." *Trópos: Journal of Hermeneutics and Philosophical Criticism* V (1): 107–124.

Loute, Alain. 2016. "La démesure du *care*: surabondance de l'amour, excédent sémantique ou contradiction?" *Cahiers du GRM* 10. http://grm.revues.org/829.

Loute, Alain. 2017. "L'imagination au coeur de l'économie de l'attention: L'optimisme sémantique de Paul Ricoeur." *Bulletin d'Analyse Phénoménologique* 13 (2). http://popups.ulg.ac.be/1782-2041/index.php?id=937.

Loute, Alain. 2019a. "Penser l'éthique de l'accueil comme une éthique de la spatialisation des soins: Le cas de la télésurveillance à domicile." *Ethica Clinica* 93: 57–66.

Loute, Alain. 2019b. "L'éthique clinique narrative porte-t-elle attention aux subjectivités? Réflexions à partir de Rita Charon." In *Valeurs de l'attention, Perspectives éthiques, politiques et épistémologiques*, edited by Nathalie Grandjean and Alain Loute, 21–44. Lille: Septentrion.

Mathieu-Fritz, Alexandre, and Gérald Gaglio. 2018. "In Search of the Sociotechnical Configurations of Telemedicine: Literature Review in the Social Sciences." *Réseaux* 207 (1): 27–63.

Mayère, Anne 2018. "Patients projetés et patients en pratique dans un dispositif de suivi à distance." *Réseaux* 207 (1): 197–225.

Oh, Hans, Carlos Rizo, Murray Enkin, and Alejandro Jadad. 2005. "What is eHealth, A Systematic Review of Published Definitions." *Journal of Medical Internet Research* 7 (1).

Oudshoorn, Nelly. 2011. *Telecare Technologies and the Transformation of Healthcare*. London: Palgrave Macmillan.

Rauly, Amandine. 2015. "Intervention publique *versus* régulation professionnelle: Conflits autour du déploiement de la télémédecine en France." *Revue de la régulation* 17. https://journals.openedition.org/regulation/11233.

Rialle, Vincent, Pierre Rumeau, Catherine Ollivet, Juliette Sabliera, and Christian Hervé. 2014. "Télémédecine et gérontechnologie pour la maladie d'Alzheimer : nécessité d'un pilotage international par l'éthique." *Journal International de Bioéthique* 25 (3): 127–145.

Ricoeur, Paul. 1964. "Vérité et mensonge." In *Histoire et vérité*, 165–193. Paris: Seuil.

Ricoeur, Paul. 1984. *Time and Narrative: Volume 1*. Translated by Kathleen MacLaughlin and David Pellauer. Chicago: The University of Chicago Press.

Ricoeur, Paul. 1985. *Time and Narrative: Volume 2*. Translated by Kathleen McLaughlin and David Pellauer. Chicago: The University of Chicago Press.

Ricoeur, Paul. 1986. "De l'interprétation." In *Du texte à l'action: Essais d'herméneutique II*, 11–35. Paris: Seuil.

Ricoeur, Paul. 1988. *Time and Narrative: Volume 3*. Translated by Kathleen Blamey and David Pellauer. Chicago: The University of Chicago Press.

Ricoeur, Paul. 1990. "Réponses." In « Temps et récit » de Paul Ricoeur en débat, edited by Christian Bouchindhomme and Rainer Rochlitz, 187–212. Paris: Cerf.
Ricoeur, Paul. 1991a. "On Interpretation." Translated by Kathleen Blamey. In *From Text to Action, Essays in Hermeneutics II*, 1–20. Evanston: Northwestern University Press.
Ricoeur, Paul. 1991b. "The Hermeneutical Function of Distanciation." Translated by John B. Thompson. In *From Text to Action, Essays in Hermeneutics II*, 75–88. Evanston: Northwestern University Press.
Ricoeur Paul. 1991c. "Imagination in Discourse and in Action." Translated by Kathleen Blamey. In *From Text to Action, Essays in Hermeneutics II*, 168–187. Evanston: Northwestern University Press.
Ricoeur, Paul. 2016. "Architecture and Narrativity." *Etudes Ricoeuriennes/Ricoeur Studies* 7 (2): 31–42.
Romele, Alberto. 2016. "Herméneutique du digital : les limites techniques de l'interprétation." https://hal.archives-ouvertes.fr/hal-01299368.
Romele, Alberto. 2020. *Digital Hermeneutics. Philosophical Investigations in New Media and Technologies.* London: Routledge.
Rouvroy, Antoinette. 2016. "'Of Data and Men': Fundamental Rights and Liberties in a World of Big Data." Council of Europe. https://rm.coe.int/16806a6020.
Simon, Pierre. 2015. *Télémédecine, Enjeux et pratiques*. Brignais: Le Coudrier.
Williatte, Lina, and Alain Loute. 2017. "La télémédecine: entre rappel du cadre de l'acte médical et déstabilisation de celui-ci." *Actualité et dossier en santé publique* 101: 44–45.

Chapter 12

The Force of Political Action in the Technological Polis

Todd S. Mei

This chapter examines the relation between political action and technology through Ricoeur's critical appropriation of Hannah Arendt's account of natality and J. L. Austin's speech act theory. This approach involves a certain bias, consistent with Ricoeur's own philosophical outlook, where political action is taken to be more primary than technology. In other words, the task of examining the role of technology in the polis presupposes understanding political action.

The easiest way to make sense of this bias is in terms of the way Ricoeur (1981) argues how the linguistic and communicative capacity of action is paramount. Political action, as a species of action, engages this communicative capacity in a distinct way that is captured by Arendt's conception of natality, or the initiation of a new beginning in the political sphere. Accordingly, understanding political action involves the task of grasping how it announces or communicates (like speech) a new beginning. So, if political action has unique conditions for its genesis that do not require or presuppose technology, then it suggests that political action not only has priority over technology but is also responsible for it in being able to explain and understand it. Ricoeur (1965b, 211) refers to this as the "culpability" or responsibility entailed in the priority of language. Indeed, this culpability will become more evident in my attempt to see how the theories of natality and speech acts illuminate technologically mediated political action.

Yet, a critic might worry that this approach commits the error of thinking technology is best understood as deriving from some more basic form of life (i.e., political action and language), when in fact life cannot be conceived or imagined apart from its technological mediation. This worry is, of course, expressed by both post-phenomenological and empirical philosophers of technology. I think some form of bias or privilege is inevitable as long as technology is discussed in terms of how it affects, enhances, or hinders human life. Some aspect of what is taken to be human is necessary for providing a critical perspective simply as a reference point. But this does not mean that empirical variation is ignored. That is more a problem for philosophies that attempt to offer a general theory about what technology is despite its variations.[1] So while the caveat of this chapter's approach has to do with its conceptual bias, the method analysis remains modest with respect to what it says about technology: namely, I propose to examine the effects of social media as a specific empirical instance of mediation operative within the technological polis.

The decision to examine social media in this context is not so spurious for reasons mainly having to do with its prevalence as a form of life (cf. Kaplan 2003, 170) facilitating political discussion irrespective of the physical proximity of interlocutors. While social media is more than simply a platform for political communication (e.g., Trottier and Fuchs 2015; Loader and Mercea 2011), my concern in this chapter is to consider how its form of mediating discussion affects how we enact and relate to political action.

In view of the aforementioned contextualizing comments, a thesis about the technological polis can now be offered. I treat social media as presenting a *risk* to the *predicatory* aims of political action. This risk primarily affects how political actions and utterances become divorced from their defining, normative context. I refer to this phenomenon as the *privation* of the conditions necessary for reasoned discussion. On this view, the force of political action, which I will relate to the speech act theoretical treatment of illocution, becomes unconstrained because the context which allows the audience to make sense of actions and utterances as reasoned statements is either absent or unobvious. Force is consequently over-determined by expressive and reactive content, thus minimalizing the rational or representational content that might inform a reasoned discussion.

One final remark: in following Ricoeur's theory of action as linguistic, I will use terms like "action" and "act," on the one hand, and "speech" and "utterance," on the other, interchangeably (Fogal et al. 2018, 6–7; cf. Mei 2018).

1. POLITICAL ACTION: SPEECH AS NATALITY

Given the priority Ricoeur gives to language, it is in one sense unsurprising that he finds Arendt's conception of action as speech highly compelling. The general debt Ricoeur owes to Arendt is well-known,[2] and the approach of this section is to exploit a distinction that some Arendt commentators take to be foundational and which some Ricoeur commentators often overlook.[3] The latter tend to focus on narrative as the basis of personal identity; but upon closer examination of Arendt's thought, the capacity to narrate appears to derive from a foundational speech act (i.e., natality). Clarifying this priority will help us to see, in the next section (§2.0), how natality advances a distinct thesis about political action that aligns amicably with Ricoeur's analysis of the pragmatic structure of speech.

When Arendt (1958, 178) states that "without the accompaniment of speech ... action would not only lose its revelatory character, but ... it would lose its subject," it seems obvious that she sees action as requiring speech. Is she advocating a kind of speech act theory? Yes, if she is implying that the "doing" of political action cannot occur apart from being linguistic in some substantial manner. Indeed, for Arendt, political action is a speech act called natality.[4] Its act, in other words, brings something new into the world:

> If action as beginning corresponds to the fact of birth, if it is the actualization of the human condition of natality, then speech corresponds to the fact of distinctness and is the actualization of the human condition of plurality, that is, of living as a distinct and unique being among equals. (Arendt 1958, 178)

The comparison between biological birth and the new initiation brought about by action is more than an analogy since Arendt suggests that like birth, and despite the fact that we do not initiate our own birth, action is an act of creation that introduces new possibilities for our political lifeworld.

In other words, the way Arendt envisages the inseparableness of action and speech allows us to see political action as achieving several things simultaneously. First, action announces motive and reason by making something happen. Second, it announces the actor as a "who" that is desiring to be recognized as both agent of the action and as one whose identity is bound up with that action. Third, the desire to be recognized as an agent opens the door to a substantial sense of identity and responsibility that can only be more fully developed by a narrative understanding (cf. Ricoeur 2000, 3, 2007, 76–79).

Given the prevalence of the role of narrative understanding in the formation of identity, it might be tempting for a reader of Ricoeur to conclude that all three points are exhausted by a narrative theoretical approach. However, the first two points can also be read in view of speech act theory, and with very different results as we will see.

On a general application of speech act theory, political action has to do with how "making something happen" in the polis announces the motive and reason of an agent publicly for consideration or even scrutiny.[5] Let us recall that following Anscombe, Ricoeur refers to the way action implicates other questions and concepts—such as goal, motive, agent, and so on—as "the language of discourse about action" (Ricoeur 2016, 90; cf. Ricoeur 2005, 94–96;

MacIntyre 1984, 209).[6] I will consider an example of this. For now, let it suffice to say that trying to understand an action as an event that presents (or as Ricoeur says, "inscribes") motive and reason sets the bar quite high. One cannot take for granted that political action is easily interpreted with regard to the conventions it affirms or challenges, something that I will examine in terms of the illocutionary and perlocutionary dimensions of meaning.

But this more detailed way of thinking about political action as a speech act is not something Arendt discusses in any detail. Instead, commentators on Arendt rely on characterizing natality as a general performative act of communication (cf. Dietz 2002, 10; Honig 1991, 99). I am not dismissing this approach. I do think, nonetheless, there is much to be gained by expressly applying speech act theory to show how political action initiates something new.

2. WHAT DOES RICOEUR ADD?

A speech act is an utterance that does something by being uttered. When it does something that is subject in some way to conventions and rules,[7] its effect or force is illocutionary. When it discloses something beyond these conventions, its effect is perlocutionary, and often involves an imaginative application in which something new is revealed about oneself, the world, or others (Mei 2018). As we will see, the way speech act theory can account for conventional and non-conventional effects will prove crucial in showing how political action predicates new meaning and how social media poses a risk to this predication.

Ricoeur's application of speech act theory offers two distinct contributions in understanding how illocution and perlocution function. The latter involves a critical use of the imaginative ability to "see as," that is to see something familiar in a new way such that it lends a critical perspective on existing political norms (hereafter, norms). The former involves a nuance: whereas illocution refers to the context of norms that inform actions, it also includes the capacity to revise these norms, or what is the social virtue of practical reasoning. As we will see, perlocutionary effects provide a significant impetus for such revision.

Let me break down these aims into three theses:

(1) Ricoeur's focus on perlocutionary effects[8] provides a distinctive way for conceptualizing how political action has a critical function for changing the way we think of and relate to others.
(2) Identity constitution via perlocutionary effects is distinct from narrative identity so often associated with Ricoeur's work. Perlocutionary identity

formation occurs by putting pressure on the illocutionary dimension of norms and conventions. So instead of privileging continuity of identity, it allows for a discordance within the narrative fabric of a polity (cf. Taylor 2015).
(3) Perlocutionary effects need to be tested. After all, there can be bad perlocutionary effects that should not be admitted into practice. Testing is performed at the illocutionary level, which uses practical reasoning to question the sufficiency of its norms. This is the illocutionary application of a process of argumentation.

To demonstrate these theses, I refer to a key feature of political action that Ricoeur takes from Arendt—namely, *suffering* (Joy 2018, 111).

At the political level, suffering can be thought as resulting from unintentional and intentional acts. Both acts involve a degree of unpredictability. The former is perhaps best expressed by Ricoeur's notion of the political paradox in which aiming for the good of political life inevitably involves isolating or disadvantaging others (Ricoeur 1965a). But even the presence of intention is no guarantor of consequences since, as Arendt observes, unpredictability lies not in the inability of the agent to foretell "all the logical consequences" but in the fact that any action begins from a point of view that can never know its outcome or end (Arendt 1958, 191). Suffering seems an inevitable consequence that makes others the victim of an action since the agent can never know exactly how her action will be understood, applied, and represented to others. Thus, Arendt (1958, 191) notes that acting and suffering are two sides of the same coin.

One of the most pressing problems about suffering is what Ricoeur calls the "difficulty, even our incapacity to bring to language the emotional, often traumatic experience that psychoanalysis attempts to liberate" (Ricoeur 1997, xxxix as cited in Joy 2018, 114–115). If action is something that is done, suffering is that which is done to another; and it is this passive role of being subject to an action that lends to suffering an ambiguity of expression and recognition. In other words, on Arendt's account, the event of suffering is one in which a public action bears upon the individual's interior sense of well-being, identity, and relations with others. So suffering, it should be noted, has a double burden. Not only is the onus of enduring harm or damage problematic, but so is finding a way to attest to it publicly. Having the courage to do so requires announcing oneself in the way Arendt envisages by the concept of natality.

Consider the action of *walking away* in a highly charged political context, that of being black in Los Angeles during the 1960s. A black man walks away from a white, male police officer in response to his question, "Where you going?" (Peralta 2008, 12:52). While the officer's question is itself an illocution (question as threat),[9] making sense of the action of walking away is problematic. Is it civil disobedience? Or, does it in some way attest to suffering? Can it be both in some sense?

Typically, analysis of the illocutionary force of action involves understanding how an action is subject to, affirms, or calls upon norms, expectations, and rules (Yalcin 2018, 402). Standard cases often refer to effects in which the relation of communication is not dramatically antagonistic, unlike the case of civil disobedience. So, for example, using a thumb's up gesture can be read as an agent's intention to signal approval that some action can commence. As long as the audience is aware of the symbolism of the gesture, then understanding is, as Austin (1975, 14; cf. 42) puts it, *felicitous*. In the example of walking away, the relation of communication is one of breaking what is expected. If the police officer's illocution does not *misfire* (Austin 1975, 16), because the black man knows he is supposed to stop and answer the question, then walking away appears to defy expectations and conventions. This defiance does not break with illocutionary norms since the police officer knows that walking away is a possible outcome to the question which then elicits another set of rules for reacting—namely, coercion, restraint, or further interrogation. In short, analyzing the action at the illocutionary level presents the action as one of non-compliance.

Ricoeur's emphasis on perlocutionary effects helps one to read the action in a different way, which highlights the issue of the suffering agent.[10] Walking away is no longer simply non-compliance but an attestation to suffering. What triggers *seeing* the act of walking away *as* something other than disobedience? It can be words surrounding the action, new awareness of socio-political information, sympathy with the agent, and so on—in short, anything perlocutionary to the context in which the action would typically be understood. What if one were able to imagine that the agent was faced with this kind of interrogation on a regular basis? Entertaining this or knowing this information enables one to see beyond the limitations of the illocutionary reading of the action and possibly opens to a revelation, to use Arendt's term—namely, the black man is tired of suffering repeatedly acts of discrimination, which undermine his sense of self-esteem and worth.[11] Kumasi, a black resident of Watts who was involved in the riots of 1965 comments on police harassment:

> Now what do you think that does to me psychologically? What does that tell me? What is message am I being fed every day? See, he [the police officer] don't understand he's feeding me a spoonful of hatred. (Perlata 2008, 13:22–13:33)

It is perhaps more feasible for an audience today to see this dimension of suffering in an action such as this due to a critically informed historical perspective. But this is only to say that an audience today is more open to considering the perlocutionary effects of walking away. The same would be possible for an audience from 1965 insofar as it is able to engage with the meaning of an action beyond conventional expectations of what an agent should be doing—that is, complying with the police officer.[12]

Another way of seeing how the change of perspective in perlocutionary effects functions is with respect to a sense of identity that is underwritten by the normative content within the illocutionary sphere. George Taylor discusses this aspect of political action under the idea of a "prospective identity," which is a future identity "not limited toward expectation of the same but toward the possibility of the new and different" (Taylor 2015, 132). Arendt's (1958, 182–183) notion of revelation is thus a kind of disrupture because it is meant to stand alone (i.e., as a new beginning) and to refigure one's perception of the act and the actor. To be sure, this process is one of critical innovation and criticism that, although challenging, has in mind the aim of greater cohesion of the plurality.

Admitting the salience of perlocutionary effects has the virtue of helping us to recognize instances where epistemic and hermeneutical injustices are operative—that is, respectively, a bias against believing someone's testimony and the misinterpretation of a statement due to some prejudice or lack of information. For this critical view and openness to be viable, it presupposes Arendt's idea that plurality is a necessary condition since it makes each individual an equal according to their own voice. This suggests that to be able to hear what another has to say involves suffering through the process of understanding the various relations and factors that inform it (cf. Mei 2015; Kaplan 2003, 72). Discovering such meaning will inevitably involve having to struggle with understanding the testimony bearing witness to what is unfamiliar to its audience, or what Ricoeur captures under the task of argumentation.

Acting upon recognition of perlocutionary effects returns individual and collective reflection to the illocutionary level at which the norms and conventions initially circumscribing the event can be critically questioned and, if needed, revised. The process of doing so is what Ricoeur captures under the term "argumentation."[13]

In the example of walking away, argumentation would require consideration of what is being affirmed by the action. The illocutionary dimension brought out the conventional meanings of non-compliance, from which perlocutionary meanings diverged. Imaginatively, at this level, we noted the suffering that the agent experienced. Argumentation is in this sense a way of investigating possible meanings according to whatever information one finds salient. Perlocutions are like wagers to read beyond the illocutionary context and therefore see a distinctive meaning to the action that was not readily apparent and could possibly be significant in view of revising norms and conventions. Could walking away be saying something about the injustice of racial and economic relations in Los Angeles? Can the injustice of non-compliance at the level of illocution be signposted a more significant injustice by means of perlocution? Of course, another option is to read the action as unacceptable. The only way to form a judgment is through the process of argumentation, which will at the very least disclose issues for the pubic to consider that may have been hidden. In this way, the new beginning of an action has its fulfillment in its being heard and considered and which may alter the normative content of a polity.

3. TECHNOLOGY AND THE FORCE OF ACTION

Herein I argue that the risk of social media is the *privation* of public reasoning—that is, the deprivation of the defining context that makes public reasoning possible. At first glance, this claim seems quite paradoxical, given that social media is about the facilitation of communication through networks, or what is often referred to as the horizontal nature of technological media. Indeed, one function of technology is to increase the availability and accessibility of virtual spaces in which plural others can articulate their views and reveal themselves. This last point suggests that social media provides a distinct kind of mediation of communication that was not possible before. Indeed, the protocols of network technology allow for spaces that are no longer merely artifactual, or located in one space, but virtual, or not bounded by any one space in particular. Yet, that which cuts one way cuts another.

What lends an adaptability and seeming ubiquity to social media communication comes with a constraint on its remit, which is oddly a lack of constraints—that is, a lack of determination of how discussion should proceed. More precisely, as we have seen in instances of everyday speech, what determines the pragmatic conditions in which speech is understandable is the illocutionary dimension in which norms and conventions provide the background knowledge for making sense of what is being proposed and performed. This dimension only functions if both the speaker and the audience are aware of the appropriate norms that allow an utterance to make sense. In other words, success of understanding depends on the integrity and recognition of norms, rules, and conventions present within the illocutionary dimension. Determining how privation manifests then needs to examine first how social media breaks down this illocutionary context.

Let us recall that Arendt describes privation as the deprivation of things that are "essential to a truly human life" (Arendt 1958, 58). The ability to be seen and heard as well as being recognized objectively as a person or member of the polis are the chief essentials since without them the words and deeds of a person do not appear, as if she did not exist. With social media, privation involves the pretense of a public space because the conventions of what it means to engage in public discussion no longer have force. Recall here that force results when an illocution achieves some effect understood by convention; and so here, I am playing on this meaning. In order for any illocution to have force, the conventions themselves must have force (or determination) for us such that it makes sense to follow them. How, then, might social media cause this force to diminish?

I offer two points drawn mostly from self-reflection, though they have been informed to some degree by empirical research: (1) lack of clear rules for discussion, which, in turn, allows for (2) the predominance of expression instead of discussion.

It is worth bearing in mind that prior to 2016, the year when the moniker "post-truth politics" became prevalent, social media was largely seen to be amicable if not essential to the project of democracy. Social media is both a means to promote democratic ideals of equal voice and is an instance of democracy at work (Feenberg 2005; cf. Sen 2009, 335–337 and Galloway and Thacker 2007, 29–31). Yet let us recall the fragile nature of plurality; democracy, equality, respect, and freedom are ideals that are only sustained by the practice of public reasoning. My use of the term "practice" is deliberate since practices involve conventions about what make an activity "counts as" a practice with respect following procedures, instructions, and so on. For it is only then that the practitioners can recognize that they are involved and invested in the same activity with a recognizable end (cf. MacIntyre 1984, 187–190, 194). To say something counts as a practice is, of course, to say that it most likely possesses some illocutionary content. But more importantly, it is to say that the practitioner recognizes that she is involved in a practice and knows what is required of her. The epistemic bar is therefore relatively high.

In general, public discussion involves conventions, or etiquette, whose burden of satisfaction is raised when discussion falls upon serious matters, such as politics. Onora O'Neill therefore notes a requirement of the public use of reason:

> What is spoken or written cannot count as a public use of reason merely by the fact that it is noised or displayed or broadcast to the world at large. Communication has also to meet sufficient standards of rationality to be interpretable to audiences who share no other, rationally ungrounded, authorities. (O'Neill 1986, 531)

To see the risk of social media, it is not necessary to go so far as having a theory about what public reason is essentially; only that it involves conventions of what it means to offer an idea for public consideration, how this idea can be tested, and that there is a mutual commitment to these criteria (cf. Murray and Starr 2018, 213). Simply offering a view indiscriminately to a public audience is not sufficient, since we expect that as public discourse, it will respect the audience as a precondition of their being a group to whom a view needs to be or ought to be articulated (MacKenzie and Porter 2019, 15).

To be sure, with social media there are user agreements relating to conduct, but these rules do not specify the terms according to which users can understand what counts as a reasoned debate and what it means to act as an interlocutor in such debates.[14] So a post offering a view about a political event may have political content, but the norms of political debate are not readily apparent. In other words, there is nothing to indicate the political post is in fact enframed by norms of debate,[15] and so it can risk appearing as a view to which one can *merely react*. What do I mean by "merely react"?

If social media risks effacing the relevance of illocutionary norms for discussion, then this absence means that speech does not operate under felicitous conditions and so risks not being able to communicate what the speaker intends—that is, not only a specific point but more generally and more importantly the idea that discussion should proceed as a public form of reasoning. Privation is therefore the risk and effect of social media *when* speech and action no longer appear as candidates for reasoned discussion due to the absence of norms indicating that it counts as such. When this occurs, actions become mere *expressions* of attitudes or preferences that, because they are not reasons, do not invoke a process of argumentation to make sense of them.

If the demand of the effort to make sense of another's words or actions is not readily apparent, there is no onus for testing the validity of a statement. Statements, in this sense, lose their locutionary basis as events attesting to something purported to be the case. So, for example, the utterance "climate change has caused the glaciers in Glacier National Park to fall from 150 to 30 in nineteen years!" will not be seen primarily as proposing a view that, by means of public discussion, should undergo argumentation. Instead, it will be taken to express an attitude with which one approves or disapproves—the so-called echo chamber effect (Williams et al. 2015, 137). Locutions, then, are not candidates for reasoned scrutiny. So how, then, are they seen?

With social media, any absence of awareness of the background knowledge required for public discussion means there is an absence of the burden of making sense and giving sense. There is a truncated hermeneutic operative here since the absence of this burden tends to generate a determinism about the discussion in which utterances are taken to be mere expressions. In such cases, expression tends to generate expression as a response. While this response can be considered a *reactive attitude*, where approval or disapproval takes the form of an emotionally informed expression, it is important to clarify that reactive attitudes in themselves are not problematic. This is because approval and disapproval, as well as the range of emotional expression, can have cognitive or representational content (Macnamara 2013). So, it is not simply a case that we should "suspend our reactive attitude" (Downie 1966, 33) as much as it is for the speaker and listener to learn how to make the cognitive content as clear as possible. But clarity is not just an onus for the one expressing something; the general norm of discussion here is one of charity—charity in interpreting what has been said and being charitable when offering a thought for consideration. Expressing a critical view such as "That's stupid" versus "I have some concerns about what you seem to be implying" takes very different communicative paths. Absent the recognition of such norms, mere or bare expression (where representational content is not clear) risks turning discussion into a thread of reactions. The worst consequence is the eventual erosion of the idea of what reasoned debate is—namely, that reasoning is synonymous with bare expressive reactions (cf. Ferrara and Yang 2015).

Empirical research can be quite illuminating here. There is evidence to suggest that with social media, participants tend not to feel constrained by norms of debate (Porten-Chee 2015; Gerbaudo 2018, 750). In other words, in face-to-face discussion, there is a normative burden having to do with sanctions on views deemed problematic or unacceptable *and* whose viability derives from the epistemic burden of having to know why one thinks what one does in order to articulate it to others and respond to their replies. Saying and doing in the public sphere comes with an accountability, whereas in the private sphere, this accountability is vague to the extent that the privative effect of social media means reasoned replies are an extravagance above and beyond bare expression, or what has been termed the harm caused by social media's "mobocratic algorithms" (Ghonim 2016). In addition, such extravagance is reinforced by the one-dimensional format of social media, the minimal allowance for characters, and how reactions are an easy alternative to discussion (e.g., an emoticon or clicking "Like"). While the last two points are self-explanatory, let me comment a bit more on what I mean by "one-dimensional." There is less opportunity to pursue in discussion necessary digressions with another person as in face-to-face instances or in forms of discussion that require interlocutors to respect the role of others. By "necessary digressions," I mean those opportunities for clarifying questions in order to better understand the position of the interlocutor, points that are rarely explicitly or transparently given due to the interpretive requirement involved in understanding language.[16] While there is no guarantee of understanding in face-to-face discussion or even reasoned debate, bare expression tends to set the bar low for public discussion by aiming at approval or disapproval (Ilyas and Khushi 2012; Kruse et al. 2017; Kwak et al. 2018).

But let us ask if too much is being made of face-to-face interaction.[17] It seems what is key to public discussion is the appearance of plural individuals. This appearing, as Levinas has argued, relies heavily on seeing the face of the person and, as Ricoeur has argued, being able to hear them. Face-to-face interlocution involves filling the distance between the interlocutors with the possibility for respect and mutuality. Arendt (1968, 241) therefore notes the Kantian direction of such ideals. On the one hand, respect involves taking seriously the process of reasoning about an assertion. The ability to speak publicly requires that one recognizes oneself as being accountable to standards of fact and truth. On the other hand, mutuality begins by respecting the other person as a rational embodied being and is also an act of the critical imagination (for Kant, an act of representation) which is defined by the struggle to apprehend and even understand the experience of others. Perhaps both respect and mutuality are possible on social media, but only as a species of what is ontologically antecedent—that is, the phenomenon of face-to-face discussion. This is only to say that social media in its current form allows for these ideals insofar as the interlocutors themselves bring the expectations and standards of public reasoning to the discussion itself.

4. CONCLUSION

So what have I argued? Internal to Ricoeur scholarship, I have shown that Ricoeur's conception of political action has a speech act dimension, as opposed to merely a narrative one, that is consistent with his debt to Arendt. External to this scholarship, I have explained in what ways Ricoeur's development of speech act theory enables us to see just how relevant illocutionary and perlocutionary dimensions are for understanding political action and its critical remit. I focused in particular on the way illocutionary content is put under pressure by perlocutionary applications and, subsequently, how the new perspectives enabled by perlocution return to the illocutionary sphere for testing by means of argumentation. Turning to the mediating role of technology, I argued that there is a risk of privation in the way social media constructs its virtual space of discussion without reference to specific norms and rules for this discussion. This privation is essentially the absence of the illocutionary information that acts as a background and resource for making judgments about how to interpret and pursue a topic of discussion. In other words, loss of this information means that there is a risk that interlocutors will fail to note that a process of testing ideas may be needed. This risk is serious enough by itself, but by way of conclusion, I would like to note how this risk appears even more perilous when couched within speech act theoretic terms.

I mentioned the example of melting glaciers to illustrate how social media can result in forgetting how locutionary statements work. I want to give a little more detail to this by saying if locution involves the assertion of truth via political action, then loss of the ability to test for the validity of such assertions—that is, the tendency to regard locutions as bare expressions—results in the attitude that there can never be any objective or factual truth to locutions. Arendt notes this problem when commenting

> Factual truth . . . is always related to other people. . . . It is political by nature. . . . Facts inform opinions, and opinions, inspired by different interests and passions, can differ widely and still be legitimate as long as they respect factual truth. Freedom of opinion is a farce unless factual information is guaranteed and the facts themselves are not in dispute. (Arendt 1968, 238)

So the constraint on public discussion is not simply that we presuppose that what each one of us has to say has some basis in what is factual, but also that what we take to constitute the sphere of facticity is agreed—that is, not any particular facts but that facts refer to things that are putatively true for everyone. If locutions are seen as mere expressions of attitudes or preferences, then either the idea of what is factual disappears or it is distorted in such a way to mean "whatever agrees with an attitude or preference."

If perlocution involves effects relating to imaginative variations of an action and their application in order to understand something in a different way, then the loss of the illocutionary sphere means such variations have no way of being tested and can become ideological or pathological. Perlocutionary effects then can be used to affirm or deny any beliefs, and instead of disrupting the identity of a group or individual, they become iterations of ideas within a closed system of signs and acts.

Finally, I should admit that while I have expressed a significant reluctance about social media, I do want to emphasize that in seeing myself as a good hermeneut, what I have offered is an analysis of its risk (of privation) as it presents itself empirically in the current historical context. All technological media and instruments have a risk by virtue of their capacity to augment some function (Mei 2016). So, on this view, there is no need, as Heidegger might lament, of a god or gods to save us. Technological design and technological use require a practical wisdom even if this wisdom cannot account for every possible consequence and application.[18] Can social media be designed mindful of the requirements and subtleties of political action? Can we use social media responsibly despite whatever tendencies it has for bare expression and reaction? In one sense, concerns over content that is uploaded or live-streamed are raising this issue. But these are explicit instances. The use of public reason is all the more subtle, elusive, and yet fragile because it announces the uniqueness of each actor for serious consideration within the plurality of many others. While I have offered an account critical of social media, a more sympathetic understanding drawing on Ricoeur might involve novel and imaginative ways for the emergence of narrative and prospective identities. I see as a necessary condition of this possibility the recognition of and adherence to specific norms and virtues that can enframe or underwrite such mediation.

NOTES

1. Cf. Lewin's (2012) response to Kaplan (2003, 164–173).
2. Joy (2018) explores the more intricate conceptual and historical linkages.
3. Arendt commentators take the reference to speech acts to be explicit (e.g., Honig 1991, 1992; Dietz 2002; Herzog 2004; cf. Keenan 1994). An exception is Loidolt (2015) and Melaney (2006). The focus on narrative in Ricoeur scholarship tends to be the default position and based on statements made by Ricoeur (e.g. 1988, 146, 246, 1992, 58, 2000, 3, 2007, 76–79). See, for example, Abel (2012, 218–219), Hoskins (2013, 96–101), Loute (2012), Truc (2011), Kaplan (2003, 60), and McCarthy (2007). Dauenhauer (1998, 135n8) notes the interest in speech between Arendt and Ricoeur as coincidental. For notable exceptions to Ricoeur scholarship with respect to political action, see Fiona Tomkinson (2012) and Iris Brooke Gildea (2018) whose theses about metaphor as integral to identity sits closely to the idea of speech act as a semantic innovation of disrupture.
4. I leave this claim undeveloped in view of the many types of speech act categories—for example, promise, threat, command, request, and instruction. My intuition is that these various types of acts can be political or nonpolitical. When the former, they are acts of natality presented in the form of a promise, threat, command, and so on.
5. Action is distinct from narrative activity since it is announcing something as opposed to telling a story about oneself, though implicitly it may be doing so when placed within a broader context. Action, in other words, announces something specifically public that is to be considered for the public and by the public and not necessarily a story about oneself. It appears Arendt sees the ability to narrate as in some sense deriving, or better yet, being made possible from the founding act of natality. Ricoeur, due to his concerns for identity, tends to elide the significance of political action as a speech act. Honig (1992, 2016) argues that the performative aspect of narrative construction through action and speech is not grounded in personal identity in the sense of identity politics since the identity Arendt has in mind is always public-facing. On the priority of action as speech act, see Dietz (2002, 10) and Honig (1991, 99).
6. For a more detailed examination of how Anscombe's analysis has significant implications with respect to reason-giving, see Ceva and Radoilska (2018).
7. I mean both social and linguistic conventions. Debates on the topic of how illocutions achieve their effects are wide-ranging (Fogal et al. 2018). For the purposes of this chapter, I do not engage with theoretical nuances and instead focus on the

pragmatics of political action; and so as unsatisfying as this may sound, I take successful illocution to be a mixture of social and linguistic conventions as well as actor intention.

8. For a fuller account, see Mei (2018). Also to note: I use the terms "effects" and "meanings" interchangeably, noting that with the latter I am employing a hermeneutical sense of the term "meaning" in which something is existentially meaningful and not the analytical sense in which meaning is specific to propositional statements.

9. A speech act that involves a double meaning is referred to as having "explicit indirection." For example, "Can you please pass the salt" is both a question and a request (Lepore and Stone 2018).

10. I am applying the pragmatics of speech act theory politically. Ricoeur comments on this specifically when reflecting on the transitions he makes with respect to theories of action in view of ethics. Moving from the formal level of analysis of action (e.g., objective structures, semantics, and syntax) toward its pragmatic dimension as being spoken (i.e., speech act theory and action as text), Ricoeur notes that he left unclear the face of impotence that goes with this ability [to say and read action], owing not only to those infirmities of every sort that may affect the human body as the organ of action, but also to the interference of outside powers capable of diminishing, hindering, or preventing our use of our abilities. In this regard, the sufferings that humans inflict on one another weigh heavily in the scale of our abilities and inabilities within the sphere of ordinary action. (Ricoeur 1997, xl as cited in Joy 2018, 115; cf. Ricoeur 2016, 94)

11. See also Ron Wilkins: "And so you feel a sense of alienation. You're culturally disoriented. You don't have a sense of identity in terms of who you really are" (Peralta 2008, 12:02–12:10).

12. In my example, I assume that the audience is not the police officer since perlocutionary effects tend to be those that outrun the original event of communication. However, this is not a necessary condition for these effects to take place. The police officer in question could have a revelation as a result of witnessing the black man walking away.

13. See also Ricoeur (1992, 174–177, 1981, 212); cf. Kaplan (2003, 70–74). For analysis of the intricacies of this process in relation to action and metaphor, see Taylor (2012), Savage (2013, 2015), and Mei (2015).

14. The website *ChangeAView* (changeaview.com) is a notable exception.

15. A further point could be developed here in which social media lacks the kind authority and reference to reputation that are involved in public reasoning (cf. Murray and Starr 2018, 230).

16. These digressions can also be considered the "dynamic force" in conversation when new information is introduced to carve up the space of reasoning. See Yalcin (2018, 403–405).

17. I leave aside the question of whether we ought to take face-to-face communication as the paradigm of language vis-à-vis Derrida on speech and Ricoeur on discourse. Suffice it to say here that because I take Arendt's notion of natality as foundational to Ricoeur's conception of political action, Arendt seems to have in mind face-to-face discussion in the public space.

18. For more on this in relation to practice and tragedy, see Carney (2018). The relation to technology may also require a tragic wisdom that has a longer historical memory.

REFERENCES

Abel, Olivier. 2012. "The Unsurpassable Dissensus: The Ethics of Forgiveness in Paul Ricoeur's Work." In *From Ricoeur to Action: The Socio-political Significance of Paul Ricoeur's Thinking*, edited by Todd S. Mei and David Lewin, 211–228. London: Continuum.
Arendt, Hannah. 1958. *The Human Condition*. Chicago: University of Chicago Press.
Arendt, Hannah. 1968. *Between Past and Future: Eight Exercises in Political Thought*. New York: Viking Press.
Arendt, Hannah. 1994. *Essays in Understanding 1930–1954: Formation, Exile, and Totalitarianism*. Edited by Jerome Kohn. New York: Schocken Books.
Aristotle. 2002. *Nicomachean Ethics*. Translated by Christopher Rowe. Oxford: Oxford University Press.
Austin, John L. 1975. *How to Do Things with Words*. Cambridge, MA: Harvard University Press.
Carney, Eoin. 2018. "Technologies in Practice: Paul Ricoeur and the Hermeneutics of Technique." PhD dissertation, University of Dundee.
Ceva, Emanuela, and Lubomira Radoilska. 2018. "Responsibility for Reason-Giving: The Case of Individual Tainted Reasoning in Systemic Corruption." *Ethical Theory and Moral Practice* 21: 789–809.
Dauenhauer, Bernard. 1998. *Paul Ricoeur: The Promise and Risk of Politics*. Lanham: Rowman & Littlefield.
Dietz, Mary. 2002. *Turning Operations: Feminism, Arendt, and Politics*. New York: Routledge.
Downie, Robin S. 1966. "Objective and Reactive Attitudes." *Analysis* 27 (2): 33–39.
Feenberg, Andrew. 2005. "Critical Theory of Technology: An Overview." *Tailoring Biotechnologies* 1 (1): 47–64.
Ferrara, Emilio, and Zeyao Yang. 2015. "Measuring Emotional Contagion in Social Media." *PLoS One* 10 (11). https://doi.org/10.1371/journal.pone.0142390.
Fogal, Daniel, Daniel W. Harris, and Matt Moss. 2018. "New Work on Speech Acts: The Contemporary Theoretical Landscape." In *New Work on Speech Acts*, edited by Daniel Fogal, Daniel W. Harris, and Matt Moss, 1–39. Oxford: Oxford University Press.
Galloway, Alexander R., and Eugene Thacker. 2007. *The Exploit: A Theory of Networks*. Minneapolis: University of Minneapolis Press.
Gerbaudo, Paolo. 2018. "Social Media and Populism: An Elective Affinity?" *Media, Culture & Society* 40 (5): 745–753.
Ghonim, Wael. 2016. Interview by Nathan Gardels. "We Have a Duty to Use Our Social Media Power to Speak the Truth." *Huffington Post*, October 26, 2016.

Gildea, Iris B. 2018. "A Poetics of the Self: Ricœur's Philosophy of the Will and Living Metaphor as Creative Praxis." *Études Ricoeuriennes/Ricoeur Studies* 9 (2): 90–103.
Herzog, Annabel. 2004. "Hannah Arendt's Concept of Responsibility." *Studies in Social and Political Thought* 10: 39–56.
Honig, Bonnie. 1991. "Declarations of Independence: Arendt and Derrida on the Problem of Founding a Republic." *American Political Science Review* 85 (1): 97–113.
Honig, Bonnie. 1992. "Toward an Agonistic Feminism: Hannah Arendt and the Politics of Identity." In *Feminists Theorize the Political*, edited by Judith Butler and Joan W. Scott, 215–235. New York: Routledge.
Hoskins, Gregory. 2013. "The Capacity to Judge and the Contours of a Political Theory of Judgment." In *Paul Ricoeur and the Task of the Political*, edited by Greg S. Johnson and Dan R. Stiver, 85–104. Lanham: Lexington Books.
Ilyas, Sanaa, and Qamar Khushi. 2012. "Facebook Status Updates: A Speech Act Analysis." *Academic Research International* 3 (2): 500–507.
Joy, Morny. 2018. "Ricoeur's Affirmation of Life in This World and His Journey to Ethics." *Études Ricoeuriennes/Ricoeur Studies* 9 (2): 104–123.
Kaplan, David. 2003. *Ricoeur's Critical Theory*. Albany: SUNY Press.
Keenan, Alan. 1994. "Promises, Promises: The Abyss of Freedom and the Loss of the Political in the Work of Hannah Arendt." *Political Theory* 22 (2): 297–322.
Kruse, Lisa M., Dawn R. Norris, and Jonathan R. Flinchum. 2017. "Social Media as a Public Sphere? Politics on Social Media." *The Sociological Quarterly* 59: 62–84.
Kwak, Nojin, Daniel S. Lane, Brian E. Weeks, Dam H. Kim, Slgi S. Lee, and Sarah Bachleda. 2018. "Perceptions of Social Media for Politics: Testing the Slacktivism Hypothesis." *Human Communication Research* 44 (2): 197–221.
Lepore, Ernie, and Matthew Stone. 2018. "Explicit Indirection." In *New Work on Speech Acts*, edited by Daniel Fogal, Daniel W. Harris, and Matt Moss, 165–184. Oxford: Oxford University Press.
Lewin, David. 2012. "Ricoeur and the Capability of Modern Technology." In *From Ricoeur to Action: The Socio-Political Significance of Paul Ricoeur's Thinking*, edited by Todd S. Mei and David Lewin, 54–71. London: Continuum.
Loader, Brian D., and Dan Mercea. 2011. "Networking Democracy?: Social Media Innovations and Participatory Politics." *Information, Communication & Society* 14 (6): 757–769.
Loidolt, Sophie. 2015. "Hannah Arendt's Conception of Actualized Plurality." In *The Phenomenology of Sociality: Discovering the "We,"* edited by Thomas Szanto and Dermot Moran, 42–55. London: Routledge.
Loute, Alain. 2012. "Identité Narrative Collective et Critique Sociale." *Études Ricoeuriennes/Ricoeur Studies* 3 (1): 53–66.
MacIntyre, Alasdair. 1984. *After Virtue*. Notre Dame: Notre Dame University Press.
MacKenzie, Iain, and Robert Porter. 2019. "Totalizing Institutions, Critique and Resistance." *Contemporary Political Theory*. https://doi.org/10.1057/s41296-019-00336-w.

Macnamara, Coleen. 2013. "Reactive Attitudes as Communicative Entities." *Philosophy and Phenomenological Research* 90 (3). https://doi.org/10.1111/phpr.12075.
McCarthy, Joan. 2007. *Dennett and Ricoeur on the Narrative Self.* New York: Humanity Books.
Mei, Todd. 2015. "Convictions and Justification." In *Paul Ricoeur in the Age of Hermeneutical Reason: Poetics, Praxis, and Critique*, edited by Roger W. H. Savage, 99–122. Lanham: Lexington Books.
Mei, Todd. 2016. "Heidegger in the Machine: The Difference Between *techne* and *mechane*." *Continental Philosophy Review* 49 (3): 267–292.
Mei, Todd. 2018. "The Poetics of Meaningful Work: An Analogy to Speech Acts." *Philosophy and Social Criticism* 45 (1): 3–26.
Melaney, William D. 2006. "Arendt's Revision of Praxis: On Plurality and Narrative Experience." In *Analecta Husserliana XC*, edited by Anna-Teresa Tymieniecka, 465–79. Dordrecht: Springer.
Murray, Sarah E., and William B. Starr. 2018. "Force and Conversational States." In *New Work on Speech Acts*, edited by Daniel Fogal, Daniel W. Harris, and Matt Moss, 202–236. Oxford: Oxford University Press.
O'Neill, Onora. 1986. "The Public Use of Reason." *Political Theory* 14 (4): 523–551.
Peralta, Stacy. 2008. *Crips and Bloods: Made in America*. Film. Balance Vector Productions.
Porten-Chee, Pablo, and Christiane Eilders. 2015. "Spiral of Silence Online: How Online Communication Affects Opinion Climate Perception and Opinion Expression Regarding the Climate Change Debate." *Studies in Communication Sciences* 58 (1): 143–150.
Ricoeur, Paul. 1965a. "The Political Paradox." Translated by Charles A. Kelbley. In *History and Truth*, 247–270. Evanston: Northwestern University Press.
Ricoeur, Paul. 1965b. "Work and the Word." Translated by Charles A. Kelbley. In *History and Truth*, 197–219. Evanston: Northwestern University Press.
Ricoeur, Paul. 1981. "The Model of the Text: Meaningful Action Considered as Text." Translated by John B. Thompson. In *Hermeneutics and the Human Sciences*, edited by John B. Thompson, 197–221. Cambridge: Cambridge University Press.
Ricoeur, Paul. 1988. *Time and Narrative: Volume 3*. Translated by Kathleen Blamey and David Pellauer. Chicago: University of Chicago Press.
Ricoeur, Paul. 1992. *Oneself as Another*. Translated by Kathleen Blamey. Chicago: University of Chicago Press.
Ricoeur, Paul. 1997. "A Response by Paul Ricoeur." In *Paul Ricoeur and Narrative: Context and Contestation*, edited by Morny Joy, xxxix–xliv. Calgary: University of Calgary Press.
Ricoeur, Paul. 2000. *The Just*. Translated by David Pallauer. Chicago: University of Chicago Press.
Ricoeur, Paul. 2005. *The Course of Recognition*. Translated by David Pellauer. Cambridge, MA: Harvard University Press.
Ricoeur, Paul. 2007. *Reflections on the Just*. Translated by David Pellauer. Chicago: University of Chicago Press.

Ricoeur, Paul. 2016. "The Problem of the Will and Philosophical Discourse." In *Philosophical Anthropology*, edited by Johann Michel and Jérôme Porée, 72–86. Cambridge: Polity.

Savage, Roger W. H. 2013. "Colonialist Ruinations and the Logic of Hope." In *Paul Ricoeur and the Task of Political Philosophy*, edited by Greg S. Johnson and Dan R. Stiver, 201–220. Lanham: Lexington Books.

Savage, Roger W. H. 2015. "The Wager of Imagination and the Logic of Hope." In *Paul Ricoeur in the Age of Hermeneutical Reason: Poetics, Praxis, and Critiquem*, edited by Roger W. H. Savage, 139–158. Lanham: Lexington Books.

Taylor, George H. 2012. "Ricoeur Versus Ricoeur? Between the Universal and the Contextual." In *From Ricoeur to Action: The Socio-Political Significance of Paul Ricoeur's Thinking*, edited by Todd S. Mei and David Lewin, 136–154. London: Continuum.

Taylor, George H. 2015. "Prospective Political Identity." In *Paul Ricoeur in the Age of Hermeneutical Reason: Poetics, Praxis, and Critiquem*, edited by Roger W. H. Savage, 123–137. Lanham: Lexington Books.

Tomkinson, Fiona. 2012. "From Metaphor to the Life-World: Ricoeur's Metaphoric Subjectivity." In *From Ricoeur to Action: The Socio-Political Significance of Paul Ricoeur's Thinking*, edited by Todd S. Mei and David Lewin, 33–53. London: Continuum.

Trottier, Daniel, and Christian Fuchs. 2015. "Theorising Social Media, Politics and the State: An Introduction." In *Social Media, Politics and the State: Protests, Revolutions, Riots, Crime and Policing in the Age of Facebook, Twitter and YouTube*, 3–38. New York: Routledge.

Truc, Gérôme. 2011. "Narrative Identity Against Biographical Illusion: The Shift in Sociology from Bourdieu to Ricoeur." *Études Ricoeuriennes/Ricoeur Studies* 2 (1): 150–167.

Williams, Hywel T. P., James R. McMurray, Tim Kurz, and F. Hugo Lambert. 2015. "Network Analysis Reveals Open Forums and Echo Chambers in Social Media Discussions of Climate Change." *Global Environmental Change* 32: 126–138.

Yalcin, Seth. 2018. "Expressivism by Force." In *New Work on Speech Acts*, edited by Daniel Fogal, Daniel W. Harris, and Matt Moss, 400–429. Oxford: Oxford University Press.

Chapter 13

Software and Metaphors
The Hermeneutical Dimensions of Software Development
Eric Chown and Fernando Nascimento

Traditionally, the most recognizable characteristics of software engineering are related to logical constructs, algorithmic thinking, and mathematical frameworks for optimizations. They are indeed at the heart of the practice. Nevertheless, an exclusive focus on these traits misses a fundamental aspect of the practice relating to its capacity to reinvent and reshape how humans act and experience the world. And this creative dimension has an intrinsically hermeneutical—interpretive—dimension.

In the first part of this chapter, we propose a phenomenological approach that describes how signs of this interpretive dimension can be recognized in the evolution of software engineering techniques. This evolutionary path may be an effect of an increasing awareness that critical aspects of software engineering cannot be reduced to a set of analytical and algorithmically driven procedures. We suggest that personal computational devices fostered an increasing realization that software development requires a hermeneutical approach to achieve its ultimate social goal of enhancing human experiences through technological artifacts.

The second part of this chapter is devoted to a philosophical consideration of the findings of the initial descriptive approach and explores the hermeneutical task required to interpret real-life contexts that will be affected by the software being developed. The software developer is the hermeneut, and the considered objects are not only user stories but also the actions and experiences humans can perform or have in the world, either actually or potentially. We argue that Paul Ricoeur's hermeneutic philosophy, particularly how he develops the concept of metaphor as semantic innovation, sheds important light to the task of software development and its implications in the world.

We conclude by suggesting how the productive imagination intrinsic to metaphorical creation plays a significant role in software development. When one considers software products as redescriptions of interpreted action, the imagination of the developer is essential in creating a new synthesis between the actions interpreted and their technological viewpoint or horizon. We also offer a few critical comments regarding the limitations of the metaphorical process as applied to software development, and some brief considerations of the social implications of the software development practice in light of its hermeneutical dimension.

1. SOFTWARE REQUIREMENTS DEVELOPMENT

The practice of software engineering, including the requirements specification phase, was derived from other engineering areas, such as civil engineering, and was initially implemented within a "waterfall" model in which each phase of the software development was rigidly self-contained and defined to provide clear unidirectional interfaces from software requirements, to software design, coding, testing, and finally deployment. The objective was to understand exactly what the customer wanted.

An important shift on the way software requirements were gathered and analyzed took place in the transition from the waterfall models in the 1970s to lifecycles using the spiral model developed by Barry Boehm in the mid-1980s (O'Regan 2016, 234), and its variations such as rapid application development (RAD) and Joint Application Development (JAD). In the spiral model, the requirements were developed throughout the software development lifecycle to rectify problems caused by changing requirements during the course of long duration projects. There was a perception of the volatile nature of requirements and the fragility of their resulting definitions, which were both driven by and impacted by human interactions.

Wiegers and Beatty propose a broader view of requirements gathering: "The features the users present as their 'wants' don't equate to the functionality they need to perform their tasks with the new product" (2013, 102). But still the requirements specification seems merely to focus on mirroring non-digital solutions to users' needs within a software environment. This approach to requirements specifications was prevalent in the software development for companies that hired software houses to create automation systems to streamline the company's activities. However, when the industry shifted to the development of applications for mobile devices, this paradigm was affected since, unlike the former situation, it was not possible for software engineers to interview the intended users of their products. Finally, it is important to note that one of the mantras of software engineering is to make a distinction between the "what" captured at the stage of requirements definition and the "how" defined in the design phase (with some possible variations to this creed). During the requirements phase, software engineers are compelled to generate a forced dichotomy between their world and the customers' world. Even if the underlying objective of deferring implementation details to the design phase is noble, many times it inhibits the developer's full engagement in the creative process of building a new and different experience born from a fusion of the developer's and the users' perspectives.

"The Agile Manifesto" (Beck 2001) published in the early 2000s marks the stage where a significant number of software developers began to adopt Agile methods such as Scrum or Extreme Programming. Among several fundamental changes, Agile methodologies propose the organization of the software lifecycle into small steps that create products based on users' stories. The emphasis on users' stories not only facilitates the engagement of the users with the development process but also points toward the linguistic shift from the previous CMM-like (Paulk 1995) "shall" statements to a model closer to a narrative that makes more explicit the interpretive nature of the requirements gathering phase. Nevertheless, in many cases, this methodology is still too focused on solely addressing the specifics of users' wants or needs rather than engaging in a creative joint exercise between users and software developers.

The advent of requirements analysis for mobile applications has brought yet another level of approximation between requirements definition and end users. As Nagappan points out (Nagappan and Shihab 2016), there are a number of recent studies for extracting software requirements from user reviews made through app stores like the Google App Store and Apple's iTunes Store. This feedback loop certainly once existed in desktop application development, but thanks to the level of immediacy that mobile devices provide for the deployment of new functionality, it is greatly enhanced and speeded up over what came before. Nevertheless, regardless of the level of automation of revision analysis, there is always a need for intense dialogue with users to understand and interpret how software should be adjusted to meet certain user demands.

We want to close these brief remarks on the requirements analysis by pointing to some features that will help us in the transition to a more philosophical inquiry into the issue. First, the evolution of requirements engineering was marked by a growing recognition of the importance of interpreting software requirements. Second, agile models introduced a narrative approach that provides a conceptual and linguistic framework more conscious of the interpretive dimension of the requirements definition. Third, formal practices in requirements specification still emphasize user intentions and thoughts. Fourth, there is an intentionally sharp separation between discussions about what users want and how these desires are realized by software. Finally, the growing area of mobile applications escapes from the traditional setting of corporate software development with clearly defined stakeholders.

2. THE NEED OF HERMENEUTICS IN SOFTWARE DEVELOPMENT

The history of software development has evolved from the early days of the advent of commercial computing (1945–1956) through the proliferation of personal computers (1972–1985) (Ceruzzi 2003) to the current trend in which significant focus is placed on mobile devices such as tablets and especially smartphones that are increasingly gaining space in our daily lives. The number of software applications for mobile devices is burgeoning exponentially each year (Nagappan and Shihab 2016), a trend driven by a demand for software artifacts that is growing ever broader and more personal.

As mobile software becomes more important and more competitive, the interpretive task becomes increasingly more critical. Building software that can be integrated into people's activities becomes, even more, the objective and challenge of software engineering. There are mighty economic forces behind software development, but these forces are seeking applications that first and foremost are capable of attracting and retaining users. Developers everywhere, especially those creating their own startups, are trying to understand how software can replace or be integrated into users' and societies' daily activities. Because of the fierce competition in this space, it is incumbent upon developers to match users' needs and desires as closely as possible.

This need to build bridges between digital artifacts and their users is the characteristic of contemporary software applications. In *Transcoding the Digital*, Marianne Van Den Boomen makes an interesting analysis of how metaphors are crucial in the way digital artifacts become used and understood. For her, "the practice of digital code exchange can only be articulated, perceived, and conceived when it is translated into metaphors" (Van Den Boomen 2014, 12–15). From the new media studies perspective, she makes a comprehensive study of the various levels at which metaphors mediate digital experiences from the level of user interface (UIs)s to the social discourse that has important repercussions for culture and society. We suggest that Ricoeur's theory of metaphor provides a rich conceptual framework to investigate and understand such mediations.

The philosophy of design has also been attentive to the hermeneutic dimension of design and technological creations. Veemas et al. describe the forms in which the bridges between design and users may fail mainly when the design framing gets disconnected from the original goal, or when the reframing proposed by designers are not acceptable for their users (Pieter et al. 2015). Jahnke, in "Revisiting Design as a Hermeneutic Practice," also explores the idea that meaning creation is fundamental to the innovation process, and it plays a dialectical role with problem solving (Jahnke 2012, 141).

In 2006, Cook et al. published an insightful article in which they highlighted that "for a software system to do something useful in the real world, its input, algorithms, and outputs must be assigned additional meanings by relating them to the real world." In a sense even closer to the conceptual framework of this article, Reijers and Coeckelbergh (2016) offer a refined perspective on how Ricoeur's narrative hermeneutics and particularly the concept of triple mimesis that we will discuss further later on in the chapter can be applied to make sense of how technologies and human experiences co-shape each other through narrative technologies.

Following Reijers and Coeckelbergh's diagnostic that current philosophy of technology approaches largely neglected the linguistics and social dimensions of new technologies, we suggest that Ricoeur's hermeneutic philosophy provides particularly relevant insights into the innovative character of software development exactly by its potential to integrate and apply our interpretative experience of the world through language through the semantic innovation created by live metaphors.

In the following section, we will investigate what aspects of a theory of metaphor, as proposed by Paul Ricoeur's hermeneutics, may foster a deeper understanding of the interpretative nature of software development. We start with a brief indication of Ricoeur's possible contributions to the philosophy of technology, and then we move on to an analysis of key characteristics of his concept of metaphors as paradigms of semantic innovation.

3. RICOEUR'S HERMENEUTICS

As David Kaplan noted in "Paul Ricoeur and the Philosophy of Technology" (2006), Ricoeur did not turn his attention to the "empirical dimensions of technology." However, Kaplan does list five themes in Ricoeur's work that are relevant to the philosophy of technology. The first is his "hermeneutic philosophy." Kaplan clarifies that for Ricoeur "hermeneutics is geared toward the interpretation of human works and other symbolically-mediated endeavors." Kaplan also highlights that during his philosophical works, Ricoeur moved from a hermeneutics that was centered on signs and symbols to a hermeneutics centered on texts and narratives, and its applications to the interpretation of human actions (Kaplan 2014, 37, 49; Ricoeur 1973).

Ricoeur agrees in many aspects with Gadamer's famous proposal of interpretation as a "fusion of horizons" (Gadamer 1975, 269–271). Understanding is the attempted fusion of the interpreter's and the writer's (or actor's) worldviews. In his comments on the differences between the Gadamerian fusion of horizon and the Ricoeurian metaphorical interpretation, George Taylor highlights the productive tension between writer and interpreter: "For Ricoeur, by contrast [with Gadamer's perspective], understanding is more a product of a tension between sameness and difference, an attempt to find similarity across, and despite, difference" (Taylor 2011, 114).

Among its several significant implications, the idea of a tensional fusion of horizons clarifies the continued vitality of interpretation in the sense that as one advances in understanding, the horizon shifts forward. It also highlights the importance of what Ricoeur calls "appropriation." "'Appropriation,'" writes Ricoeur, "means 'to make one's own' what was initially 'alien.' According to the intention of the word, the aim of all hermeneutics is to struggle against cultural distance and historical alienation. Interpretation brings together, equalizes, renders contemporary and similar" (Ricoeur 1981, 183).

As we keep at our sight the concepts of interpretation as a tensional fusion of horizons and appropriation that will play an essential role in the application of Ricoeur's theory to software development, we want to turn our attention to metaphors within Ricoeur's hermeneutic framework for three main reasons. First, as it provides an extremely insightful conceptual background to explore the application of hermeneutics to software development since, as we already suggested, metaphors are a common component of software applications. Second, because Ricoeur explored key hermeneutic concepts through metaphors in several books and articles. For instance, in "Metaphor and the Central Problem of Hermeneutics" (1973a) Ricoeur discusses the key dialectics of explanation and understanding and sense and reference based on the assumption that metaphors can be seen as works in miniature. And third, because for Ricoeur, metaphors are paradigms of semantic innovation (Ricoeur 1973b), which is a feature that interests us in relation to the innovations brought up by software development.

4. METAPHORS AS A MODEL OF SEMANTIC INNOVATION

When Aristotle proposed the importance of metaphors both in his *Poetics* and *Rhetoric*, he had in mind several dimensions of the metaphorical utterance, which he expressed by defining metaphors as the work of the genius who has an eye for resemblances (Aristotle 1995, 1459a 5–10). *To metaphorize well is to "see-as,"* to "perceive the similarity in dissimilars," remarked Aristotle. It is a creative process that builds approximations between an apparent irreducible distance. Along with Ricoeur, we want to retain Aristotle's insight of metaphors as "seeing-as" for our subsequent analysis of metaphors in software development.

Nevertheless, as we advance in the history of the concept, this initial productive aspect of metaphors fades away, and its reputation becomes questionable, as the concept of metaphor was to become reduced during the Middle Ages and early Modern times to a simple figure of speech that turns metaphors into minor stylistics and ornamental artifact (Ricoeur 1978).

The resurgence of interest in the study of metaphor (Gibbs 2008) is marked by contemporary studies from two main perspectives in linguistics: semiotics and semantics. While the former deals with metaphor in its isolated manifestation as a sign within a closed linguistic system, the latter, represented by scholars with whom Ricoeur dialogues extensively such as I. A. Richards, Max Black, and Monroe Beardsley, considers metaphor in the context of sentence and discourse.

This semantic perspective is what most interests us in this chapter as it offers the conditions for an analysis of metaphor as a rich cognitive process for the creation of new meanings. Ricoeur suggests that this semantic innovation of metaphor is born of the exercise of an intense cognitive activity that he calls, from the Kantian epistemology, "creative imagination." The metaphor, therefore, can *redescribe our experience in the world*, opening a fissure in the predetermined structures of established linguistic lexicons.

A key aspect in the contemporary semantic theory of metaphor, exemplified by Max Black's (1977a) work, is that the semantic innovation of metaphors moves from the level of the word to the level of semantic predication. "Seeing as" is no longer merely seeing one word through or as another, but is an exploration of the possibilities of the semantic field to give meaning to reality by varying the usual predication associated with the words. Ricoeur calls this sentence-level perspective a tension theory as opposed to a word-level approach that relates to a substitution theory (Ricoeur 1978). This new way of seeing allows for an exploration of how a new concept can be created out of the effort to reconcile the apparent paradox engendered by an *impertinent predication* (Cohen 1967).

Within this movement from a mere substitution of words to a discourse-level impertinent predication, Max Black suggests that the secondary subject of a metaphor, what I. A. Richards calls the "vehicle," is better understood as a system rather than just as an individual thing. For example, in "society is a sea," Black considers not only "sea" as an individual isolated object, but also as a system of ideas that are linked to "sea." He calls this semantic system the "implicative complex" (Black 1977b). It is the set of "associated implications" that could be attributed to the secondary subject. What Black pointed out about the secondary subject of a metaphor is also true for the primary subject. We do not restrict our understanding of metaphors based on the isolated lexical entry for "society," but the meaning of the metaphors implies and impacts the implicative complex attached to "society."

It is within this deeper framework of *impertinent predications involving implicative complexes* that Ricoeur explores the hermeneutical potential of metaphors. Such enriched hermeneutic analysis of metaphors that moves away from an impoverished rhetorical use of metaphors within a substitution theory to a much denser and richer system of tensional and productive associations is what grants metaphors its potential of *semantic innovation*, of proposing new meanings to worldly experiences and actions, or as Beardsley calls it, "emergent meanings."

Ricoeur also explores the creative and imaginative aspects of metaphors through the dialectics of sense (what is said) and reference (that about which something is said). The reference is the projection of language toward the world of meanings. For Ricoeur, metaphors open up a new form of metaphorical reference, as they do not point toward something that is already there, but they point to some new possible meanings. They suggest an expansion of one's current set of meanings, as they "redescribe reality inaccessible to direct description" (1984, xi).

Both in *Metaphor and the Central Problem of Hermeneutics* (1972) and in *Time and Narrative* vol. 1 (1978), Ricoeur approximates this creative power of metaphor to the idea of mimesis, which provides us an interesting perspective on metaphors through the analysis of the moments of the narrative triple mimesis: prefiguration, configuration, and refiguration. The implicative complexes that are part of a metaphorical statement exist in a prefigured world of meaning characterized by their established literal meanings. A new metaphor is a twist of the prefigured meaning in a configured impertinent predication that references an emergent meaning. Readers of the metaphor refigure their world through the tensional fusion of horizons between the metaphorical reference suggested by the metaphor and their prefigured literal meanings involved in the metaphor. We could call this process the metaphorical cycle.

5. SOFTWARE DEVELOPMENT AS A METAPHORICAL PROCESS

We now test the applicability of Ricoeur's theory of metaphor with three examples of software development, each more applicable than the last. We start with short message service (SMS) messages, an engineering-driven implementation of an existing metaphor. We then transition to social networks where the metaphor is more strongly embedded in the applications. Finally, we present what may be the ultimate example of software as a metaphorical process—the development of the touch UI software layer.

SMS and Text Messages

A strength of mobile applications is that they provide the ability to break spatial and temporal boundaries. The fact that mobile devices are still referred to as "phones" gives primacy to their role as phones that are no longer tethered to a single location. Given the phone's job as a communication device, it is interesting then to consider other forms of communication that have spun out of the mobile industry, such as texting. The calling apps on cell phones removed space as a constraint on communication, but not time. The prefigurations of text messages are simple—telegrams and letters. Both remove spatial and temporal restrictions on communicating; one need not be close either in time or space to communicate with someone else. That these things are such strong antecedents to texting can even be seen in the iconography of many text applications, which are meant to resemble envelopes for letters.

Texting dates back to 1984 and a system called "SMS" (Milian 2009). The original version of SMS was shaped almost entirely by a combination of the pre-existing forms of communication that removed temporal and spatial restrictions, along with the technological limitations present at the time (e.g., up to 160 characters per message). Thus, SMS did not start out as a reimagination of communication, rather it began as a fairly straightforward marriage of technology to an existing metaphor. The original metaphor "SMS messages are letters" is a kind of second-order metaphor in that a letter is a metaphor itself for communication, thus it does not resonate as easily as a metaphor like "selecting is touching," which we will examine in a subsequent example.

The strong ties between the original formulation of SMS and technology hampered its adoption. To a user without expert knowledge, the 160-character limit is completely unnatural with respect to familiar forms of communication. Then there were the problems of actually creating a message. Early phones did not feature full keyboards, and typing in a message was laborious. In other words, the technology itself was a barrier for nontechnical users that prevented them from fully acquiring the metaphor.

Over time many of the original technological limits on SMS and its usage were dropped—keyboards were introduced, autocomplete technology was added, texts could be longer than 160 characters, messages could eventually include audio and video, and so on, a kind of constant reformulation of the technology. As this happened, developers worked to incorporate the changes into their systems in order to improve the software and make it more useful. These changes naturally broke the original, clumsy, metaphor. An important step in this process being when SMS transitioned to "instant messaging." While letters remove one kind of temporal restriction on communication, they are anything but instant. By emphasizing this speed, developers were able to reformulate the original metaphor.

The new metaphor became "texting is conversation." This is an improvement over the original metaphor in many regards, not the least of which is that it is no longer a second-order metaphor. Unlike a letter, a text message thread between two people online at the same time can be in real-time the way an actual conversation is. There are even a number of ways that texting improves the original experience. Aside from removing spatial and temporal constraints, text conversations reduce attentional demands, allowing, for example, multi-tasking (or even having multiple conversations simultaneously). Further, these conversations leave a concrete trace; they do not have to be remembered—they can be scrolled back to. Finally, they do not require speaking, meaning they can happen even in places where speaking out loud would bother other people nearby. Thus, texting has expanded the meaning of what a conversation is.

By 2010, SMS had become the most widely used application in the world (Gayomali 2012; Erickson 2012), ubiquitous to the point where calling mobile devices "phones" does not really make sense anymore. The conversation metaphor has become embedded so deeply that many teenagers text in favor of talking even when they are in the same location.

Social Networks

Most of the money made by developers in the modern app economy comes from advertising. To succeed then, the software development of an app must gain and hold a user's attention. Software developers have to figure out how to keep people engaged (online) as much as possible. From this point of view, this type of software is attempting to supplant previous entertainment industries such as movies and television. Indeed television itself is built on the metaphor "a window to the world" (Hutchinson 1950). Apps as a whole are effectively trying to replace television as that window. The prefiguration, as embodied by the television industry, was for each industry itself to produce as much content as possible. Of course, creating such content is expensive and risky—after spending large amounts of money developing a show television networks would essentially hope that the show was what the consumers wanted, and many such shows were dropped after only airing a few episodes. Thus, the temporal dimension of the old model was long—the time from creating a new show, to writing it, producing it, and finally airing it could take a year or more.

With social networks, the content comes from the users. This solves two major problems simultaneously—the app developers are no longer responsible for the creation of all of the content required to keep users engaged, and since users are creating the content themselves, it is much more likely to be interesting and engaging to them. In other words, the fusion of horizons comes directly from making the users part of the process. For this to work, these networks need as many people on and connected to each other as possible to keep the supply of content high. Thus, it is in the best interest of developers to design their software to encourage people to connect with each other even if they are not in relationships in the real world. Thus, in this reconfiguration, users of social networks are encouraged to form relationships with people they may never have met. "Conversations" with celebrities and people on the other side of the world are equally possible. Further, the temporal dimension of content creation has been shrunk to almost nothing—content is consumed almost immediately after it is created and the consequences of poor content are minimal.

Social networks go beyond the simple "screen as entertainment" metaphor though they are built around the semantic context of relationships and thus refine the general "entertainment" metaphor to be more specifically about relationships, be they friendships, love, or other. These relationships are expressed somewhat differently in different networks, but in general, social networks attempt to capture the spirit of checking in on, or catching up with, friends. In the real world, friends interact by meeting—which requires being in the same place at the same time. When people meet with friends, relatives, and acquaintances, they inevitably spend time trading information, or "catching up," telling each other what has happened in their respective lives since the last time that they met. In a social network, a "post" is a metaphor for this process. A post, be it a tweet, a picture, or a video, informs the world what the poster has been doing. In person, our friends respond to our news, creating a conversation and providing support. In a social network, the metaphor turns these conversations into comments on the post, and support is expressed and collected in the form of "likes." Thus, the full metaphor is that "catching up with friends is posting, commenting and liking."

A given social network contains "feeds," which again build on the foundation provided by the basic software list construct. The simplest organizational structure for such feeds, and one most easily implemented with a list structure, is a list ordered by time—a timeline. And indeed, many social networks are organized around this concept. However, a raw timeline would break the catching up metaphor. Instead, UIs are generally conceived as a combination of time and conversations. The result is a system that is more coherent and meaningful. This additional structure is a necessity because the technology removed the previous restrictions on conversations that they are fixed in time and space. The extra structure provides the user with a mental model that allows them to make sense of what otherwise might be a string of seemingly unrelated conversational snippets and to overcome the semantic distance from the concept of conversations face-to-face. It also allows them the illusion of seeing people they are connected with as friends. Thus, the refiguration of the relationship concept under this metaphor breaks temporal and spatial boundaries, and even stretches the definition of relationship to include people that we may never have met.

The "friends catching up" metaphor is a useful starting place for social networks. It is familiar to everyone, and with the concept of a feed organized by a combination of time, people, and conversations, the metaphor is easily grasped by even nontechnical users—they are not required to learn specialized commands or to use complicated interfaces. Meanwhile, social networks have expanded the semantic field of what it means to catch up. On the one hand, people can connect to an arbitrary number of other people whether they directly know them or not. On the other hand, the information that they choose to share with others can be carefully curated if desired. The intimacy of close sharing with a few friends is shifted to public-facing sharing with as many other people as wanted. This weakening of what it means to be a friend directly benefits the social networks as it helps them solve their content creation problem.

Touch User Interfaces

Pre-2007 the cell phone industry was dominated by companies that used hardware keyboards, this was the prefiguration that faced Apple as it developed the iPhone. Up until that time, the ubiquitous UI for cell phones was a combination of a list, for example, a list of phone numbers in an address book or a list of applications to choose from, and the arrow keys on the hardware keyboard. Users used the arrows to navigate to the item in the list that they wanted, and then selected their choice by pressing the "enter" key. Cell phone keyboards themselves were essentially miniaturized versions of the keyboards used on personal computers. This had the advantage that many people already had experience and a mental model of how keyboards worked. As recounted by Steve Jobs in his keynote introducing the iPhone (Wright 2015), Apple saw this model as a problem that needed to be solved. The problem Jobs focused on was that keyboards are fixed and unchangeable. Modern software, by contrast, is dynamic. But changing software on a cell phone was hamstrung by the existing mapping to the keyboard. If new software functionality was added, for example, there might not be any buttons available that weren't already mapped to other things.

Once Apple decided to get rid of the hardware keyboard, its developers had many problems to solve, but certainly among the most important was simply how to let the users choose what to do next. The operating system that Apple was developing as part of this process, iOS, is built as a series of layers. At the top of this hierarchy is where user interactions are handled. In early versions of iOS, this layer was called Cocoa Touch. Apple's engineers had to create this layer as a bridge for app developers who want their applications to be interactive. Since the developers writing Cocoa Touch could not anticipate every possible type of interaction that developers might want to support, they had to create a system that was as general as possible. The simplest version of such an interface is to create a button for every possible interaction. For example, with a fixed number of applications, an operating system could assign a button to each one. An improvement over this system, and the one employed by cell phones up until then, was to create a list display and to let the user navigate by using arrow keys. Once the user had navigated to the appropriate action, they could then take an action by hitting the appropriate selection key. The metaphor employed by this system, such as it was, was "actions are taken by typing." Such a system is easy to implement as it can be repurposed for applications or actions within an application. Once the keyboard was taken away, it required a new way of thinking about this layer.

The obvious thing for the engineers to do was to look at other models. At the time, Apple had reinvented itself largely on the back of iPods, a successful product that did not rely on keyboards, one that instead used a more intuitive interface based on a wheel. The scroll wheel could be used either for the choosing applications or selections within an application, which is exactly what it did on the iPod (note: Apple actually made mockups of iPhones using this design) but the wheel had the major drawback that it interfered with the screen. If the screen were to be a central feature of the phone, then that left the engineers with little choice but to use the screen itself for the user interface. As it happens, around that time an enabling technology had become available—the capacitive touch screen. With such a screen, it was possible to detect a user's fingers. Thus, the question became how to use this information to allow users to make choices. An obvious option, and one Apple explored, was to make a virtual scroll wheel—Apple even patented this technology. In other words, in several different ways, Apple could have easily relied on the prefiguration of either cell phones at the time, or its own iPod. Instead, it went user-centric and looked at how people choose things in the world. The answer is that they use their hands for grabbing, and their fingers for pointing. As Steve Jobs said in the keynote introducing the iPhone, "We're going to use the best pointing device in the world. We're going to use a pointing device that we're all born with—we're born with ten of them. We're gonna use our fingers" (AppleInsider 2012). Putting the technology together with how users do things in the world meant that a user could "point" at something on the screen and the capacitive screen could identify where on the screen they touched it. The reason this layer is called Cocoa Touch is because the act of touching the screen became a driver for the entire interface.

Previously hardware buttons were slaved to specific actions. With touch, it was possible to create virtual buttons for an unlimited number of actions. This necessitated a new development paradigm in Cocoa Touch to let app developers put their own buttons on the screen, and also to let those buttons be connected to arbitrary actions. In software terms, this paradigm is called event-driven programming. Cocoa Touch monitors the screen and can process what the user does and where on the screen the user touches, for how long, and so on. App developers, in turn, can grab the results of this processing and use it to determine the user's intent. Thus, Steve Jobs's idea that software should be dynamic and upgradeable was now realizable because the touch interface was general purpose enough to support any sort of user interactions that app developers might dream up. Apple's choice to look to the user for inspiration, instead of engineers, led to a new software paradigm and, not coincidentally, the most successful consumer product in history.

The tenor of the touch metaphor is "selecting" or "choosing," and the vehicle is "touching." The power of this metaphor comes from the fact that it almost isn't a metaphor at all. As Jobs noted in his keynote, fingers are the greatest selection devices ever created. The richness of the metaphor does not come from, nor does it require, a major cognitive reorganization on the part of the receiver, it comes from the ways that the technology was able to employ it. All of the layers in iOS, and now Android, work to hide the user from all of the details necessary to make this happen. In terms of touch UI, this led to a series of predications starting with "selection is touching" and including others like "grabbing is touching and holding," "swiping is moving" and others. This metaphor is so powerful and successful that stories of such children trying to touch and swipe other devices like televisions are legion as users increasingly have become unaware of the layers of technology required to make it work.

6. CONCLUSIONS

Let us conclude with a summary that highlights a few points of the ways in which metaphors play an important role in software development, and how the recognition of its hermeneutical dimension clarifies some social implications of the practice.

The exponential growth and sophistication of software platforms and the ubiquity of mobile devices increase the importance of a well-thought-out software requirements process. Some of the traditional approaches that focused on an accurate and direct reproduction of tasks and activities are not suitable to cope with this current scenario. The interpretive dimension of software development has become more apparent, and the need to understand not only user stories but also the real-life experiences that will be affected by the software is increasingly a crucial component of successful software development that holds important social implications. To think about this hermeneutical dimension of software development, Ricoeur offers a model in which interpretation can be applied to both stories and actions. His approach also highlights the dialectic between the world of the action and the world of technology, and it invites the developer to use productive imagination to create metaphorical impertinent predictions rather than merely a recreation of an experience. It encourages moving the criteria for good software development from mere adherence to what already exists in the world to a metaphorical interpretation of those realities that create new meaning and sense to human experiences. The hermeneutic approach also points to the openness of metaphors implemented in software that may be applied and used in ways that were not contemplated by their developers.

We suggested that Ricoeur's theory of metaphors enables us to look at the software creation process through the lens of the semantic innovation model created by metaphors. Ricoeur suggests that the impertinent predication of metaphors gives rise to a new meaning through the semantic clash between two implicative complexes. Through the examples of text messages, social networks, and touch interfaces, we sought to show how software development involves the creation of an impertinent predication between the implicative complexes of the technological artifact (the world of the developer) and those with which users have contact outside the technological environment (the world of the user). Software developers create an impertinent predication that redescribes reality (chatting is sending typed texts to device) and invites users to an interpretive operation that fuses their "pre-technological" horizons with the redescription offered by the application (catching up is sharing / reading posts in a social network). The software requirements specification should propose a fruitful fusion of the technological horizon of the software developer with the user's horizon, bridging the semantic gap. The software itself is the realization of the reinterpretation of the world that emerges from a meaning-producing encounter between these two worldviews. Software developers are bridging the gap between the technical and the user's horizon. So, a pair of cryptographic keys becomes the "signature" that proves one's identity. One could expand the list of examples of historical, and cultural, "alienations" with which interpretation must struggle proposed by Ricoeur to include technological alienation, or the distance between the meaning of software algorithms and data structures and their relationships to the real-world entities and actions. To overcome such distance in the process of creating new software, developers must create a metaphorical interpretation of reality.

Software development, especially for mobile devices applications, must redescribe our experience in the world. Similar to the way a Shakespearean metaphor would invite us to see time as a beggar, a mobile application invites us to see a click on an icon as a token of friendship and appreciation. Mobile applications may redescribe trust and excellence with the number of stars users attribute to a ride service or a virtual store. Software developers need to see software entities as physical or social ones that will be redescribed by the use of the software. Most applications reflect the tension between similarity and difference in metaphors. Emoticons are representations of emotions, but they are impoverished by the lack of immediate contact between interlocutors. This metaphorical dimension of software development requires a hermeneutical skill from developers and designers, and this is as important as the technical skills in the creation of reliable and efficient software.

The intricacy of the software requirements definition phase becomes clear as one realizes that software metaphors are embedded into a multidimensional relationship of implicative complexes. So, a metaphor of a letter brings with it a set of associations (message, communication, post office, mail carriers, delivery time, and privacy) that may enhance or complicate its understanding. Developers also need to consider how implicative complexes evoked by metaphors are different among their potential users. For applications that are increasingly targeting a global audience, it may mean that there may be a need to "translate" metaphors if implicative complexes lead to undesired interpretations within a given cultural sphere.

Although software development is indeed metaphorical, there are, however, crucial differences in how Shakespeare's "time is a beggar" and the technological "selecting is touching" metaphors operate. Such differences are linked to distinct underlying intentionalities and materialities of poetry and technology that we should briefly explore as we conclude this article and open the discussion to further explorations.

The concept of live metaphors proposed by Ricoeur helps us contrast the life cycle of poetic and technological metaphors. Live metaphors have not yet been transformed into another polysemic entry of the linguistic lexicon and maintain their character of openness to new interpretations and, therefore, of revealing new meanings through fusions of different horizons. Metaphors have a life cycle. They are proposed as an eruption of new meanings, they are born intriguing and defying as impertinent predications. Over time, a metaphor's use may get sedimented into a new closed meaning that becomes a semantic overloading of an existing or new entry of the vocabulary. So, the computer "mouse" became a new meaning of the word originally used mainly to describe a small rodent. Mouse as a metaphor rested in peace the day it became a new entry in our collective vocabulary (or Wikipedia for that matter). "Time is a beggar" remains defiant, open to new interpretations, no single new word or polysemic entry for time or beggar in the vocabulary captures the meaning pointed to by this live metaphor.

While poetic metaphors seek to maintain their vitality by opening themselves to different interpretations, technological metaphors, despite using the same underlying semantic innovation mechanism, are intended to be incorporated into the common lexicon in the easiest and fastest way possible. Technological metaphors want to speed up the metaphorical life cycle so that the possibilities of meaning converge to their intended functionality as fast as possible. The impertinent predication must quickly become the usual sense of predication, "touching" shall become a synonym of "selecting." The materiality of the software as metaphor and the rapid spread of software applications through application stores facilitates this cycle by pushing the use of these metaphors into large communities of users and communication media rapidly and extensively.

The metaphorical cycle that we suggested inspired by Ricoeur's triple mimesis seems to be a privileged way of reflecting on the social implications of software metaphors. The impact of metaphors configured in the form of applications for mobile devices is particularly expressive due to the immense speed of their dissemination through application stores and increasingly efficient over-the-air installation mechanisms. The metaphor of number of stars for service evaluation configured in ride applications quickly reconfigures the semantic horizon of users who start to see the stars as the quality of the service provider. The metaphorical cycle makes it evident how software metaphors have important social implications when redescribing horizons of meaning at the personal, interpersonal, and institutional levels.

As we contemplate the intrinsic hermeneutic nature of the software development process through software metaphors, we deepen our understanding of its intricacies and challenges beyond the logical and coding skills, and unveil the profound social implications, which is increasingly playing a prominent role in our contemporary societies.

REFERENCES

AppleInsider. 2012. "Virtual Scroll Wheel Patent Shows Alternate iOS Input Method." *AppleInsider*, August 7, 2012. https://appleinsider.com/articles/12/08/07/virtual_scroll_wheel_patent_shows_alternate_ios_input_method.

Aristotle. 1995. *Poetics*. Translated by Demetrius Longinus. Cambridge, MA: Harvard University Press.

Beck. 2001. "The Agile Manifesto. Agile Alliance." http://agilemanifesto.org/.

Black, Max. 1977a. "More About Metaphor." *Dialectica*. https://doi.org/10.1111/j.1746-8361.1977.tb01296.x.

Black, Max. 1977b. "More About Metaphor." *Dialectica* 31 (34): 431–457.

Ceruzzi, Paul E. 2003. *A History of Modern Computing*. Cambridge, MA: MIT Press.

Coeckelbergh, Mark, and Wessel Reijers. 2016. "Narrative Technologies: A Philosophical Investigation of the Narrative Capacities of Technologies by Using Ricoeur's Narrative Theory." *Human Studies* 39 (3): 325–346.

Cook, Stephen, Rachel Harrison, Meir M. Lehman, and Paul Wernick. 2006. "Evolution in Software Systems: Foundations of the SPE Classification Scheme." *Journal of Software Maintenance and Evolution: Research and Practice* 18 (1): 1–35.

Erickson, Christine. 2012. "A Brief History of Text Messaging." *Mashable*, September 21, 2012. http://mashable.com/2012/09/21/text-Messaging-History.

Gadamer, Hans Georg. 1975. *Truth and Method*. Translated and edited by Garrett Barden and John Cumming. New York: Seabury Press.

Galvis Carreño, Laura V., and Kristina Winbladh. 2013. "Analysis of User Comments: An Approach for Software Requirements Evolution." In *Proceedings of the 2013 International Conference on Software Engineering*, 582–591. ICSE '13. Piscataway, NJ, USA: IEEE Press.

Gayomali, Chris. 2012. "The Text Message Turns 20: A Brief History of SMS." *The Weekly Law Reports*, December 3, 2012. https://theweek.com/articles/469869/text-message-turns-20-brief-history-sms.

Gibbs, Raymond W., Jr. 2008. *The Cambridge Handbook of Metaphor and Thought*. Cambridge: Cambridge University Press.

Hutchinson, Thomas H. 1950. *Here Is Television, Your Window to the World*. New York: Hastings House.

Iacob, Claudia, and Rachel Harrison. 2013. "Retrieving and Analyzing Mobile Apps Feature Requests from Online Reviews." In *2013 10th Working Conference on Mining Software Repositories (MSR)*, 41–44.

Jahnke, Marcus. 2012. "Revisiting Design as a Hermeneutic Practice: An Investigation of Paul Ricoeur's Critical Hermeneutics." *Design Issues* 28 (2): 30–40.

Kaplan, David. 2014. *Readings in the Philosophy of Technology* (2nd ed., Vol. 37). Lanham: Rowman & Littlefield Publishers.

Kaplan, David M. 2006. "Paul Ricoeur and the Philosophy of Technology." *Journal of French and Francophone Philosophy* 16 (1/2): 42–56.

Leffingwell, Dean. 2010. *Agile Software Requirements: Lean Requirements Practices for Teams, Programs, and the Enterprise*. Boston: Addison-Wesley Professional.

Milian, Mark. 2009. "Why Text Messages Are Limited to 160 Characters." *The Los Angeles Times*, May 3, 2009. https://latimesblogs.latimes.com/technology/2009/05/invented-text-messaging.html.

Nagappan, Meiyappan, and Emad Shihab. 2016. "Future Trends in Software Engineering Research for Mobile Apps." In *2016 IEEE 23rd International Conference on Software Analysis, Evolution, and Reengineering (SANER)*. https://doi.org/10.1109/saner.2016.88.

O'Regan, Gerard. 2016. *Introduction to the History of Computing: A Computing History Primer*. Cham: Springer.

Paulk, Mark C. 1995. *The Capability Maturity Model: Guidelines for Improving the Software Process*. Boston: Addison-Wesley.

Pieter, Vermaas, Dorst Kees, and Thurgood Clementine. 2015. "Framing in Design: A Formal Analysis and Failure Modes." In *Proceedings of the 20th International Conference on Engineering Design (ICED15)*, July 27–30, 2015, Politecnico di Milano, Italy. Glasgow: The Design Society.

Ricoeur, Paul. 1973a. "Metaphor and the Central Problem of Hermeneutics." *Graduate Faculty Philosophy Journal* 3 (1): 42–58.

Ricoeur, Paul. 1973b. "Creativity in Language." *Philosophy Today* 17 (2): 97–111.

Ricoeur, Paul. 1973c. "The Model of the Text: Meaningful Action Considered as a Text." *New Literary History* 5 (1): 91–117.

Ricoeur, Paul. 1978. *The Rule of Metaphor: Multi-Disciplinary Studies of the Creation of Meaning in Language*. Translated by Robert Czerny. Toronto: University of Toronto Press.

Ricoeur, Paul. 1981. *Hermeneutics and the Human Sciences: Essays on Language, Action and Interpretation*. Edited by John B. Thompson. Cambridge: Cambridge University Press.

Ricoeur, Paul. 1984. *Time and Narrative: Volume 1*. Translated by Kathleen MacLaughlin and David Pellauer. Chicago: The University of Chicago Press.

Taylor, George H. 2011. "Understanding as Metaphoric, Not a Fusion of Horizons." In *Gadamer and Ricoeur: Critical Horizons for Contemporary Hermeneutics*, edited by Francis J. Mootz, III, and George H. Taylor, 104–118. London: Continuum.

Van Den Boomen, Marianne. 2014. *Transcoding the Digital: How Metaphors Matter in New Media*. Amsterdam: Institute of Network Cultures.

Wiegers, Karl Eugene, and Joy Beatty. 2013. *Software Requirements*. Redmond: Microsoft Press.

Wright, Mic. 2015. "The Original iPhone Announcement Annotated." *TNW*, September 9, 2015. https://thenextweb.com/apple/2015/09/09/genius-annotated-with-genius/.

Chapter 14

Narrating Artificial Intelligence
The Story of AlphaGo
Esther Keymolen

From self-driving cars to voice assistant Alexa, artificial intelligence (AI) is currently perceived as one of the most promising technologies that is driving innovation. While the field of AI has already been up and running for more than sixty-five years, recent breakthroughs in foundational techniques such as machine learning, the use of neural networks, and natural language processing have put it in the spotlight (Crawford and Whittaker 2016, 2). AI is defined as "the theory and development of computer systems able to perform tasks normally requiring human intelligence" (Jobin et al. 2019, 389). The European Commission refers to AI as "systems that display intelligent behaviour by analysing their environment and taking actions—with some degree of autonomy—to achieve specific goals." The commission further adds that "AI-based systems can be purely software-based, acting in the virtual world (e.g. voice assistants, image analysis software, search engines, speech and face recognition systems) or AI can be embedded in hardware devices (e.g. advanced robots, autonomous cars, drones or Internet of Things applications)."

What these definitions show is that AI can become an integral part of both the online and offline world, if such a distinction can still be made (The-Online-Initiative 2015). In other words, AI is on its way to becoming an ontological part of the objects and environments we live our lives with and in. And while there is debate on how and to what extend technologies mediate and shape human beings, their relations, and the world they inhabit, there does appear to be some rare consensus among philosophers that technologies are not merely "mute" instruments. Technology shapes, it matters, and it deserves our attention. This inherently makes AI an utterly important research subject for philosophers of technology.

One important prerequisite for such an endeavor is that philosophers have some kind of access to the technology. Located after the so-called empirical turn (Achterhuis 2001), many approaches such as postphenomenology (Ihde 1990), critical constructivism (Feenberg 2009, 2017), and Actor-Network Theory (Latour 1993) have been geared to investigating real-life examples and case studies. In particular, postphenomenology, with its emphasis on the first-person perspective and the mediating relations opening up everyday life, has increasingly manifested itself as a case-based, empirically informed approach (Rosenberger and Verbeek 2015; Aagaard et al. 2018). One of the basic assumptions in postphenomenology is that technologies "open up" the world for human beings in specific, concrete ways and that we can describe these different variations or stabilities in order to understand how human beings perceive of themselves, of others, and of the world around them.

1. HOW AI CHALLENGES POSTPHENOMENOLOGY

At first sight, one could think that this focus on the contextual, everyday use of technology ensures that postphenomenology and AI are a perfect fit. After all, if AI is well on its way to becoming a ubiquitous technology, then investigating how it mediates everyday-life practices seems not only highly relevant but also quite feasible. Nevertheless, AI developments impose certain challenges on the postphenomenological approach which need careful consideration.

First of all, as postphenomenology predominantly focuses on how technologies mediate *everyday life*, the technologies that are central to such analyses are oftentimes mundane objects or consumer products. However, not many AI applications already made it into consumer products and the few that exist are not yet highly sophisticated. Oftentimes they are still in a beta-phase, have high error rates, and only possess limited functionality. This obviously does not make them unworthy of investigation. We can still learn a lot from these first-generation commercial AI applications. However, they might also blindside us, as they provide only a limited and maybe even discarded image of what AI can bring about.

This immediately brings us to the second challenge, as the fact that everyday-life AI is not yet highly advanced does not mean that the progress and developments that are currently being made behind the scenes do not warrant philosophical reflection. In other words, don't we run the risk of ending up with an outdated and toothless philosophy of technology if we only focus on the everyday context of technologies-in-use?

Thirdly, there is the so-called black box problem (Pasquale 2015). Even if we would be able to gain access to relevant AI applications, that does not imply that we also understand what we perceive. This can have different causes. Sometimes, the AI application is shielded for proprietary reasons. For instance, when the companies behind the applications do not want to provide too much information, it could negatively impact their market position. It might also be that the AI application is opaque due to inherent, technical complexity. Moreover, in general, AI applications operate in intricate networks of human and nonhuman actors. Behind a sleek interface reside software, hardware, databases, servers, AI companies, advertisement companies, intelligent services, hackers, and so on.

Considering these challenges, it becomes clear that if postphenomenology already in an early stage wants to engage with these upcoming technologies, new ways of gaining access to these technological practices are needed.

It is important to note that postphenomenology is not a static approach and that the scholars in the field certainly cannot be accused of conservatism. Doing justice to its pragmatism roots, there definitely is room to explore and reinterpret the postphenomenological toolbox as it has been initiated by founding father Don Ihde (1990). Over the past couple of years, several researchers have come up with new methodological means to further develop the postphenomenological approach.

Some scholars have combined postphenomenology with Actor-Network Theory to better address the socio-technical context in which technology mediates everyday life (Rosenberger 2018; Keymolen 2017). Others have incorporated elements of the philosophy of Foucault to bring the idea of power more to the foreground (Verbeek 2011). Another very promising direction that currently is being explored is the intensification of the empirical approach, for instance, by building on ethnography (Kudina and Verbeek 2019) and investigating the relation between postphenomenology and existing empirical methods (Aagaard et al. 2018).

While some of the listed challenges connected to investigating AI might be mitigated by turning to one of these approaches—for example, ANT might be a fruitful way for investigating the networks in which AI applications are embedded and empirical research in the form of focus groups or in-depth interviews with experts might be a way of bridging the knowledge gap, I want to take the opportunity to explore another avenue.

This chapter will aim at broadening the scope of what "empirically informed" generally comes down to in postphenomenology, by investigating to what extent Ricoeur's narrative discourse might be useful to interpret and gain access to new technological developments (Ricoeur 1991). Instead of focusing on the first-person experience of everyday life, mediated by technological artifacts, it will take, what Ricoeur would refer to as, a necessary "detour" to explore the stories we tell about AI and how, in these stories, we imagine AI to become part of our everyday life.

This narrative approach is not meant to replace the more "traditional" postphenomenological analysis or the empirical approaches described earlier. Rather, it should be seen as a new addition to the continuous expanding methodological toolbox of postphenomenology. In particular, it could serve well as a strategy to engage with the increasing number of stories that surface on the development of AI. AI is presented to us in movies, TV series, books, performances, and—as will be the case in this chapter—documentaries. These stories generally fall out of the scope of the other approaches as they are not perceived as "empirical," "phenomenological," or "focused on everyday-life." This chapter will show that engaging in a hermeneutical interpretation of these stories is however not only valuable in itself but can also better equip us to undertake a postphenomenological analysis.

Central to this chapter will be the documentary "AlphaGo" (Kohs 2017), which narrates one of the greatest breakthroughs in the field of AI. AlphaGo is the name of an algorithm that learned how to play the very complex and ancient Chinese board game Go and was developed by DeepMind, an AI company later acquired by Google. By engaging with this story, we should become able to get a better understanding of state-of-the-art AI and read ourselves into a context where we interact with its transformative power, before it is even part of everyday life (Kaplan 2006). Moreover, as the documentary's goal is to share information on AI in an accessible manner, it might prove to be a good way to open at least some parts of the AI black box.

This chapter will first provide an overview of some of the key concepts of Ricoeur's narrative discourse. This will be followed by a short methodology section and a synopsis of the documentary. Next, the actual analysis of the documentary will take place, by making use of Ricoeur's key concepts and highlighting the relevance for a postphenomenological analysis. The chapter will end with a concluding section, summarizing the most important outcomes.

2. RICOEUR'S NARRATIVE DISCOURSE

One key starting point in Ricoeur's work—which it actually shares with postphenomenology—is that human life cannot be directly known or understood, but that it is always mediated. And for Ricoeur, it is particularly mediated through *language*. However, as spoken language can be short-lived, Ricoeur predominantly turns to written texts or analogue forms that capture language, such as, in our case, a documentary. By focusing on "captured" language instead of spoken language, how the text becomes interpreted—its meaning—can become detached from the intentions of the actor who initiated the words.

Although Ricoeur did not come up with a systematic, general theory of interpretation, in his work, he did develop a variety of fruitful concepts and techniques that can enable a rich, hermeneutic interpretation of texts. Particularly, his narrative discourse that focuses on the understanding of actions appearing in a certain place and timespan mediated by texts (both actions and texts understood in a broad sense) is relevant to the analysis of AlphaGo.

Key Concepts

How the reader reads herself into the texts, and interprets its structure and genre can all be seen as a form of "appropriation" (Pellauer 2016). By making the text "ours," we are not just interpreting text, we are trying to make sense of the world (*referentiality*), to relate to others (*communicability*), and to get to know ourselves (*self-understanding*) (Ricoeur 1991, 27). This subjectivity also entails that there is no such thing as a finite interpretation. Interpretations need to be checked and contrasted with other interpretations and over time, interpretations might lose their explanatory force, making room for new ones (also see: Gadamer 1972).

A Story

"A story," Ricoeur (1981, 239) claims,

> describes a sequence of actions and experiences of a certain number of characters, whether real or imaginary. These characters are represented in situations which change or to the changes of which they react. These changes, in turn reveal hidden aspects of the situation and the characters, give rise to a new predicament which calls for thought or action or both. The response to this predicament brings the story to its conclusion.

What "makes" a story is however not the mere succession of events leading to an acceptable conclusion. A story is just as much about the expectations one has while engaging with texts (both as a narrator and reader), actively bringing together bits and pieces to make a meaningful whole. A story is characterized by an active coming together of a chronological and a non-chronological dimension. The former, Ricoeur refers to as the *episodic dimension*, the latter is the *configurational dimension*. Both in *the art of narrating a story* and in *the art of following a story* these dimensions play an important role and actually are intrinsically intertwined (Ricoeur 1981, 241). After all, just as even the simplest narrative is more than merely the succeeding episodes leading up to a conclusion, demanding some act of interpretation, of finding a closure, also constructing meaningful totalities out of a story cannot be done if one would completely abolish its narrative structure.

This two-dimensional distinction actually resonates—to a certain extent—with the idea of multistability, one of the core, postphenomenological concepts that refers to the ontological openness of artifacts to different interpretations and their ability to furnish different relations. Just as a story is more than its chapters, scenes, or episodes, a technology is also more than the sum of its screws and bolts, of its software and hardware. Just as people have to read themselves into the text to interpret and give meaning, people co-shape the mediating relations with their devices. Just as a text can have a different meaning to different people, so can technology.

Genre

It goes without saying that not all stories are the same: a theater piece is not the same as a book, a comic is different from a novel. Also, a documentary as AlphaGo is situated in a specific genre domain. It is nonfiction, which generally is understood as "representing reality." However, that it refers to reality does not make it a mere copy of reality either. Every nonfiction is also in a certain way fiction. The documentary in itself is an audio-visual artifact adhering to independent rules and structures which are inherent to the genre (therefore, fiction); however, it is simultaneously also referring to something outside of itself, to actions taking place in the real world (therefore, nonfiction). In other words, both fiction and nonfiction as *productive imaginations and representations* prescribe a new reading of reality, making the life-world appear under specific circumstances.

The fact that we are able to pinpoint the genre of a text is because it is embedded in a tradition that is located between *innovation* and *sedimentation*. We recognize characteristics that allow us to categorize a piece of work as a documentary, thriller, or tragedy. However, this is never an exhaustive description. It always remains a singular work, which allows for new relations with the tradition (Ricoeur 1991, 23–24). It is the variation between the two poles of innovation and sedimentation that in the case of the documentary will result in a productive representation.

Emplotment

A plot is a key part of a story. First, it "serves to make *one* story out of the multiple incidents" (Ricoeur 1991, 21). Second, as a plot also gathers all kinds of heterogeneous components, from actors who perform actions to those who suffer them, from well-planned interactions to unintended consequences, it is both "concordant and discordant" in nature (Ricoeur 1991, 21). Finally, the plot mediates the episodic and configurational dimension of the story. It is where the competition between succession and configuration becomes explicit.

Because we are, through culture and tradition, familiar with different types of plots, we can also learn from them. We can relate a plot to certain virtues or "forms of excellence" (23). Ricoeur speaks of "phronetic understanding" to emphasize its practical character (23). It is not so much about understanding the techniques or rules, which a plot follows. Rather, the significance of a text lies in the "intersection of the world of the text and the world of the reader" (26). It relates more to lived experience than to theoretical knowledge.

It is not merely in the text, but also in the reader, in the act of reading that the *emplotment* takes place. It is where hermeneutics and phenomenology come together and "where narrative and life can be reconciled with one another" (26), as "reading is itself already a way of living in the fictive universe of the work." Or in other words, "stories are recounted but they are also *lived in the mode of the imaginary*" (27).

> A text is not something closed in upon itself, it is the projection of a new universe distinct from that in which we live. To appropriate a work through reading is to unfold the world horizon implicit in it which includes the actions, the characters and the events of the story told. As a result, the reader belongs at once to the work's horizon of experience in imagination and to that of his or her own real action. (Ricoeur 1991, 26)

3. RICOEUR AND ALPHAGO

AlphaGo is a 2017 documentary directed by Greg Kohs (Kohs 2017). In this movie, viewers follow the developments of the AI application AlphaGo: a computer program developed by the AI start-up DeepMind (owned by Google) to play the ancient board game called Go.

In the AI community, the board game Go has long been considered as one of the biggest challenges for AI to master. While the rules of the game might be rather straightforward—two players take turns placing either black or white stones on a 19 by 19 grid board with the aim to capture the opponent's stones or surround empty space to make points of territory (Silver and Hassabis 2016)—the game itself is "mind-bogglingly complex—far more complex than chess. A game of 150 moves (approximately average for a game of Go) can involve 10^{360} possible configurations" (Granter et al. 2017, 619). Playing Go, there are more possibilities "than there are atoms in the universe," the DeepMind blog emphasizes (Silver and Hassabis 2016). Because of this infinite possibilities, Go is—much more than chess—associated with creative and imaginative thinking, something that is thought of as being difficult, if not impossible, for AI to achieve (Bory 2019, 631).

In order to nevertheless successfully try to reduce this complexity, AlphaGo utilizes two deep neural networks: a policy network to provide the probability of certain moves and a so-called value network that delivers a position evaluation (Silver et al. 2017).

First, AlphaGo engages in supervised learning to train the policy network. It uses human expert games to learn how Go is played. Next, by reinforcement learning, the algorithm also plays millions of games against an instantiation of itself to improve its value network (Li and Du 2018, 78). Based on the combination of these trained networks, AlphaGo searches for the proper moves by anticipating possible future states (a so-called lookahead search) and predicts the most-likely successful state.

> AlphaGo looks ahead by playing out the remainder of the game in its imagination, many times over—a technique known as Monte-Carlo tree search. But unlike previous Monte-Carlo programs, AlphaGo uses deep neural networks to guide its search. During each simulated game, the policy network suggests intelligent moves to play, while the value network astutely evaluates the position that is reached. Finally, AlphaGo chooses the move that is most successful in simulation. (Silver and Hassabis 2016)

After training on 160,000 recorded games of professional human Go players and then playing more than 30 million games against versions of itself, AlphaGo was ready for its first match against a human professional player. In October 2015, we witness AlphaGo beat Fan Hui, a European Go champion with 5-0. After recovering from this defeat, Hui joints the DeepMind team to help further train AlphaGo. Subsequently, in March 2016, a new historical match is organized: AlphaGo will play Go against Lee Sedol, one of the world's best Go players. Now, this match, again consisting out of five games, receives wide media attention and ends—much to the surprise of both the Go and the AI community—in a victory for AlphaGo. The AI wins with 4-1.

The Art of Reading

For my analysis of the documentary, I proceeded as follows: I watched the documentary itself four times. The first time was one year ago. At that time, I did not anticipate writing about the documentary. This viewing was therefore, to a certain extent, pre-theoretical. Obviously, it is not possible to put all knowledge and experience between brackets, but while watching, I did not actively investigate how the content of the documentary resonated with my background knowledge. This first viewing of the documentary was, nevertheless, important as it "set the scene" for the analysis. It allowed me to familiarize myself with the overarching narrative and storyline, to gain a first understanding of the plot and the actors involved. While watching but also after the documentary ended, I remember experiencing feelings of excitement, wonder, empathy, and also resistance and sadness. The documentary touched upon topics that were meaningful to me and sparked my interest; it became clear that there was something at stake.

The second viewing had the explicit goal of identifying themes, ideas, starting points to engage in a hermeneutic analysis of the documentary. I made notes, paused the documentary frequently, played back some parts, and so on. The third viewing took the form of multiple close-reading sessions of the transcript of the documentary. In this way, I was able to further elaborate my preliminary analysis of the second viewing. The fourth viewing was again a total viewing of the documentary in order to loosely validate and further fine-tune my analysis that came out of the previous rounds.

In addition to the multiple rounds of viewing (and reading), I also engaged with critical texts, comments, and analyses of the documentary and academic literature on AI. This took place after the second viewing and enabled me to develop a richer interpretation of the documentary.

Emplotment in AlphaGo: Game as a Structuring Narrative

Following Ricoeur, we know that the way in which a story is structured, is not neutral. It pre-sorts our understanding and interpretation of a specific subject. The dominant narrative of the documentary is without any doubt shaped by the game Go itself. By using the game as an overarching structure, actors immediately are assigned specific and recognizable roles: they are participants, opponents, winners and losers, judges, commenters, and supporters. This game narrative also allows for a clear delineation of the action space by introducing specific rules, attributes, and expectations.

Choosing for the game as organizing structure of the documentary is telling on different levels. First of all, game and play are often seen as a key aspect of human life (Frissen et al. 2015). As Huizinga (1955, 173) in his classic book *Homo Ludens* states: human culture "arises in and as play, and never leaves it." Following Huizinga, play is characterized by the fact that it is an activity, human beings freely engage in. It is different from their ordinary, everyday life, as it is not spurred by the need to fulfill basic, everyday needs such as finding food or shelter. Rather, one engages in play for the sake of play itself. Taking place within a specific timeframe and on a specific location, with its own rules and order, people who play are absorbed by the game, they are completely dedicated to it, often experiencing profound feelings of excitement, joy, fear, and tension (also see Gadamer 1972). All in all, to play a game is to enter a "sacred sphere" (Huizinga 1955, 9).

Huizinga explains that many forms of living—from developing rituals, poetry, to religion and warfare—are built up on patterns of play. As play is such a fundamental form of social life, it is decisive that in the documentary, it is in the context of a game, human and AI "meet," and "get to know" each other. Following Huizinga, this element of play will, however, not just vanish when the game is over. It will put a long-lasting, decisive mark on future human-AI interactions and the societal practices that will be established within the AI domain.

While Huizinga emphasizes the distinction between everyday life and play—he refers to the "magical circle" people step into when playing—I do not perceive this distinction to be an impermeable hiatus. Rather, from a first-person perspective, we can be immersed in play, while also becoming aware of that "immersiveness." In other words, we can be in-play while also being aware of how this play relates to the world around us. It is our double, eccentric position (cf. Plessner 1975), our reflexivity that gives room to emplotment in a game. It is where the act of play and life come together, where the role we play and the life we live touch each other and a new light becomes shed on our perceptions, beliefs, and actions. At that moment, game truly mediates life. In the documentary, Frank Lantz, director of NYU Game Centre, explains:

> Go is putting you in a place, where you're always at the very farthest reaches of your capacity. There's a reason that people have been playing Go for thousands and thousands of years, right? It's not just that they want to understand Go. They want to understand what understanding is. And maybe that is truly what it means to be human. (Kohs 2016, 00:01:05)

Second, play and game are not merely something closely connected to human life, they are also inherently intertwined with technology, particularly with AI. In order to build AI, one needs input of the real world, so-called training data. The real world, however, is a messy place and as a result training data can be biased, multi-interpretable, or too complex to distill meaningful patterns from. This can make it hard to train AI in a successful way. By focusing on games, one takes a very specific, well-defined snippet of real life. On the one hand a game is rule-based, making it relatively easy to familiarize AI with the action-space in which it has to perform. On the other hand, a game like Go gives access to a specific form of human intelligence, bringing together logic and creativity, both seen as essential to—in the end—develop strong AI. The game as a specific representation of the world is therefore instrumental in building AI.

Third, there is also a commercial interest in choosing for the game narrative. As a game is distinct from everyday life, it provides a safe space to experience the power of AI. AI can "win" without immediately threating the familiar world as it does not *directly* have a tangible impact on that world. By presenting AI's progress to the world as a game, DeepMind is able to: "gain trust in its new products, turning the old imaginary of a new intelligence as a potential Frankenstein's monster into the narrative of togetherness, in which AI is an essential partner for the progress of humankind" (Bory 2019, 639).

4. SEDIMENTATION AND INNOVATION IN ALPHAGO

Now that we have established that "game" is the overarching structure of the documentary's narrative, we now move to the interpretation of the documentary itself. Following Ricoeur, we know that in the art of reading, we are always maneuvering between the poles of sedimentation and innovation. Our unique and meaningful understanding of a story cannot materialize without the background knowledge guiding our reading. In reading we are not just focused on the story itself but on the web of familiar concepts and traditions in which the story is embedded. Such a first important element of sedimentation can be found in the historical sequence of games between man and machine, in which the documentary implicitly is embedded. In 1997, there was the chess game between IBM's Deep Blue and the then world champion Kasparov. Deep Blue won. In 2011, IBM's Watson defeated two champions—Ken Jennings and Brad Rutter—in the questions-based game Jeopardy. And in 2016, there is the match of AlphaGo vs. Lee Sedol.

Man versus Machine Games

All these games between technology and human beings are to display, prove even, that in a domain, which is generally thought of as being dominated by human superiority, technology can outclass the human intellect. This antagonistic setting brings forth the frame of the human player and the technological player as rivals, opponents, and enemies. With every lost game, beliefs about human capacities, human uniqueness, and the place of the human in the world are put under pressure. With Ricoeur, we can wonder how much of this framing will impact our future perceptions and interactions with AI.

In addition to this dominant, antagonistic framing, also some other shared characteristics can be identified, particularly in the AlphaGo and the Deep Blue match. On both occasions, there was: a leading Western tech company involved, a non-Western world champion as human opponent, and both games attracted a lot of media attention (Bory 2019, 631). This media attention plays an important role in the AlphaGo documentary. The recurring press conferences and TV and internet commentators, which appear in the documentary serve as an important source of interpretation. As many people might still be new to the game of Go, the media literally tell them what they are looking at.

Research into the overlap and differences between Chinese and American coverage of the event indicates that overall, when AlphaGo was framed as human by, for instance, describing it by making use of human qualities such as "intuitive" or "creative," this often brought along questions on what it actually means to be human. Covering and co-experiencing the unexpected performance of the AI opened up a conceptual space to reflect on the porous boundaries between human and machine. This fundamental discussion took place both in the American and the Chinese media (Curran et al. 2019).

A difference between the American and Chinese media interpretation, however, was that the Chinese press made use of the "non-threat" frame much more than the American press. While the Chines press acknowledged the impact of AlphaGo's performance on the assumptions of what it means to be human and what it means to play Go, they overall did not assess this as a danger. The researchers suggest that this might be traced back to different cultural backgrounds—or macroperceptions—in postphenomenological terms. The Chinese public might be more familiar with Go, as the game is part of their education. They might also be more open to AI because of a less apocalyptic, religious tradition—such as Christianity in the Western world—and a different history of science fiction (Curran et al. 2019, 5–6).

This kind of findings emphasizes the importance of connecting macroperception and microperception in postphenomenological research in AI. While the dominant frame of man versus machine may significantly impact the mediating relations that can be developed with AI tools and services, the macro-context in which these relations are constructed, should not be overlooked. An overarching framing like the man versus machine perspective can still have different meanings and normative connotations due to different macroperceptions. While Ihde has always emphasized that microperception and macroperception are inherently intertwined, he does seem to suggest that they should be analyzed in different ways: postphenomenology is best suited to look into the microperception, whereas the macroperception asks for a cultural hermeneutics approach (Ihde 1990, 30). In interpreting the AlphaGo documentary, it becomes clear that a phenomenological analysis can highly benefit from such a hermeneutical approach. This raises the question if a postphenomenological analysis should not be executed in tandem with a hermeneutic analysis.

Next to the elements of sedimentation, there are also elements of *innovation* to be found in the documentary. While AlphaGo is embedded in this canonical sequence of man versus machine games, it also differs from its predecessors. The documentary reinterprets the man versus machine framing in three distinct, innovative ways (Bory 2019, 632). First, whereas Deep Blue was a hardware-based machine with enormous calculating force, AlphaGo was presented as a body-less, self-learning intelligence. Second, where the functioning of Deep Blue was hidden from its opponent and the public, the tech team behind AlphaGo went to great lengths to describe the way the AI operates. Third, where Deep Blue was predominantly described as mimicking human intelligence, AlphaGo is characterized as going "beyond its human guide," coming up with "something new, and creative, and different" (Kohs 2017, 00:50:48).

The Tragedy of AlphaGo

Another important element that helps us to interpret the documentary is its *tragic-like emplotment*. Resonating with the old Greek tragedies—as another form of sedimentation—we can recognize Lee Sedol as a tragic hero. He enters the game freely, cheered along by the whole Go community, together with the media who watch and comment on this historic game, much as if they take on the role of the choir in old Greek tragedies. Sedol is—overly—confident that he is going to win, but then fate strikes.

Also, the antagonistic setting of man versus machine confrontations in which the Alphago-Seedol game is embedded resonates with the Greek tragedies of grand conflicts. With every lost game, he comes to realize that it was *hubris*, which led him to accept the challenge. He ignored the signs: the previous man-machine games, which were all won by the machine (including the most recent game of AlphaGo against European champion Fan Hui), the overall disruptive developments in the AI domain, and the excellent reputation of the DeepMind AI team, specifically. After entering the magical circle of the game, there is, however, no way back. He is captured by the rules of the game and has to play until the end. In AlphaGo versus Sedol, fate and freedom come together, enforcing the game's tragical character. Sedol reflects on his daunting position: "I can't believe this is happening. Regardless of your opponent's level, to be defeated not by three-zero, but five-zero? Losing to AlphaGo by five-zero would really hurt my pride" (Kohs 2016, 1:03:06).

After every game, Sedol has to face the press. The first time, he stated to be surprised by the high playing level of AlphaGo, the second time he said to be speechless, the third time he felt he had to apologize to the audience. Sedol:

> I think I have to express my apologies first. If I had been able to play better or smarter, the results might have been different. I think I disappointed too many of you this time. I want to apologize for being so powerless. I've never felt . . . this much pressure, this much weight. I think I was too weak too overcome it. (Kohs 2016, 1:02:05)

Experiencing self-doubt and guilt, the responsibility of playing on behalf of "humanity" becomes an enormous burden for Sedol to bare. However, nobody really doubts his efforts. The fact that Sedol actually manages to win the fourth match only underlines his determinacy. As De Mul describes the tragic hero, he stresses: "When they go down in their struggle, this is generally not the result of a lack of determination, but rather of the inhuman grandness of their efforts. They lose themselves in the catastrophic course of events" (De Mul 2014, 45).

Phronetic Understanding

As viewers of the documentary, we can empathize with Sedol as we all have firsthand experiences of being captured between freedom and necessity, between making a choice and not being able to escape its consequences. Aristotle states that watching a tragedy unfold before your eyes invokes feelings of compassion and fear with the audience, liberating "the intense emotions that are called forth by the tragic events" (De Mul 2014, 43).

There is something to be learned from the suffering of Sedol. Or as Ricoeur states it, we can move from tragic wisdom to practical wisdom (Ricoeur 1994). The one-dimensional perspective on, for instance, "the human" or "the machine" or "intelligence" when confronted with the complexity of life can lead to sometimes insoluble conflicts. Deciding on the best action in these situations is not so much about coming up with a univocal solution, as it is about developing a judgment that acknowledges and responds to the tragic character of the conflict (also see: Nussbaum 2002).

A tragedy, such as the AlphaGo game, does not only force us to face the limitations of our human condition, it also gives us the opportunity of undergoing what De Mul calls an "aesthetic experience" (De Mul 2014, 50). Actually, on two occasions in the documentary, an outspoken aesthetic experience occurs. The first aesthetic experience is move 37 in the second game, made by AlphaGo, the second one is move 78 in the fourth game by Sedol.

During the second match, while Sedol takes a break to smoke, AlphaGo plays move 37. Almost all professional commentators agree that this is a move, no human would play. AlphaGo actually agrees with this judgment as the AI states that there is a 1 in 10,000 probability that this move would have been played by a human player. When Sedol comes back into the room and sees move 37, the camera focuses on his face and we witness his confusion. Sedol comments:

> I thought AlphaGo was based on probability calculation and that it was merely a machine. But when I saw this move, I changed my mind. Surely, AlphaGo

is creative. This move was really creative and beautiful (Kohs 2016, 00:52:17) . . . . This move made me think about Go in a new light. What does creativity mean in Go? It was a really meaningful move. (Kohs 2016, 00:53:35).

On several occasions in the documentary, the word "beautiful" is employed to describe move 37. However, one can argue that beautiful is a too limited expression of what happens here. Move 37 is not something we merely desire or that sparks our admiration. It is both a stunning and devastating move (cf. Yu 2016). Stunning, because it opens up the possibility of understanding and playing Go in a completely new way. Devastating, because creativity and intuition seemingly no longer solely belong to human beings. Move 37 sparks *sublime* beauty (on tragedy and the sublime, see: De Mul 2014, 56–59). We can feel the agony of Lee Sadol, while we simultaneously also share in his wonder for this extraordinary move. A move, which on the one hand is *non-human*, in the sense that a human player would probably never had played it and on the other hand is utterly *human*, as it displays a capacity for creativity and intuition, we until now had only ascribed to humans.

In the fourth match, with move 78, Sedol is able to play a wedge move at the center of the board, making the game so complex that AlphaGo is no longer able to evaluate its position correctly. The AI starts to play strange and even silly moves and resigns after a while. Sedol wins the match. At the press conference, move 78 is called a "god's move." Asking Sedol, what he was thinking when he played that move, he replies: "At that point in the game, move 78 was the only move I could see. There was no other placement. It was the only option for me, so I put it there." AlphaGo, however, estimated—once more—that only 1 out of 10,000 humans would actually play that move. In a sense, move 78 came to mirror move 37. Twin moves, but not identical twins as they were both extraordinary, both beautiful, but not both sublime.

With move 78, as one DeepMind scientist explains, we are left: "a little bit in awe of the human brain's power, in particular, Lee's amazing ability to cause AlphaGo problems and find something seemingly out of nothing. And so, we really want to understand what had happened" (Kohs 2016, 1:14:47). In other words, at least for the DeepMind crew, move 78 was not merely beautiful, it also quickly became another opportunity to further improve their AI, commodifying move 78. People watching the victory, on the other hand, were completely overwhelmed by excitement, resulting in an intense and all-consuming joyful experience. However, therefore also their experience did not have a sublime character, as it lacked the ambiguity of going through conflicting emotions as a whole, which is a typical aspect of the sublime. Rather, these feelings of joy and excitement, at least for a short period of time, smothered the feelings of fear and loss that had been growing with every lost game. Sedol (Kohs 2016, 1:12:19) explains:

I heard people shouting in joy when it was clear that AlphaGo had lost the game. I think it is clear why. People felt helplessness and fear. It seemed like we humans are so weak and fragile. And this victory meant . . . we could still hold our own. As time goes on, it will probably be very difficult to beat AI. But winning this one time, I felt like it was enough. One time was enough. I heard from so many people saying they were running out in the street. They were so happy. They were chanting, they were celebrating.

By interpreting the tragic-like emplotment of the documentary, we can identify some basic characteristics of human-AI interactions, which could be useful for a postphenomenological analysis. First, hubris leads human beings to mistakenly believe that they can completely control AI and that it is merely an instrument to be played with or to have at your disposal. This hubris is not merely in Sedol, who takes on the challenge of competing with AlphaGo, but also in the computer scientists of DeepMind, who are convinced that they contain AlphaGo and can explain its actions. In other words, hubris might blind people for the mediating workings of the technology. Second, this loss of control is not merely a bad thing. As we saw, there is also beauty in a technology, which accomplishes things we thought of as belonging solely to the human domain. The complexity of experiencing the sublime in our interactions with technology demands from postphenomenology a sensitivity for those conflicting feelings that exist in tandem. It is not just that different people can have different mediating relations with an AI application, but within a specific mediation, different, conflicting meaning-given movements can co-exist.

5. TRANSFIGURATION: WHEN DOCUMENTARY AND LIFE INTERSECT

In this section, the art of narrating a story and of following a story come together. In the previous sections, we have established the game-like structure of the documentary, and noticed how it moves between the poles of sedimentation and innovation and appeals to our phronetic understanding. Now we are able to explore how this story mediates our relation to ourselves, others, and the world around us.

Self-understanding

The documentary explains that Go is being thought of not just as a game, but as having "a very deep philosophy. The Go board reflects the individual who's playing. The truth is gonna show itself on the board. You won't be able to hide it" (Kohs 2016, 00:16:45). Or as Fan Hui explains, "Because all the things I learned in my life is [*sic*] with Go. It looks like a mirror. I see Go, I also see myself. For me, Go is real life" (Kohs 2016, 00:04:20).

Playing Go is also a form of self-expression, a way of distinguishing oneself from others. Sedol: "It has become a type of creation of mine. I want my style of Go to be something different, something new, my own thing. Something that no one has thought of before" (Kohs 2016, 00:18:25). So, if Go is a form of self-expression and a way to get to know oneself, it is interesting to investigate if AlphaGo mediates this relation with the self in a new way.

In the first part of the documentary, we get a glimpse of such mediation through Fan Hui who just lost, as a professional Go player, from a computer program for the first time in history. During the game, we see him go through a kaleidoscope of emotions: surprise, irritation, fear, resignation. After the final game, he comments: "I feel something very . . . strange. I lose with [*sic*] a program. And I don't understand myself anymore" (Kohs 2016, 00:10:16). He leaves and it takes more than an hour for him to return to the DeepMind offices. To the question if he is okay, he replies to experience mixed feelings. He is not happy to have lost the game, but he is happy to be part of this historic moment. After going through the emotions of losing the game, of losing himself, he seemingly has found a new part of his identity: the first human Go professional to have lost from a computer program, making him part of an historical event. Later in the documentary, he will join the DeepMind team to help them improve AlphaGo.

Lee Sedol too comes to see himself in a new light through his interaction with AlphaGo. Notwithstanding, or actually maybe because of the four games that he lost, he declares: "I feel thankful and feel like I have found the reason I play Go. I realize it was really a good choice, learning to play Go" (Kohs 2016, 01:25:36).

All in all, both Go players in their interaction with AlphaGo, go through a *personal transition*. Losing from AlphaGo invokes some sort of existential anxiety at first but also provides room for a new way of understanding themselves. They change through their interaction with AlphaGo. By focusing on these first-person experiences, we can nuance the earlier discussed antagonistic framing inherent in the man versus machine narrative. The human Go players are not diminished but transformed by their interaction with the machine.

Communicability

In the documentary, AlphaGo becomes "the other." From a postphenomenological point of view, we could speak of an—augmented—alterity relation. The Go players interact with AlphaGo as another actant (Latour 1993).

What immediately stands out in this interaction between players and AlphaGo is the lack of intersubjectivity. Yes, AlphaGo plays moves, which are creative and beautiful, but it does not play Go with its opponent. Playing is not just following the rules of the game—which AlphaGo does perfectly—more importantly, it is a coming together of actors who co-experience the event by looking at each other, reading emotions, smelling, hearing, and talking. These parts of the game are lost when AlphaGo plays. Fan Hui explains that normally when playing Go, he can feel "many, many things. But with AlphaGo, you can feel nothing. So, when you can feel nothing when you play, you have more and more questions about yourself" (Kohs 2016, 00:36:39). Here, we see an immediate connection between the second-person and first-person experience. As the former lacks intersubjectivity, the subjective experience becomes much more encompassing, overwhelming even.

In the documentary, we see Lee Sedol looking at Aja Huang, who is sitting opposite of him, placing the stones on the board on behalf of AlphaGo. As a player, it comes naturally to look at one's opponent's face. However, Aja Huang is not Sedol's opponent. Huang too feels how unusual this interaction is: "I can actually feel the spirit and courtesy of a great Go player like Lee Sedol.... It was the first time he faced a strange opponent, I think. It is non-human, has no emotion, it's cold. But he stayed very calm. And I can feel his mental strength" (Kohs 2016, 00:33:08). The limited role of Huang and the absence of AlphaGo in the interaction makes that after a couple of games, Sedol brings along friends to go through the moves once more when the game is over. This is something Go players usually do together, but with AlphaGo has become impossible. All in all, the *lack of intersubjectivity* makes *loneliness* the fundamental mood of this interaction.

This need for understanding the other and building a meaningful connection is also reflected in the active search of the audience to find ways to interpret AlphaGo. They are looking for "tells" of AlphaGo, which might give away something of its modus operandi and many journalists interview DeepMind scientists to explain to them how AlphaGo "thinks."

It is striking how far the DeepMind crew goes to demystify AlphaGo in the documentary. There are quite a few scenes shot in the control room where we witness the scientists follow, comment, and interpret the moves of AlphaGo. Seemingly, the muteness of the artificial intelligence is balanced with the human networks in which AlphaGo is embedded.

Although, DeepMind wants to stay far away from anthropomorphizing AI, they also stress its humanness, even without its ability for intersubjectivity. One of the DeepMind-scientists states:

> AlphaGo is human-created, and I think that's the ultimate sign of human ingenuity and cleverness. Everything that AlphaGo does, it does, because a human

has either created the data that it learns from, created the learning algorithm that learns from that data, created the search algorithm. All of these things have come from humans. So, really, this is a human endeavor. (Kohs 2016, 00:33:08)

Referentiality

Finally, the documentary also tells us something about the world and how it changes because of AlphaGo. First of all, it puts the world of Go itself in a different light. When people start to realize that AlphaGo is not a "normal," mediocre program and that what at first sight might look like a wrong move, a mistake, actually turns out to be a game-changing move, they acknowledge Go has entered a new paradigm because of AlphaGo. "The lessons that AlphaGo is teaching us are going to influence how Go is played for the next thousand years," one commentator claims (Kohs 2016, 01:20:05). Sedol: "What surprised me the most was . . . that AlphaGo showed us that moves humans may have thought are creative, were actually conventional" (Kohs 2016, 01:21:03).

However, that Go has shown people that they might be less creative than they initially thought did not make them stop play. On the contrary, one of the effects of the AlphaGo game and the documentary was a rush on Go boards and stones.

The documentary wants to show us also another world, which reaches far beyond the game of Go. Particularly at the end of the documentary, the narrative of the man versus machine framing is abandoned to sketch a future where AI and human beings empower each other. Here the mutual shaping vision, which is key to postphenomenology and other approaches after the empirical turn, becomes apparent. Reporter Cade Metz reflecting on the game makes the link between Sedol's transformation and how this might happen to all of us:

> At least in a broad sense, move 37 begat move 78, begat a new attitude in Lee Sedol, a new way of seeing the game. He improved through the game. His humanness was expanded after playing this inanimate creation. And the hope is that, that machine, and in particular, the technology behind it, can have the same effect with all of us. (Kohs 2016, 01:24:42)

6. CONCLUSION

The aim of this chapter was to investigate in which way the documentary AlphaGo can give us access to AI technology, which is not yet part of everyday life, so we can come to understand AI's mediating qualities in a timely manner and inform a postphenomenological analysis. By adopting key elements of Ricoeur's narrative discourse, we engaged in a step-by-step analysis of the documentary, bringing forth some valuable findings.

Among others, we established the importance of actively connecting micro and macroperception. While theoretically interlinked, in practice, they are often addressed separately. The cultural embeddedness of microperceptions can however lead to different and even conflicting human-AI-world relations, as the framing of U.S. and Chinese media outlets show. Moreover, this macro-perception is also important to understand the tragic emplotment that resonates in the documentary, referring back to the narrative of the ancient Greek tragedies. Finally, the macro-perception is also important to grasp the power relations that play a role in the set-up of the documentary. By choosing a game-setting to display their AI's strength, DeepMind (Google) creates a safe action-space where AlphaGo never becomes too threatening.

When it comes to the first-person experience, we saw that the initial hubris of the human actors led them to believe they could control the AI, blinding them for the mediating workings of AlphaGo. Or, loosely referring to Ihde (1990): they wanted the transformation of the AI without its mediation. This loss of control can, however, go together with a feeling of awe. The complex, layered experience of the sublime is something we might expect to perceive more often in our future interactions with AI. This asks of postphenomenology to invest in crafting a new sensitivity for those conflicting feelings coming together in such an experience.

As all mediating technologies bring some aspects of reality to the fore and hide others in the background, we learn from our analysis that one of the crucial aspects that gets lost in AI-interactions is intersubjectivity, resulting in feelings of strangeness and loneliness. That people look for "tells," look for togetherness with AlphaGo is not mere anthropomorphizing, it is a fundamental need of people to interpret and to be interpreted. Recent investigations in explainable AI, in developing methods to explain what happens in the black box, predominantly focus on explanations that enable engineers to make the system more robust or to comply with legal requirements. However, our analysis indicates that explainability might also be *crucial to live a flourishing life together with AI*. What kind of explanation is needed, or to put it differently, what kind of *mediation* is desired from an existential point of view, might be a valuable question postphenomenology could actually contribute to if it makes room for new understandings of what it means to operate as an "empirically informed" approach. In our analysis of AlphaGo, Ricoeur's narrative discourse has proven to be a valuable way to interpret and gain access to AI.

REFERENCES

Aagaard, Jesper, Jan Kyrre Berg Friis, Jessica Sorenson, Oliver Tafdrup, and Cathrine Hasse, eds. 2018. *Postphenomenological Methodologies: New Ways in Mediating Techno-Human Relationships*. Lanham: Lexington Books.

Achterhuis, Hans. 2001. *American Philosophy of Technology: The Empirical Turn*. Indianapolis: Indiana University Press.

Bory, Paolo. 2019. "Deep New: The Shifting Narratives of Artificial Intelligence from Deep Blue to AlphaGo." *Convergence* 25, no. 4: 627–642.

Crawford, Kate, and Meredith Whittaker. 2016. *The AI Now Report: The Social and Economic Implications of Artificial Intelligence Technologies in the Near-Term*. New York: AI Now.

Curran, Nathaniel Ming, Jingyi Sun, and Joo-Wha Hong. 2019. "Anthropomorphizing AlphaGo: A Content Analysis of the Framing of Google DeepMind's AlphaGo in the Chinese and American Press." *AI & Society* 1–9.

De Mul, Jos. 2014. *Destiny Domesticated: The Rebirth of Tragedy Out of the Spirit of Technology*. Albany: SUNY Press.

Feenberg, Andrew. 2009. "Democratic Rationalization: Technology, Power and Freedom." In *Readings in the Philosophy of Technology*, edited by D. M. Kaplan, 139–155. Lanham: Rowman & Littlefield Publishers.

Feenberg, Andrew. 2017. *Technosystem: The Social Life of Reason*. Cambridge: Harvard University Press.

Frissen, Valerie, Sybille Lammes, Michiel De Lange, Jos De Mul, and Joost Raessens. 2015. "Homo ludens 2.0: Play, Media, and Identity." In *Playful Identities: The Ludification of Digital Media Cultures*, edited by Valerie Frissen, Sybille Lammes, Michiel De Lange, Jos De Mul, and Joost Raessens, 9–50. Amsterdam: Amsterdam University Press.

Gadamer, Hans-Georg. 1972. *Wahrheit und Methode*. Tübingen: Mohr.

Granter, Scott R., Andrew H. Beck, and David J. Papke, Jr. 2017. "AlphaGo, Deep Learning, and the Future of the Human Microscopist." *Archives of Pathology & Laboratory Medicine* 141, no. 5: 619–621.

Huizinga, Johan. 1955. *Homo Ludens: A Study of the Play-Element in Culture*. Boston: Beacon Press.

Ihde, Don. 1990. *Technology and the Lifeworld: From Garden to Earth*. Bloomington: Indiana University Press.

Jobin, Anna, Marcello Ienca, and Effy Vayena. 2019. "The Global Landscape of AI Ethics Guidelines." *Nature Machine Intelligence* 1, no. 9: 389–399.

Kaplan, David M. 2006. "Paul Ricoeur and the Philosophy of Technology." *Journal of French and Francophone Philosophy* 16, no. 1/2: 42–56.

Keymolen, Esther. 2017. "Trust in the Networked Era: When Phones Become Hotel Keys." *Techné: Research in Philosophy and Technology* 22, no. 1: 51–75.

Kohs, Greg, Gary Krieg, Josh Rosen, and Kevin Proudfoot. 2017. *AlphaGo*. United States: Moxie Pictures.

Kudina, Olya, and Peter-Paul Verbeek. 2019. "Ethics from Within: Google Glass, the Collingridge Dilemma, and the Mediated Value of Privacy." *Science, Technology, & Human Values* 44, no. 2: 291–314.

Latour, Bruno. 1993. *We Have Never Been Modern*. Harvard: Harvard University Press.

Li, Fangxing, and Yan Du. 2018. "From AlphaGo to Power System AI: What Engineers Can Learn from Solving the Most Complex Board Game." *IEEE Power and Energy Magazine* 16, no. 2: 76–84.

Nussbaum, Martha C. 2002. "Ricoeur on Tragedy: Teleology, Deontology, and Phronesis." In *Paul Ricoeur and Contemporary Moral Thought*, edited by John Wall, William Schweiker, W. David Hall, and David Hall, 264–276. New York: Routledge.

Pasquale, Frank. 2015. *The Blackbox Society: The Secret Algorithms That Control Money and Information*. Harvard: Harvard University Press.

Pellauer, David, and Bernard Dauenhauer. 2016. "Paul Ricoeur." In *The Stanford Encyclopedia of Philosophy*. https://plato.stanford.edu/archives/win2016/entries/ricoeur/.

Ricoeur, Paul. 1981. *Hermeneutics and the Human Sciences: Essays on Language, Action and Interpretation*. Edited and translated by John B. Thompson. Cambridge: Cambridge University Press.

Ricoeur, Paul. 1991. "Life in Quest of Narrative." In *On Paul Ricoeur: Narrative and Interpretation*, edited by David Wood, 20–33. London: Routledge.

Ricoeur, Paul. 1994. *Oneself as Another*. Chicago: University of Chicago Press.

Rosenberger, Robert. 2018. "Why It Takes Both Postphenomenology and STS to Account for Technological Mediation: The Case of LOVE Park." In *Postphenomenological Methodologies: New Ways in Mediation Techno-Human Relationships*, edited by Jesper Aagaard, Jan Kyrre Berg Friis, Jessica Sorenson, Oliver Tafdrup, and Cathrine Hasse, 171–198. Lanham: Lexington Books.

Rosenberger, Robert, and Peter-Paul Verbeek. 2015. "A Field Guide to Postphenomenology." In *Postphenomenological Investigations: Essays on Human-Technology Relations*, edited by Robert Rosenberger and Peter-Paul Verbeek, 9–42. Lanham: Lexington Books.

Silver, David, and Demis Hassabis. 2016. "AlphaGo: Mastering the Ancient Game of Go with Machine Learning." *Google AI Blog*, July 2, 2016. https://ai.googleblog.com/2016/01/alphago-mastering-ancient-game-of-go.html.

Silver, David, Julian Schrittwieser, Karen Simonyan, Ioannis Antonoglou, Aja Huang, Arthur Guez, Thomas Hubert, Lucas Baker, Matthew Lai, and Adrian Bolton. 2017. "Mastering the Game of Go Without Human Knowledge." *Nature* 550, no. 7676: 354–359.

The-Online-Initiative. 2015. "Background Document: Rethinking Public Spaces in the Digital Transition." In *The Onlife Manifesto*, edited by Luciano Floridi, 41–48. Cham: Springer.

Verbeek, Peter-Paul. 2011. *Moralizing Technology: Understanding and Designing the Morality of Things*. Chicago: The University of Chicago Press.

Yu, Haofeng. 2016. "From Deep Blue to DeepMind: What AlphaGo Tells Us." *Predictive Analytics and Futurism* 13: 42–45.

Conclusion

Hermeneutic Responsible Innovation

Wessel Reijers, Alberto Romele,
and Mark Coeckelbergh

Technological change is at the forefront of human thinking, across the world. More than ever before, intellectuals, politicians, and the public at large are grappling with the wonders and the terrifying dangers of technology. This comes clearly to the fore in widely read books such as Yuval Noah Harari's *Homo Deus* (2013) and Soshana Zuboff's *Surveillance Capitalism* (2019), in political movements thematized around technology, such as Anonymous, the Pirate Party, or Cyber Anarchism, and in increasingly formalized government strategies to tackle the normative impacts of technology; think of the "High Level Expert Group on Artificial Intelligence" in the European Union. We started this book by emphasizing the importance of the activity of interpretation in this context, and by outlining a new research program under the heading of Hermeneutic Philosophy of Technology (HPT). This program has literally and figuratively unfolded in the subsequent pages, which have led the reader through fourteen chapters that have shaped it in a multitude of directions.

Overall, *Interpreting Technology* has demonstrated the strength of HPT in three ways. First, it has illustrated its versatile nature, in line with the "empirical turn," by interweaving discussions of a great plenitude of technologies: ranging from AlphaGo and generic software code, via social media and quantified-self technologies, to technologies applied in particular domains such as healthcare and education. Second, it has for the first time provided a comprehensive connection between the Ricoeur-inspired HPT and established theories of technology, notably Postphenomenology, Actor-Network Theory, and Critical Theory of Technology. This interlinking will form the basis of many fruitful dialogues between these approaches and HPT. Third, the book has opened up a robust connection between philosophy of technology and ethics of technology, two "fields" that have thus far not interacted much; think, for instance, of the few connections between established theories in the latter, such as "ethical impact assessment," or "value sensitive design," and the first. That HPT can provide such a connection is partly because of the strength of Ricoeur as a philosopher, who has consistently linked his philosophical hermeneutics to ethics (*Oneself as Another*), as well as to core debates in legal and political theory (*The Just*). Whatever a Ricoeurian philosophy of technology might look like, it will have strong ethical and political aspects.

The chapters of this book offer the reader a number of case studies on how to theorize technology from Ricoeur's hermeneutic perspective. These case studies have introduced "tools" for interpreting technology. Bruno Gransche, in his highly original contribution (chapter 5), mentions what these tools might look like in a post-Ricoeur HPT, for example, putting to the fore notions like *texture* and *objective spirit* into the field as powerful means for analysis. We can also think of Annemie Halsema's lucid engagement with Ricoeur's model of *mimesis* (chapter 6), establishing it as a powerful hermeneutic model of analysis that might complement the models offered by post-phenomenology and Actor-Network Theory. Or, consider Noel Fitzpatrick's detailed effort to make Ricoeur's "Little Ethics" (chapter 7) and the *ethical aim* (the good life, with and for others, in just institutions) relevant for ethics of technology. Heading toward the realm of politics, Guido Gorgoni and Robert Gianni (chapter 10) show convincingly how Ricoeur's notion of *responsibility* offers a valuable perspective for discussion in RRI.

These studies, which demonstrate the potential of hermeneutic tools coming out of HPT, are "put to work" through applications of Ricoeur's oeuvre to fields other than philosophy of technology. Perhaps the most "narrative" and surprising contribution of the book, by Esther Keymolen (chapter 14), defies categorization altogether and at the same time illustrates how a new ethnographic (or "netnographic") method arises out of Ricoeur's hermeneutics. "Weaving" (or texturing, as Gransche might put it) the story of AlphaGo into a structured interpretation of the capacities of Artificial Intelligence, Keymolen takes the reader on a journey that mixes academic analysis with an engaging storyline. David Lewin eloquently shows Ricoeur's relevance for the modern classroom (chapter 8), in which technologies are increasingly mediating the educational experience, thereby opening a corridor to educational studies. Geoffrey Dierckxens takes us from the classroom to the hospital ward (chapter 9), making a powerful case for introducing Ricoeur's hermeneutics to understand the technological mediation of prostheses. Alain Loute also engages in an important discussion of Ricoeur's hermeneutics in the context of e-health (chapter 11) and does so while also illustrating the relevance of exploring the limits of HPT, how technologies affect the activity of interpretation and in a way destabilize the hermeneutical effort.

Finally, a book on Ricoeur cannot be finished without putting the hermeneutic tools of HPT and their application in *context*, by engaging in a mediation with other approaches in philosophy of technology. Eoin Carney puts everything at work to construct what one might call a "re-encounter" between Ricoeur and postphenomenology (chapter 2), two strands of theory that have gone apart but now might come together again. Another long awaited re-encounter, though one that has been more "covered up" is forcefully brought to light by de Bas de Boer and Jonne Hoek, generating a dialogue between two philosophical giants: Ricoeur and Latour (chapter 3). Speaking in clear terms to Ricoeur's neo-Marxist sympathies, David M. Kaplan builds a bridge to critical theory of technology, and most notably Andrew Feenberg (chapter 4). Ernst Wolff has drawn these efforts together in a rich and engaging discussion of Ricoeur's main contributions to our thinking concerning technology (chapter 1). Finally, Todd Mei touches upon a highly important contextualization that reaches beyond philosophy of technology, into political theory, by turning to Ricoeur's notion of political action that is deeply inspired by Hannah Arendt (chapter 12).

It is on this last note that we turn to a new "horizon" of HPT, namely the introduction of a philosophical hermeneutic—and political—perspective in the nascent field of responsible innovation (RI). *Interpreting Technology* has largely occupied itself with the stringent task of interpreting technology, as the title strongly suggests. That is, it has offered tools for interpretation, case studies in which these tools were applied, and dialogues between the HPT perspective and other approaches. As such, we have come to be aware of different technological mediations, related ethical and political impacts, as well as principles to understand these impacts. To somewhat echo Marx, these philosophical efforts have led to an interpretation of our technological world; yet, the point is *also* to change it. Responsible innovation, insofar as it is distinct from philosophy and ethics of technology, harbors a strong practical dimension, one that asks for guidance in taking *political action*; the theme of Mei's chapter. We take our cue as well from Gorgoni and Gianni, and Ricoeur's forward-looking notion of responsibility, yet also want to go beyond the notion of responsibility as such and toward a notion of responsible *practice*. What follows is merely a first sketch of this new horizon of HPT, called *Hermeneutic Responsible Innovation* (HRI).

1. FROM THEORY TO ACTION

The challenge is this: How to translate HPT as a theoretical program in philosophy into *action* that gives meaning to the notion of responsible innovation? Even though Ricoeur did not occupy himself at all with the question of responsible innovation as we now understand it, the question of responsible *technical* practice (in the sense that Wolff ascribes to it, as a constant engagement with *techne*) has been increasingly on the forefront of his thinking in his later years (Reijers and Coeckelbergh 2020). We have to understand responsibility foremost as a *political* concept in this context; for Ricoeur followed Arendt in seeing responsible action as acting in public, under the condition of plurality. One book that has touched upon a similar thematic is Dries Deweer's engaging *Ricoeur's Personalist Republicanism* (2017), which discusses how Ricoeur has made a move throughout his career, from being politically engaged in the personalist movement of Maritain and Mounier, to turning away from explicitly political questions in favor of gaining a new understanding of the human condition from a phenomenological and hermeneutic perspective, in order to finally return to the question of the "political paradox," as discussed in Kaplan's chapter (chapter 4), and to a new understanding of the republican political project.

Let us now turn to the field of responsible innovation, which has witnessed significant expansion in the past decade, partly due to the growing impact of technological innovations on society and partly due to increasing institutional efforts, for instance, by actors like the EU, to render innovation practices more responsible—in the sense of anticipating potential problematic impacts of innovation and resolving those. At the time of its inception, responsible innovation was strongly related to utilitarian risk/benefit analysis and forms of institutional impact assessment—with a focus on administrative implementation. Yet, in recent years, philosophically informed critiques have moved the field more in the direction of Ricoeur's hermeneutic project. First, authors have questioned the limited notion of "innovation," for it is largely understood in techno-economic terms, and also questioned whether innovation can be "responsible" at all, for responsibility as prudence often invokes a backward-looking attitude while innovation is volatile and uncertain (Blok and Lemmens 2015; Blok 2019). Second, there has been an explicit hermeneutic strand of research, as explored by Armin Grunwald (2014), which explicates the role of narrative visions in our understanding of innovation practices. A link has been made between hermeneutic analysis and virtue ethics, for instance, by proposing to educate innovators in accordance with interpretations of innovation that can be found in narratives like Mary Shelley's *Frankenstein* (see Sand 2018; Grinbaum and Groves 2013). Third, responsible innovation has been moved in the direction of political philosophy, by acknowledging that innovation as an activity has not only ethical dimensions but also political ones, having to do with power, deliberation, and resolution of conflict (Himmelreich 2020; Wong 2019).

The Ricoeur-inspired research program of HPT promises to contribute to the field of responsible innovation in mainly three ways. First, it would reinforce the interpretation of innovation as an eminently political activity, thereby making it sensitive to the challenge posed by the *political paradox*. Second, it would introduce not only a hermeneutic perspective to understand innovation, as arguably has been done by existing research, but also a hermeneutic theory of *technical practice*, which enables an understanding of innovation according to the dynamic of "ascending complexification and descending specification" (Ricoeur 1992, 158); a dynamic that also features in Ricoeur's discussion of the practices of interpretation and argumentation in legal contexts (Ricoeur 2000, 125). Third, Ricoeur sheds light on the nature of *responsible judgment* in innovation practices, most notably through his discussion of Arendt's notion of political action as modeled on Kant's concept of aesthetic judgment (Ricoeur 2000, 94). The question here is to what extent an activity of making (innovation) could be open to political action, given that a judgment of beauty is only accessible to those with a backward gaze, and action disperses while innovation leaves its traces in material things. In what follows, we will not fully develop these three elements of a possible integrated research program of HRI but rather sketch the outlines of what a research agenda could look like, also touching upon telling examples of recent technological innovations.

2. THE POLITICAL PARADOX

Foremost, perhaps, the political paradox reveals that Ricoeur is an often-unacknowledged thinker of *cybernetics*, or what Heidegger designated as the mode of thinking that replaces philosophy. The political paradox arises out of the fact that modern political entities progressively behave as "the whole *and* the part, as container *and* the contained" (Ricoeur 2000, 93. emphasis added). As such, Ricoeur's thinking resonates with that of contemporary philosophers of technology like Yuk Hui, who recognizes that "the giant force of technology is autosystematising at all orders of magnitude" (Hui 2019, 250). When we consider responsible innovation as a technical practice, we should acknowledge that in the current time of capitalist production and globalization, it is indeed a practice that leads to systemization at all levels of society—from the household (e.g., Amazon's Alexa), via the factory floor (e.g., automated warehouses) to the political realm. Especially in the context of the current COVID-19 pandemic, more energy is spent on finding new ways to organize and systematize political discourse, make political decisions, and implement policies; for it has raised the need for remoteness, or political action at a distance.

"The political" in this understanding is not limited to *politics* as it is commonly understood, as what goes on in parliaments, governmental committees, or the whims of dictators. Rather, it is linked to the institution: to the acting and speaking in public that is necessary to sustain human (public) forms of living together. It concerns strengthening and sustaining power in common while putting checks on the inevitable power over (Ricoeur 1992, 220). The most eminently political innovation practices in this regard are those that aim at bringing about new institutional forms and consequently new forms of deliberation, decision-making, and enforcement. Examples of such institutions are platforms leveraging blockchain technology, such as Ethereum, but also rather loose systems of interrelated technologies as the one implemented by activists like Aurdey Tang in Taiwan, which are lauded as bringing a "digital democracy" (Nabben 2020). Development of such "institutional innovations" or perhaps "techno-institutions" has accelerated significantly in the past years, aided by improvements in large-scale deliberations platforms, electronic voting, and game-theoretical incentive mechanisms, and, consequently, deserves serious scrutiny.

How would the political paradox shed light on institutional innovations? Let us first try to understand the meaning of the political paradox more precisely. Ricoeur puts this forward as the tension between the promise and the risk of political activity (Deweer 2013). On the one hand, the promise of political activity is to fulfill the *telos* of humanity, showing progress in developing a constitution under which all humans are equals; to enable co-decision and self-determination of peoples. In this regard, also the distanciation involved in the design of new techno-institutions contributes to further empowerment of human cooperation. For instance, online platforms like Discord enable the Ethereum community to self-organize from all corners of the world, enabling some form of democratic co-decision of developers in constructing a new techno-institution. On the other hand, the risk of this same political activity is to constitute an inevitable threat of domination of some over others. To stay with the same example: in the Ethereum community, some states of exception like the notorious "DAO Attack" (Dupont 2017) (an exploit of a newly created "decentralized autonomous organization" in which an attacker used established rules as reflected in the software code to practically steal millions of dollars' worth of cryptocurrency) have triggered non-democratic, fairly centralized responses that have created fractures in the community. Those who felt dominated by an imposed decision to "hard fork" the system (split the system into two independent versions, one in which the "theft" was effectively made undone) left the community and continued their own "fork." Some argue that Ethereum has given rise to an adversarial "dark forest" (Robinson and Konstantopoulos 2020): its organizing powers also generate attempts for harmful exploitation.

This brief discussion shows two things: how innovation itself can indeed be turned into a political activity, and how the outcomes of this type of innovation give rise to the political paradox. Accordingly, our research agenda faces a theoretical and a practical challenge. Theoretically, the challenge is to reconcile the activity of innovation with the activity of political action. Connecting to the HPT program outlined in the introduction, this involves investigating the politics of the "grammar" or "conditions of possibility" of technologies. We will further discuss this point, under "responsible judgment." Practically, the challenge is to critically approach innovations that promise to strengthen our techno-institutions, even those that seem completely benevolent. As discussed throughout this book, Ricoeur is on the side perhaps of the hermeneutics of suspicion when it comes to technology, but at the same time considers our technological condition not only negatively; rather perhaps as a *pharmakon* (poison as well as cure)—in the words of Bernard Stiegler (1998). Concretely, this means that we should critically scrutinize new tools such as platforms for online participatory budgeting (e.g., "Decide Madrid"), for making collective, democratic decisions across borders, cultures, and languages (e.g., "Democracy Earth") and for organizing deliberations online (e.g., "Discourse"). Yet, we should also consider them as potential allies in the struggle for global justice and equality.

Note that techno-institutions figure as an illustration here. The point made stands for innovation—and responsible innovation—in general, insofar innovations have a bearing on the human capacity to act and speak in concert. So far, we have discussed the political paradox in relation to the outcomes of innovation (e.g., the digital platforms), which brings us to the question of the dynamic of innovation practices.

3. TECHNICAL PRACTICE

If *responsible* innovation is possible (i.e., if innovation can be responsible), it raises the question of how discreet, fine-grained actions (e.g., coding a piece of software) can in fact be related to those high-level principles such as justice and equality that characterize responsibility in public life. It is one thing, as a practitioner, to understand the ambivalent or problematic nature of technical innovation practices—for instance, by engaging with the Frankenstein narrative—but it is quite another thing to be aware of the way in which one's everyday actions in a way contribute to forms of living together. In other words, how can one understand which impact an innovation practice will have on the whole when it is only recognizable at the granular level? For instance, one little mistake in the coding of a financial innovation ("fintech") might lead to an exploit that could harm millions of people. What constitutes responsible conduct in this context?

We find a promising answer to this question in philosophical work on practice. Consider Bourdieu's theory of practice that focuses on *habitus*, which constitutes "systems of durable, transposable dispositions" that "generate and organise practices and representations" (Bourdieu 1992, 53). According to this notion of practice, discreet actions could be explained by means of *habitus*, the already-present, sedimented social "repertoire" that each individual possesses. Hence, individual actions are linked with the social, the larger whole. Similarly, consider MacIntyre's (1992) hermeneutic theory of practice, which makes explicit how narrative extends individual practices through standards of excellence and life plans to the moral community or tradition as a whole. That is, practitioners can explain the actions they engage in by referencing the totality of relevance in which these actions are embedded; which includes the life choices they made and the standards of excellence that define practices for them. Yet, with scholars like Blok (2019), we might ask critical questions regarding the applicability of these accounts of practice in the realm, or ontology, of "innovation." First, if innovation is concerned with the "new," how can it be informed by standards of excellence that derive from an already existing "tradition" or *habitus*? Second, if the narrative mode that mediates practice is only a *historical* mode, derived from "what really happened," how can it account for the imaginative variations, which lie closer to the mode of *fiction*, that mediate practices of innovation (e.g., ideation, brainstorming, "out of the box" thinking)? And third, do the views of Bourdieu and MacIntyre not confront us with a certain determinism, that leaves some but only little room for the unexpected, which—for Arendt—is the basis of (responsible) action?

Ricoeur's theory of (technical) practice (1992), which leans much more toward Arendt's ontology of action than Bourdieu's and MacIntyre's, to some extent solves this conundrum. There is no space here to do right to the richness of Ricoeur's theory, so let us dwell on one of its central elements: the dynamic of ascending complexification and descending specification. On the one hand, this dynamic introduces a structure very similar to MacIntyre's, namely one that departs from basic actions and actions chains (e.g., coding a piece of software in Python), moving toward practices as global actions that stretch through time (e.g., developing a financial technology), which are, in turn, nested in life plans (e.g., being a computer scientist) and a narrative unity of life, and standards of excellence. On the other hand, it produces a movement from distant ideals that are incorporated in life plans and, in turn, inform practices and their associated action chains and basic actions. In this way, technical practices are not only shaped by historical traditions (upwards movement) but can also be ruptured by new insights and ideals coming from fiction, which mediate fine-grained actions in everyday life (downward movement). For instance, a software engineer might read the fictional story in

Orwell's *1984* and gain new insights concerning privacy and security (distant ideals), reflect on her life plan and standards of excellence (what does it mean for me to be a good computer scientist) and adjust her daily practices, action chains, and basic actions (e.g., applying principles of data minimization).

In this way, Ricoeur's theory of technical practice leaves room for both sedimentation (tradition) and innovation. In the introduction chapter, we have insisted on the transcendentals, grammars, or conditions of possibility in which technologies and technological mediations are embedded. However, it would be wrong to see these conditions of possibility as deterministic. In his reflections on the social imaginary, Ricoeur has opportunely insisted on the articulation between ideology and utopia, the former related to the justification of the status quo, the latter to its evolution. Similarly, we here argue that innovation practices should be understood in light of their conditions of possibility. And yet, these conditions are not static; they are dynamic.

As an example of how this might play out in innovation practices, one could, for instance, think of the introduction of COVID-19 apps in different countries. The innovators of these apps would on the one hand engage in technical practices that derive from a professional tradition, one that ideally incorporates design practices that align with established standards of excellence for privacy and security, some of which are captured in legislation like the General Data Protection Regulation (GDPR) in the EU and in international standards. At the same time, developers face the open uncertainty of an innovation challenge, of an unprecedented collection and analysis of personal, health-related data that can have severe societal consequences (e.g., people who had been near to a COVID-19 positive person might have to go in quarantine without actually being infected, on a large scale, with serious consequences for work, income, family life, etc.). This challenge requires them to pass beyond the sedimentation of a historical tradition, and engage with imaginative variations of probable scenarios to envision the potential impact of a non-existent app. This exercise would, in turn, feed back into their innovation practices, coming up with new solutions to not-yet-existing societal problems.

Ricoeur's notion of technical practice and its central movement of ascending complexification and descending specification captures this dynamic in the fullest way. At the political plane, as Kaplan has shown (chapter 4), it is translated in the notions of ideology (representing sedimentation) and utopia (representing innovation). Hence, responsible innovation has to take both these aspects of the movement internal to its practices into account. It is both "closed" as a tradition, and at the same time "open" toward the future. This remark leads us to the third potential contribution of HRI, namely a revisiting of the dynamic between determinative and reflective judgment in *responsible* judgment.

4. RESPONSIBLE JUDGMENT

For Hannah Arendt (1958), responsible, political action is radically open toward the future, and harbors the possibility of the unexpected. It actualizes virtue in the public realm, unfolding in acting and speaking in concert. Ricoeur has been thoroughly influenced by Arendt's theory of the Vita Activa, most notably in his later writings on ethics and politics. Yet, through his appraisal of Arendt's work appears a nuanced (as always) critique (Ricoeur 1983, 2000). This critique opens up a vast problematic that stretches across our thinking concerning responsible innovation; which has to do with the distinction between forward-looking and backward-looking responsibility.

As a human activity, responsible innovation resides at the "in between" in Arendt's ontology of labor, work, and action. Namely, it belongs to *work* insofar it is concerned with the production of durable objects of use (technologies, systems, but also models). However, at the same time, it belongs to (political) *action* insofar it is concerned with the political virtue of responsibility; of taking responsibility for things we have not ourselves done (e.g., climate change). It could therefore be conceptualized as "work in the mode of action" (Reijers 2020). As such, responsible innovation ought to bring forth a hybrid between a durable world and plurality (i.e., a dynamic between closedness and openness as discussed earlier); between durability and fragility (as temporal modes). And yet, this seems problematic in Arendt's work. As soon as action is transformed into making (e.g., a political decision into a constitution), it seizes to give rise to the intangible web of stories that constitutes the public sphere. As part of the fabricated world, a narrative is no longer capable of constituting genuine responsible action.

The root for Arendt's perspective on responsible action lies in her adoption of Kant's model of aesthetic judgment (of the beautiful) for responsible (political) judgment, for it too knows no end (no telos) and establishes a sensus communis based on plurality, which establishes the examplarity of the particular and hence the agents of action to distinguish themselves from all others, to enact the unexpected. This radically open-ended notion of responsible action is questioned by Ricoeur, who assigns both a role to aesthetic, reflective judgment *and* to teleological judgment (Ricoeur 2000, 106). As often with Ricoeur, it is not either/or, but neither/nor. Ricoeur therefore tentatively disagrees with Arendt and opens up a perspective according to which responsible innovation can both involve reflective and teleological judgment. Such an understanding of responsible innovation would allow for an acting in concert among innovators and stakeholders to be transformed into the world that is made, without losing its character of responsibility. This means that like written constitutions and laws, the products of innovation can "coauthor" responsible practices: the fabricated world does not only alienate but can also emancipate and participate.

Yet, it seems that Ricoeur did not finalize his work on a synthesis between reflective and teleological judgment. His remarks on Arendt clearly offer a critique and point at an alternative, but this alternative remains inconclusive. Perhaps the key can be found in Ricoeur's characterization of the narrative unity of life as the "unstable mixture of fabulation and actual experience" (Ricoeur 1992, 162). Yet, this narrative unity of life hits upon its limits in tragic events, which offers a non-philosophical "mixture of constraints of fate and deliberate choices" (Ricoeur 1992, 242). Perhaps the challenging task that Ricoeur puts before us, reflecting on the discussions above, is how to generate utopias that are not mere phantasmagories, but rather realizable projects. In other words, utopian notions of responsible innovation that are open toward the future but at the same time embedded in concrete situations.

Taking up the project of reconciling the two modes of judgment in responsible innovation might be a primary research objective of HRI. To illustrate the importance of this thematic, we return to cybernetics and systems engineering. The global application of complex systems engineering, either for good (e.g., the digital democracy project in Taiwan) or for ill (e.g., the use of Cambridge Analytica to sway voters for Brexit and the election of Trump as president of the United States), shows an increasing animosity toward the *unexpected* in Arendt's political action. Yet, in order to not throw the baby out with the bathwater, we have to follow Arendt in defending the unexpected in public life without putting innovation practices and their products on a second rank. Rather, the innovative has to become political, and the political innovative.

5. CONCLUSION

This concluding chapter has provided a reflection on the fourteen chapters that have kick-started the hermeneutic philosophy of technology (HPT) research program, which is highly indebted to the work of Paul Ricoeur. Additionally, it has opened up a new horizon of HPT, which extends the program from creating a theoretical understanding of technological hermeneutics to a practical application of the program in the form of a politically oriented HRI. The discussion has shown that a Ricoeur-inspired program for responsible innovation can add to the current state of the art in at least three ways. First, it can insert the problematic of the "political paradox" in responsible innovation, which makes the political character of innovation explicit. Second, it captures the dynamic of responsible innovation through a theory of technical practice that pays heed to the movements of ascending complexification and descending specification. Finally, it could work toward a synthesis of reflective and teleological judgment in responsible innovation.

REFERENCES

Arendt, Hannah. 1958. *The Human Condition*. Vol. 24. Chicago: University of Chicago Press. https://doi.org/10.2307/2089589.

Blok, Vincent. 2019. "Towards an Ontology of Innovation: On the New, the Political-Economic Dimension and the Intrinsic Risks Involved in Innovation Processes." In *Routledge Handbook of Philosophy of Engineering*, edited by Diane P. Michelfelder and Neelke Doorn. London: Routledge.

Blok, Vincent, and Pieter Lemmens. 2015. "The Emerging Concept of Responsible Innovation. Three Reasons Why It Is Questionable and Calls for a Radical Transformation of the Concept of Innovation." In *Responsible Innovation 2: Concepts, Approaches, and Applications*, edited by Bert Jaap Koops, Ilse Oosterlaken, Henny Romijn, Tsjalling Swierstra, and Jeroen van den Hoven, 1–303. Heidelberg: Springer. https://doi.org/10.1007/978-3-319-17308-5.

Bourdieu, Pierre. 1992. *The Logic of Practice*. Edited by Richard Nice. Stanford: Stanford University Press. https://doi.org/10.1080/0046760X.2017.1384855.

Deweer, Dries. 2013. "The Person and the Political Paradox: The Personalist Political Theory of Paul Ricoeur." *Appraisal* 9, no. 4.

Deweer, Dries. 2017. *Ricoeur's Personalist Republicanism: Personhood and Citizenship*. Lanham: Lexington Books.

Dupont, Quinn. 2017. "Experiments in Algorithmic Governance: A History and Ethnography of 'The DAO,' a Failed Decentralized Autonomous Organization." In *Bitcoin and Beyond*, edited by Malcolm Campbell-Verduyn. London: Routledge.

Grinbaum, Alexei, and Christopher Groves. 2013. "What Is 'Responsible' About Responsible Innovation? Understanding the Ethical Issues." In *Responsible Innovation*, edited by Richard Owen and John Bessant, 119–143. London: John Wiley & Sons, Ltd.

Grunwald, Armin. 2014. "The Hermeneutic Side of Responsible Research and Innovation." *Journal of Responsible Innovation* 1, no. 3: 274–291.

Harari, Yuval Noah. 2013. *Homo Deus: A Brief History of Tomorrow*. New York: Harper Collins Publishers.

Himmelreich, Johannes. 2020. "Ethics of Technology Needs More Political Philosophy." *Communications of the ACM* 63, no. 1: 33–35.

MacIntyre, Alasdair. 2007. *After Virtue: A Study in Moral Theory*. Third Edition. Notre Dame, Indiana: University of Notre Dame Press. https://doi.org/10.1017/CBO9781107415324.004.

Nabben, Kelsie. "Hacking the Pandemic: How Taiwan's Digital Democracy Holds COVID-19 at Bay." *The Conversation*, September 11, 2020. https://theconversation.com/hacking-the-pandemic-how-taiwans-digital-democracy-holds-covid-19-at-bay-145023.

Reijers, Wessel. 2020. "Responsible Innovation Between Virtue and Governance: Revisiting Arendt's Notion of Work as Action." *Journal of Responsible Innovation*, 1–19. https://doi.org/10.1080/23299460.2020.1806524.

Reijers, Wessel, and Mark Coeckelbergh. 2020. *Narrative and Technology Ethics*. London: Palgrave MacMillan.

Ricoeur, Paul. 1983. "Action, Story and History: On Re-Reading the Human Condition." *Salmagundi*, no. 60: 60–72.

Ricoeur, Paul. 1992. *Oneself as Another*. Edited by K. Blamey. Chicago: University of Chicago Press.

Ricoeur, Paul. 2000. *The Just*. Chicago: University of Chicago Press.

Robinson, Dan, and Georgios Konstantopoulos. 2020. "Ethereum is a Dark Forest." *Medium*, August 28, 2020. https://medium.com/@danrobinson/ethereum-is-a-dark-forest-ecc5f0505dff.

Sand, Martin. 2018. "The Virtues and Vices of Innovators." *Philosophy of Management* 17, no. 1: 79–95. https://doi.org/10.1007/s40926-017-0055-0.

Stiegler, Bernard. 1998. *Technics and Time 1*. Stanford: Stanford University Press.

Wong, Pak-Hang. 2019. "Democratizing Algorithmic Fairness." *Philosophy & Technology* 33: 225–244.

Zuboff, Shoshana. 2019. *The Age of Surveillance Capitalism*. New York: Public Affairs.

Index

accountability, 119, 130, 131, 175, 176, 181, 219
Achterhuis, Hans, xiii, 27, 43, 143, 190, 250
actant, 43, 45–48, 50–57, 265
actor-network, 53–57
Actor-Network Theory (ANT), xvi, 250, 251, 271, 272
Alexa, 120, 249, 275
algorithm, 79, 87, 89, 90, 120, 121, 126, 131, 196, 201, 202, 219, 229, 233, 243, 252, 256, 267
AlphaGo, ix, xix, 252–72
alterity, 32–34, 86, 265
Amazon, 162, 275
ambiguity, xvi, 4–7, 10, 13, 28, 29, 32, 120, 213, 263
Apple, 126, 231, 240–42
application, 5, 17, 37, 67, 91, 153, 202, 204, 211–13, 220, 221, 233, 234, 272, 273, 281; AI, 250, 251, 255, 264; hermeneutic, 30, 31, 33, 35, 38, 40; prosthetic, 155, 156, 158, 159, 161, 163, 166, 168; software, 230–32, 236–38, 240, 241, 243, 244, 245, 249
appropriation, 4, 18, 30, 31, 40, 71, 209, 234, 253

Arendt, Hannah, xviii, 99, 111, 113, 136, 181, 209–16, 220, 221, 273–75, 278, 280, 281
Aristotle, x, 32, 79, 83, 102, 104, 105, 114, 129, 165, 166, 195, 234, 262
artifact, x, xvi, 3, 13–15, 18, 20, 28, 36, 37, 40, 44, 47–52, 56, 57, 64, 72, 81, 85, 89, 160, 172, 229, 232, 235, 243, 251, 254
artificial intelligence, ix, xviii, 87, 147, 249, 266, 271, 272
ascription, 122, 124, 125, 131, 132, 173
Athena, 79
attestation, 103, 104, 112, 132, 140, 142, 146, 173, 183, 214
Austin, J. L., xviii, 209, 214
autonomy, 62, 63, 87, 136, 137, 155, 156, 159, 162, 164, 165, 167, 181, 193, 194, 249

Benjamin, Walter, 166, 191, 199
Big Data, 201, 202
bioethics, 153, 156, 159, 167, 168
black box problem, 49, 126, 250, 252, 268
blockchain, 276
body, xxi, 13, 36, 38, 64, 99, 117, 119, 122–25, 127, 129, 154–68, 261; biography, 154, 159, 161–63, 166

Index

Bourdieu, Pierre, xiv, 278
Butler, Judith, 100, 110–12

capability, xvi, 3, 15, 31, 104, 139, 141, 167
Carney, Eoin, xvi, 27–41, 224, 273
Cassirer, Ernst, 76–78, 80–84, 88, 89
Charon, Rita, 192, 193, 197
Chown, Eric, xviii, 229–45
civilization, 4–9, 13, 65
clinical ethics, 191, 192, 197
Coeckelbergh, Mark, ix–xxi, 44, 154–56, 162, 233, 271–82
communication, 10, 12, 15, 17, 45, 63, 68, 70, 79, 80, 101, 102, 108, 190, 193, 194, 196, 197, 200, 210, 214, 216, 217, 237, 244, 245
conditions of possibility, x, xiii, xiv, xv, xix, 78, 277, 279
configuration, 30, 44, 76, 77, 79, 82, 83, 89–91, 93, 104–6, 236, 253–55
COVID-19, ix, 119, 275, 279
critical theory of technology, xvi, 61, 66, 69, 72, 271, 273
cybernetics, 275, 281

data, 80, 81, 86, 89, 90, 117–32, 140, 153, 190, 201, 202, 243, 258, 267, 279; data self, xvii, 118, 119, 126, 128, 131, 132
de Boer, Bas, xvi, 43–58, 273
DeepMind, 252, 255, 256, 259, 261, 263–66, 268
deep neural networks, 256
democracy, 9, 70–72, 200, 217, 276, 277, 281
democratic rationalization, 66, 68
de Mul, Jos, 16, 262, 263
deprivation, 46, 216
detour, xvi, 29, 31, 88, 89, 110, 118, 147, 251
Dierckxsens, Geoffrey, 153–68, 173
digital, xvii, xxi, 17, 18, 77, 79, 80, 82, 86, 90, 99, 102, 106, 118, 119, 120, 122, 123, 125, 189, 190, 196, 200, 201, 203, 230, 232, 277; democracy, 276, 281; health, 191, 193, 194, 196, 197, 202, 204; hermeneutics, xvii, 18, 117, 121; identity, xvii, 100, 101, 102, 105, 106, 108–13, 118, 123
discourse, 27–29, 40, 41, 44, 47, 71, 72, 75, 76, 79–81, 85, 100, 109, 140, 192, 194, 197, 199, 200, 218, 235, 251–53, 268, 275–77
distanciation, xi, 34, 37, 70, 118, 119, 194

education, xvii, 5–9, 90, 135–48
e-Health, xviii, 189, 190, 193–94, 203–5, 272
Ellul, Jacques, 62, 144
embodiment, xvii, 34, 38, 39, 67
empirical turn, xiii, xvi, xix, 27, 29, 43, 75, 143, 190, 250, 267
emplotment, xix, 46, 47, 58, 78, 102, 104, 105, 205, 254, 255, 257, 258, 261, 264, 268
enframing (*Gestell*), 135, 144, 146, 147
environment, 9, 12, 13, 15, 17, 18, 36, 44, 66, 99, 100, 148, 176, 180, 184, 196, 230, 243, 249
Ethereum, 276
ethics of technology, xv–xvii, 11, 18, 118, 271, 273
explanation, 44, 46, 47, 49, 54, 55, 58, 64, 70, 86–88, 90, 91, 202, 234, 268

Facebook, 90, 99, 100, 107–10, 122, 126, 131
face-to-face, ix, 219, 220, 240
Feenberg, Andrew, xvi, xix, 48, 61, 65–67, 69–72, 143, 217, 250, 273
fintech, 277
Fitzpatrick, Noel, xvii, 117–32, 272
Floridi, Luciano, 117
Foucault, Michel, 119, 122, 123, 136, 190, 200, 203, 251
Frankenstein, 259, 274, 277
Frankfurt School, xvi, 61, 62
freedom, 8, 10, 31, 62, 64, 67, 68, 71, 101, 217, 221, 261, 262

Freedom and Nature, 64, 67
Freud, Sigmund, 29, 30, 36, 67, 70

Gadamer, Hans-Georg, xi, xii, 30, 31, 37, 38, 68, 70, 85, 88, 194, 233, 234, 253, 257
game, ix, 101, 102, 156, 180, 199, 252, 255–68, 276
General Data Protection Regulation (GDPR), 131, 279
Gianni, Robert, xviii, 171–83, 272, 273
gift, 181, 182
good life, 11, 12, 71, 128, 129, 272
Google, xix, 90, 118, 120, 121, 126, 231, 252, 255, 268
Gorgoni, Guido, xviii, 171–83, 272, 273
grammar, x, xiii, xv, xix, 14, 81, 179, 277, 279
Gransche, Bruno, xvii, 17, 75–91, 155, 272
Greimas, Algirdas, xvi, 43, 45–47, 51–53, 57, 102

Habermas, Jurgen, 11, 27, 61, 63, 65–71
Halsema, Annemie, xvii, 99–113, 272
health, 10, 66, 69, 164, 190–93, 196–98, 201–4, 279
healthcare, 17, 153–68, 191–92, 271
Hegel, Georg Wilhelm Friedrich, 66, 76–77, 80, 88
Heidegger, Martin, xi, xiii, xv, 16, 27–30, 37–41, 62, 135–36, 138–39, 143, 148–49, 194, 221, 275
hermeneutic idealism, 194
Hermeneutic Philosophy of Technology (HPT), ix, x, xiii–xv, xix–xx, 271–72
Hermeneutic Responsible Innovation (HRI), x, xv, xx, 271–75, 277, 279, 281
hermeneutics of suspicion, xvi, 27–28, 30, 32, 36, 39, 277
hermeneutics of trust, xvi, 28, 30, 36–38, 40
hermeneutic turn, 29, 190

High Level Expert Group on Artificial Intelligence, 271
History and Truth, 64, 199
Hoek, Jonne, xvi, 43–58, 273
Homo interpretans, 121, 124
Homo ludens, 113, 257
Huizinga, Johan, 113, 257–58
human-technology relation, 16, 19, 33, 40, 44, 86, 136
Husserl, Edmund, 38–39, 163
hybrid, 47–48, 51, 53–54
hybrid intentionality, 160–61, 168

idealism, 81, 92, 190, 194
idem/ipse, xvii, 103, 107, 119, 121–24, 127–28, 138, 140, 192
ideology, xvi, xix–xx, 11, 61, 64, 68–72, 155, 159–62, 167, 279
Ihde, Don, xii–xiii, 27, 33–39, 41, 67, 86, 250–51, 260, 268
illocution, 179, 183, 210, 212–18, 220–23
imagination, xviii–xx, 17, 54, 68–69, 155, 163, 165–68, 191–200, 204, 220, 230, 235, 237, 243, 254–55; productive, xviii–xix, 17, 200, 230, 243, 254
imputability, 128, 130–31, 181
industrial revolution, 7–8
information, 10, 34–35, 79, 81, 86, 108, 117, 119, 124, 132, 144, 164, 189–93, 201, 204, 214–15, 220–23, 239–41, 251–52; technology, 16, 196
innovation, xv, xviii–xx, 5, 18, 35, 171–75, 177, 180–84, 195, 198–201, 222, 229, 233–36, 243, 245, 249, 254, 259–60, 264, 273–82
inscription, 49–50, 117–18, 123, 126–27, 190
instant messaging, 237
institution, xx, 5–6, 11, 16, 19, 128, 131, 161, 178–83, 192, 199, 200, 203, 272, 276–77
intentionality, 35, 128, 160–61, 168

interface, 31–32, 86, 230, 232, 240–43, 251
internet, 99, 101, 106, 108, 113
interpretation, ix, xi–xii, xvi, 7, 8, 12, 15, 28–31, 37–40, 44, 55, 64–68, 71–72, 76, 80, 82, 84, 86–93, 110, 113, 117, 120, 126, 129, 137, 140–44, 148, 189, 194, 202, 233–34, 243–45, 252–53, 257, 259, 260, 271–75
intersubjectivity, 56, 266, 268

Jobs, Steve, 240–42
The Just, 131, 172, 272
justice, 18, 47, 71, 114, 128–29, 131, 182, 184, 201, 251, 277

Kant, Immanuel, xv, 36, 58, 62, 66–67, 128, 195, 197–98, 220, 235, 275, 280
Kaplan, David, xvi, 16, 27, 44, 61–72, 92, 155, 171, 190–91, 210, 215, 222–23, 233, 252, 273–74, 279
Keymolen, Esther, xix, 249–68, 272

labor, 4, 8–9, 14, 280
language, x–xiv, xx, 8–9, 17, 29, 31, 38, 65, 71, 75–85, 88–93, 99, 103–4, 108, 112, 117–20, 126, 130, 140, 147, 178–80, 190, 194–200, 204, 209–10, 213, 219, 223, 233, 236, 249, 252, 277
Latour, Bruno, xvi, xxi, 43–58, 250, 265, 273
learning, 135–49
Lewin, David, xvii, 16, 31–32, 135–48, 155, 222, 272
liability, 175–76
life-stories, 102, 104, 106–7, 111, 153–54, 158, 165
little ethics, 114, 118–19, 128–31, 272
live metaphors, 233, 244
lived body (*Leib*), 163
lived experience (*Erlebnis*), 17, 29, 88, 90, 153–54, 157, 255
Loute, Alain, xviii, 189–205, 222, 272

machine learning, 33, 89, 121, 132, 249, 261, 265, 267
MacIntyre, Alasdair, 212, 217, 278
mapping, x, xv, xxi, 114
Marcuse, Herbert, 27, 61–66
Marx, Karl, 29–30, 61, 64, 67–68, 70, 273
material hermeneutics, xii, xvi
medial turn, 77, 91
mediarchy, 196
mediation, xiv, xviii, xx–xxi, 6, 15, 18, 43, 46, 56, 67, 77, 88–89, 93, 110, 114, 117, 118, 121, 123, 139, 177, 183, 192, 194, 196–97, 199, 202, 210, 216, 232, 264–65
medical practice, 153, 157, 159, 192, 203
medium, 32, 38, 70, 76–84, 91
Mei, Todd, xviii, 18, 209–22, 273
metaphor, xii, xviii, 67–68, 79, 92, 195, 222–23, 229–45
micro-perception, 260, 268
mimesis, 17, 104, 114, 203, 233, 236, 245, 272
Modern Constitution, 47, 50–51, 53, 55–56, 58
modernity, 3, 4, 6, 9, 11, 43, 58
multistability, xii–xiii, 35, 254
Murdoch, Iris, 165

narration, xvii, 15, 53, 55, 87, 106, 111, 114, 122, 128, 147, 165, 199
narrative, x, xii, xvii, xix, xx, 15, 17, 46, 52–58, 68, 100–28, 137–38, 145–46, 148, 156, 158, 161, 165–66, 191–92, 194–95, 197, 199, 202, 205, 211, 222, 231, 233, 255, 268, 272, 274, 278, 280–81; clinical ethics, 189, 192, 197; competence, 193; identity, xvii, 16, 99–14, 118, 121–28, 136–38, 145–47, 153–54, 158–59, 165, 178, 192, 212; medicine, 189, 205; technologies, 153–59, 163, 168, 233
narrativity, xviii, 15, 45, 108, 189, 191–96, 203, 205

Nascimento, Fernando, xviii, 229–45
natality, xviii, 209–13, 222–23

objective spirit, 77, 80, 82, 84–85, 88, 91, 272
Oneself as Another, 11–12, 15, 67, 100, 105, 118, 121, 132, 155, 166, 272

pedagogical relation, 135–38, 140–42, 144, 147, 149
pedagogy, 138, 140, 146
perlocution, 212–16, 220–21
personalized medicine, 202
pharmakon, 126–27, 277
philosophy of technology, ix–xvi, xviii–xxi, 3–4, 16, 19, 27, 38, 43–45, 61–63, 66, 75–77, 143, 149, 153–56, 159–60, 168, 190–91, 233, 250, 271–73
phronesis, 162, 165–66
play, 257–58
plot, 45–47, 52–53, 58, 101, 104–6, 112, 114, 254–55, 257
poiesis, 76, 83–84
political action, xviii, 209–15, 220–23, 273, 275, 280–81
political paradox, 11, 19, 70, 213, 274–77, 281
politics, 4, 6–8, 11–12, 14, 19, 52, 71–72, 119, 181, 217, 222, 272, 275, 277, 280
postphenomenology/post phenomenological, xx–xxi, 27–28, 32, 37–40, 250–52, 260, 264, 268, 273
power, xxi, 4, 6–7, 11, 14, 19, 29–30, 62–63, 66, 68–69, 71, 119, 144, 161, 164, 177, 179, 182, 196–97, 199, 223, 251, 259, 268, 274, 276
pragmatic/pragmatism, 27–28, 40, 82, 178–79, 200, 211, 216, 223, 251
praxis, 8, 21, 76, 180
precautionary principle, 172, 175–76
prefiguration, 83, 91, 104, 236–38, 240–41

privation, 210, 216–18, 220–21
prosthesis, 154, 156, 158–63, 168

quantified self, xvii, 117–20, 122–25, 128–32, 271
question/questioning, x, xv, xviii, 4, 8, 11, 16, 27, 30, 35, 37, 39–41, 68, 70, 76, 88–89, 99, 103, 117, 121–32, 140, 143, 147–49, 154, 158–61, 163, 166, 168, 190–94, 199–201, 203–4, 213–14, 223, 241, 260, 265, 268, 274–75, 277–78

rationality, 19, 48–49, 61–62, 65–66, 69–70, 201, 217
rationalization, 61, 63, 66, 68, 204
recognition, xiv, 15, 38, 136–38, 140, 145–48, 177–78, 181–83, 191, 203, 213, 215–16, 219, 222, 231, 242, 249
reconfiguration, 15, 145, 239
referentiality, 253, 267
refiguration, 83, 90–91, 104, 236–38, 240–41
reflective judgment, 279–80
Reijers, Wessel, ix–xx, 17–18, 44, 56, 92, 154–56, 162, 190, 233, 271–82
responsibility, xviii, xix, 6, 11–12, 16, 71, 130, 132, 172–83, 209, 211, 262, 272–74, 277, 280
responsible innovation, x, xv, xx, 177, 183, 273–75, 277, 279–82
responsible judgment, 275, 277, 279–80
Responsible Research and Innovation (RRI), 171
responsible subject, 173, 177, 183
responsiveness, 176, 178
Ricoeur, Paul ix–xiii, xv–xx, 3–20, 27–31, 34, 37–41, 43–47, 53–58, 63–72, 75–81, 83–85, 87–89, 91–93, 99, 100, 102–14, 117–33, 136–37, 139–40, 142, 145–49, 153–59, 161–62, 165–68, 172–74, 176–83, 189–92, 194–95, 197–202, 204–5, 209–15, 220, 222–23, 229, 232–36,

243–45, 251–55, 257, 259, 262, 268, 271–81
Romele, Alberto, ix–xx, 17–18, 20, 86, 92, 132, 137, 156, 190, 194–95, 271–82
Rouvroy, Antoinette, 201–2
Rule of Metaphor, 67

schema/schematization, 46, 68, 76, 81, 84, 86, 91–93, 195, 197–201, 204
science and technology studies (STS), xiii, 43
science, ix, xiii, 4, 9, 29–31, 39, 43, 49, 57, 61–64, 67, 69–70, 92, 117, 132, 140, 171, 196
sedimentation, 4, 85, 190, 198, 200, 254, 259–61, 264, 279
self-esteem, 128–31, 214
self-exposure, xvii, 100, 110–11, 113
self-expression, 100–101, 108, 110–11, 113, 265
self-understanding, 7, 17, 28, 77, 85, 88–89, 110, 118, 173, 176, 253
semantic innovation, xviii, 195, 201, 222, 229, 233–36, 243, 245
semantic optimism, 190, 198–200, 202
semantic productivity, 201
semiotics, xii, xvi, 43, 45–50, 53–58, 235
Simondon, Gilbert, 204
SMS, 120, 236–38
social imaginary, xvii, xix, 154, 161, 163, 279
social imagination, 155, 166–68, 198
social media, xviii, 99–102, 106–14, 123, 130–31, 148, 153, 210, 212, 216–23, 271
social networks, 79, 196, 236, 238–40, 243
social object, 197
software, xviii, 101, 229–45, 249, 251, 254, 271, 276–78
software requirements, 230–31, 242–44
solicitude, 128–31

space, xviii, xxi, 8, 10, 15, 58, 62, 76, 81–84, 108, 111, 180–81, 196, 201, 203–5, 216–17, 232, 237, 239, 255, 257–58, 260, 268, 278
spatiality, 102, 106–7, 114, 190, 202–4
speech, xi, 8–9, 19, 34, 99, 111–12, 195, 198, 209, 216, 218, 222–23, 235, 249, 261; act, xviii, 18, 93, 209–12, 220, 222–23
Stiegler, Bernard, xxi, 118, 127, 129, 132, 143, 277
story, xix, 38, 45–47, 57–58, 99, 102, 104–12, 114, 122, 125–26, 137, 141, 154, 157, 159, 165–67, 191–92, 197–200, 222, 253–55, 257, 259, 264, 272, 278
subject, 6, 10, 18, 29–30, 36, 43–48, 55–57, 61, 63–64, 71, 89, 124, 135–39, 145–46, 148, 172–74, 176–78, 181, 183, 192–94, 196, 200–201, 211–14, 235, 249, 257
subjectivity, 45, 56–58, 103, 136–38, 181–82, 192, 253, 266, 268
sublime, 263–64, 268
suffering, 10, 62, 128, 146, 192, 213–15, 223, 262
symbol, 28–31, 40, 93
symmetry, xiv, 43–45, 47–48, 56
systems engineering, 281

techne, xvi, 3, 13, 15, 19–20, 76, 78–80, 85, 88–90, 274
technical object, 196–98, 204
technical practice, xvii, 274–75, 277–79, 281
technique, xvii, 3–5, 11, 13–15, 19, 61, 64, 79, 119, 146, 196, 202, 229, 249, 253, 255–56
technological mediation, x, xii–xiv, xix, xx–xxi, 18, 32–33, 37–38, 139, 209, 272–73, 279
technological relation, 5, 136, 138
technological thinking, xvii, 136, 138–39, 146

Index

technologies of the self, 122–23, 126, 132
technology, xi–xxi, 3–11, 13–20, 27–41, 43–45, 53, 56, 58, 61–72, 75–93, 99, 107, 117–18, 121–22, 127, 129, 135–36, 138, 140, 143, 146, 148–49, 153–56, 158–63, 168, 171, 183, 189–93, 200–203, 205, 209–10, 216, 220, 224, 233, 237, 239, 241–44, 249–51, 254, 258–59, 264, 267, 271–73, 275–78, 281
telemedicine, 193, 198, 203–5
telemonitoring, 198
teleological judgment, 280–82
teleology, 54, 57
telos, 128, 138, 276, 280
temporality, 6, 45, 102, 106–7, 110, 114, 121, 190, 202, 204
text, xvii, 8, 15, 18–19, 20, 34, 45–46, 48–49, 53–56, 75–78, 80–81, 83, 85, 88–89, 92, 101, 109, 113, 117, 127, 147, 194–95, 197, 199, 223, 252–55; messages, 237, 243
texting, 237–38
texture, xvii, 75–85, 88–92, 272
TikTok, 120, 122
Time and Narrative, 15, 17, 67, 100, 104, 126, 128, 199–200, 202, 204, 236
tool, xvii, 4–5, 13, 19, 39–40, 61, 63–64, 75–76, 78, 89, 91, 146, 192–93, 197, 203, 204, 260, 272–73
trace, ix, 39, 41, 57, 86, 117–18, 123, 193, 204, 238
tracks, 76, 82–86, 90–91, 93
tradition, 12, 70–71, 77, 103, 141, 144, 148–49, 180–81, 198, 200, 254–55, 260, 278–79
tragedy, 224, 254, 261–63
trails, 51, 53, 76, 82–83, 85–86, 90–91, 93
transcendental, x, xiv–xv, xix–xxi, 62, 85

transfiguration, 264
translation, xii, xv, 7, 13, 19–20, 52–54, 79, 172, 191–93, 196–97, 199–205
truth, x–xii, xix, 48–49, 66, 68, 217, 220–21, 264
Twitter, 99

understanding, x–xi, xv–xvii, 6–8, 11–20, 28–30, 32, 36–38, 40, 43–47, 52–58, 61, 68–70, 75–78, 80, 85–89, 91–93, 102, 105–6, 110, 112, 117–18, 122, 127, 132, 137, 147, 153–54, 156, 164–68, 172–73, 176, 181–83, 192, 195, 209, 211–12, 214–16, 219–20, 222, 233–35, 244–45, 252–53, 255, 257–59, 262–66, 274–75, 280–81
universal, 5, 19, 55, 65–66, 69, 178, 200, 204
urban/urbanization, 10, 15
utopia, xvi, xx, 31, 61, 68–72, 161–62, 279, 281

Verbeek, Peter-Paul, 35–37, 67, 75, 143, 149, 160, 250–51
virtue, 18, 36, 55–56, 143, 165, 176, 212, 215, 221–22, 255, 274, 280
Vita Activa, 280
voice assistant, 249

weaving, 75, 78–79, 82, 91, 93, 271–72
Wolff, Ernst, xvi, 3–19, 273–74
work, xvi, xviii, 3, 7–9, 11, 13–20, 29, 31, 33–34, 37–38, 40, 44–48, 54, 61, 64, 66, 70–71, 76, 78–82, 84–85, 100, 111, 117–18, 126–27, 132, 138, 156–59, 165, 190–91, 193, 195–202, 204, 212, 217, 221, 233–35, 239, 242, 252–55, 272–73, 278–82

YouTube, 118, 120

About the Contributors

Bas de Boer is a postdoctoral researcher at the philosophy department at the University of Twente, Netherlands. He recently defended his PhD dissertation on the role of technologies in scientific practice, with a focus on imaging and stimulation technologies in the neurosciences. His current research focuses on how technologies in healthcare shape the understanding and experience of health.

Eoin Carney received his PhD in philosophy from the University of Dundee (Scotland) in 2018. His thesis, Technologies in Practice: Paul Ricoeur and the Hermeneutics of Technique, was a study of Paul Ricoeur's work in light of recent approaches to the philosophy of technology. His primary research areas are hermeneutics, the philosophy and ethics of technology, and, more recently, the personal identity tradition in analytic and continental philosophy.

Eric Chown is the Sarah and James Bowdoin Professor of digital and computational studies at Bowdoin College. Besides teaching computer science courses in artificial intelligence, cognitive architecture, and computer programing, Eric Chown also enjoys researching the learning in humans and machines. He was awarded a five-year National Science Foundation Faculty Early Career Development Grant that was used to buy specialized robots for his project—Computational Models of Space in Navigation and Other Domains.

Mark Coeckelbergh is a philosopher of technology. He is professor of philosophy of media and technology at the Department of Philosophy of the University of Vienna and president of the Society for Philosophy and Technology. He also has an affiliation as professor of technology and social responsibility at the Centre for Computing and Social Responsibility, De Montfort University, UK. He is the author of *Liberation and Passion* (2002), *The Metaphysics of Autonomy* (2004), *Imagination and Principles* (2007), *Growing Moral Relations* (2012), *Human Being @ Risk* (2013), *Environmental Skill* (2015), *Money Machines* (2015), *New Romantic Cyborgs* (2017), *Using Words and Things* (2017), *Moved by Machines* (2019), *Introduction to Philosophy of Technology* (2019), and *AI Ethics* (2020).

Geoffrey Dierckxsens (PhD) is head of the Interdisciplinary Research Lab for Bioethics (IRLaB) at the Department of Contemporary Continental Philosophy in the Institute of Philosophy of the Czech Academy of Sciences (CAS) in Prague. He recently received the Lumina Quaeruntur Award for prospective researchers to establish IRLaB. Dierckxsens specializes in French phenomenology and hermeneutics, in particular as applied to bioethics and cognitive theory.

Noel Fitzpatrick (doc ès lettres, Paris VII) is professor of philosophy and aesthetics and the Dean of Graduate School of Creative Arts and Media (GradCAM). He is also the academic lead of the European Culture and Technology Lab (ECt Lab+) of the European University of Technology based at Technological University Dublin (TU Dublin). He is a leading member of the European Artistic Research Network (EARN), a member of Ars Industrialis, and a founding member of the Digital Studies Network at the l'institut de recherche et innovation (IRI) at the Pompidou Centre in Paris. His most recent publication is collective with Bernard Stiegler entitled *Bifurcate: There Is No Alternative*, open humanities press, 2020.

Robert Gianni is a research fellow at Maastricht University and investigator for Ethics and Responsible Research and Innovation at the BISS Institute. As a political philosopher, Robert works on the relationship between democracy and the role of emerging technologies. His research interests focus on understanding how these two poles interact and how democratic principles can be reinvigorated by strengthening inclusiveness.

Guido Gorgoni is assistant professor in legal philosophy at the University of Padova (Department of Political Sciences, Law and International Studies), where he also teaches legal informatics; and he is Chercheur Associé at the Séminaire interdisciplinaire d'études juridiques, Université Saint-Louis, Bruxelles. His research activity focuses on the evolutions of the legal idea of responsibility on the evolution of legal sources and regulation, on new technologies and on responsible research and innovation. His interest on

Ricoeur's thought focuses on the legal philosophy of Ricoeur, namely on topics such as responsibility, the subject of rights, and recognition.

Bruno Gransche has been a philosopher at the Institute of Advanced Studies FoKoS at the University of Siegen since 2017. He works as a researcher and lecturer in the fields of philosophy of technology and ethics, socio-technical cultural techniques, and future-oriented thinking. He is a research fellow at the Fraunhofer Institute for Systems and Innovation Research ISI in Karlsruhe, where he worked as a philosopher and Foresight expert until 2016.

Annemie Halsema is associate professor at the Department of Philosophy of the Vrije Universiteit in Amsterdam, Netherlands. Her research specialization is in French phenomenology and hermeneutics and in feminist philosophy. She wrote two books on Luce Irigaray, and together with Fernanda Henriques edited *Feminist Explorations of Paul Ricoeur's Philosophy* (Lexington 2016).

Jonne Hoek is a PhD candidate at the philosophy department at the University of Twente, Netherlands. He teaches philosophy of law at the Radboud University, Netherlands. His research interests are metaphysics, philosophy of technology, and existential philosophy.

David M. Kaplan is professor in the Department of Philosophy and Religion at the University of North Texas. He is author of *Ricoeur's Critical Theory* (SUNY Press), and editor of *Reading Ricoeur* (SUNY Press), *Philosophy, Technology, and the Environment* (MIT Press), and *The Philosophy of Technology* (Rowman and Littlefield). He has published several articles on hermeneutics, narrative theory, and technology—particularly genetically modified food, artificial ingredients, and junk food. He also manages the Philosophy of Food Project.

Esther Keymolen is assistant professor in ethics, law, and policy of new data technologies at the Tilburg Institute for Law, Technology, and Society (Tilburg University). She has a background in philosophy of technology and postphenomenology. Her research focuses on the role of trust and privacy in networked and smart environments. In her book *Trust on The Line*, she developed a conceptual trust model, which she applied to analyze different topical cases, such as: Airbnb, smart cities, smart toys, digital hotel keys, and personalized online advertisement.

David Lewin is lecturer in philosophy of education at Liverpool Hope University (UK). His current research addresses the religious and philosophical implications of modern technology. In addition to publishing several articles on the philosophy of technology and its relation to religious philosophy, his first monograph *Technology and the Philosophy of Religion* was published in 2011 (Cambridge Scholars Publishing).

Alain Loute is assistant professor at the Center for Medical Ethics (ETHICS EA 7446) of the Catholic University of Lille (France). He is co-holder of the "Chaire Droit et éthique de la santé numérique" (Catholic University of Lille). His research interests are hermeneutics, care ethics, technology ethics, and digital health. He is also the author of a book *La création sociale des normes* (Olms, 2008) on Ricoeur and a dozen articles on the philosophy of Paul Ricoeur. Recently, he has coedited *Donner, Reconnaître, Dominer, Trois modèles en philosophie sociale* (with L. Carré, Septentrion, 2016) and *Valeurs de l'attention, Perspectives éthiques, politiques et épistémologiques* (with N. Grandjean, Septentrion, 2019).

Todd S. Mei is senior lecturer in philosophy at the University of Kent. From 2008 to 2012, Mei was a lecturer in the philosophy and religious studies departments at the University of Kent, and from 2012 to 2015, he was a lecturer in philosophy at the University of Dundee. His current research interests include a phenomenological understanding of the meaning of land and how it relates to questions of well-being, dwelling, and economic justice; understanding the notion of work according to analogies to discourse and metaphor; and the development of a hermeneutical theory of truth in relation to the work of Paul Ricoeur.

Fernando Nascimento is assistant professor in digital and computational studies at Bowdoin College teaching courses on philosophy of technology and hermeneutics. His research is organized in three interconnected academic axes of ethics, hermeneutics, and digital technologies and has Paul Ricoeur as its main theoretical reference. Prior to his academic positions, he worked for almost twenty years in the telecommunication industry developing software for mobile devices worldwide. He is currently codirector of the Digital Ricoeur project, and director of the Society for Ricoeur Studies since 2018.

Wessel Reijers is a postdoctoral research associate at the Robert Schuman Centre, European University Institute, Florence. His fields of expertise include philosophy and ethics of technology, philosophical hermeneutics, responsible innovation, citizenship studies, and science and technology studies. In his current project, he focuses on the impacts of social credit systems on the institution of citizenship. He has published multiple journal articles on hermeneutic philosophy of technology, responsible innovation, and distributed governance. Together with Mark Coeckelbergh, he published a monograph titled *Narrative and Technology Ethics* (2020) with Palgrave MacMillan.

Alberto Romele is associate researcher at the IZEW, the International Center for Ethics in the Sciences of the University of Tübingen, Germany. He has been associate professor of philosophy of technology at the Catholic University of Lille. He is the author of *Digital Hermeneutics: Philosophical Investigations in New Media and Technologies* (Routledge 2019) and co-editor of Towards a Philosophy of Digital Media (Palgrave Macmillan 2018).

Ernst Wolff is professor of philosophy at the KU Leuven in Belgium and extraordinary professor in the philosophy department of the University of Pretoria in South Africa. His research is mainly in social and political philosophy, action theory, hermeneutics, and African philosophy. He has published extensively on Ricoeur's work in the context of technology, politics, and responsibility. His upcoming book focuses on the technicity of action.

www.ingramcontent.com/pod-product-compliance
Lightning Source LLC
Chambersburg PA
CBHW022009300426
44117CB00005B/105